The World Food Problem

FOURTH EDITION

The World Food Problem

Toward Ending Undernutrition in the Third World

Howard D. Leathers
Phillips Foster

Viva Books
New Delhi | Mumbai | Chennai | Kolkata | Bengaluru | Hyderabad | Kochi | Guwahati

First Indian Edition 2011

VIVA BOOKS PRIVATE LIMITED

- 4737/23 Ansari Road, Daryaganj, New Delhi 110 002
  E-mail: vivadelhi@vivagroupindia.net, Tel. 42242200
- 76, Service Industries, Shirvane, Sector 1, Nerul, Navi Mumbai 400 706
  E-mail: vivamumbai@vivagroupindia.net, Tel. 27721273, 27721274
- Jamals Fazal Chambers, 26 Greams Road, Chennai 600 006
  E-mail: vivachennai@vivagroupindia.net, Tel. 28290304, 28294241
- B-103, Jindal Towers, 21/1A/3 Darga Road, Kolkata 700 017
  E-mail: vivakolkata@vivagroupindia.net, Tel. 22836381, 22816713
- 7, Sovereign Park Apartments, 56-58, K. R. Road, Basavanagudi, Bengaluru 560 004
  E-mail: vivabangalore@vivagroupindia.net, Tel. 26607409, 26607410
- 101-102 Mughal Marc Apartments, 3-4-637 to 641, Narayanguda, Hyderabad 500 029
  E-mail: vivahyderabad@vivagroupindia.net, Tel. 27564481, 27564482
- Beevi Towers, First Floor, SRM Road, Kaloor, Kochi 682 018, Kerala
  E-mail: vivakochi@vivagroupindia.net, Tel: 0484-2403055, 2403056
- 232, GNB Road, Beside UCO Bank, Silpukhuri, Guwahati 781 003
  E-mail: vivaguwahati@vivagroupindia.net, Tel: 0361-2666386

www.vivagroupindia.com

Published by arrangement with

Lynne Rienner Publishers, Inc.
1800 30th Street, Suite 314
Boulder, Colorado 80301
USA

ISBN : 978-81-309-1531-9

Published by Vinod Vasishtha for Viva Books Private Limited, 4737/23, Ansari Road, Daryaganj, New Delhi - 110 002. Printed and bound by Raj Press, R-3, Inderpuri, New Delhi - 110 012.

S0000 / S0810

Contents

Tables and Figures

Tables

Figures

Preface

The first half of 2008 saw dramatic increases in the price of food. Suddenly, the world food problem was news, moving from the agate print of the financial section to the headlines of the front page. Did this mark the end of decades of progress in the effort to end undernutrition in the world—and the beginning of a new trend toward ever increasing worldwide hunger? This fourth edition of *The World Food Problem* revises earlier editions to consider some of the implications of these recent developments.

The 2008 experience draws our attention to the increasing interconnectedness of rich and poor countries; thus, we have expanded our discussion of recent research into globalization and economic growth. The 2008 experience also points to nonfood uses of agricultural output as a potentially important driver of world food prices, which is reflected in our expanded coverage of biofuels.

Our understanding of the causes of and policy solutions to the world food problem continues to evolve. Economic research is shedding new light on relevant issues: for example, economic historians are gathering new evidence about the impact of nutrition on economic growth, and economists are using historical patterns of temperature change to project the impact of climate on agricultural production. Policy initiatives give reasons for both hope and caution: for example, agricultural research programs have shown success in increasing cassava yields in western Africa, but substantial foreign aid has failed to make a difference in places like Zambia and Eritrea. This new edition also incorporates up-to-date statistical information.

Despite these changes, the principal messages of the first edition are as valid today as they were in 1992. Undernutrition remains a problem for hundreds of millions of people in developing countries. Poverty, income inequalities, population growth, and illness continue to be important causes of undernutrition. And general economic growth and improved agricultural productivity will be integral parts of any strategy to reduce world hunger.

The Plan of the Book

We begin with an emphasis on definitions and facts. As the material develops, our emphasis shifts to behavioral models of society (for example, economic and demographic) and how these models relate to undernutrition. Finally, we discuss how these models can be applied in evaluating nutrition policy alternatives.

In Part 1, malnutrition is revealed as a leading killer throughout the world, with undernutrition in the developing world the main nutrition problem. Before considering the causes of undernutrition and policy alternatives to alleviate it, we examine the facts and provide answers to questions such as: What is malnutrition? How do we measure it? Who is malnourished? What are the trends? In Part 2, we look at the main causes of undernutrition and attribute these causes mainly to economic, demographic, agronomic, and health variables. In Part 3, we explore public policies that will have an impact on undernutrition. The final chapter offers some speculations about the future.

We have attempted to integrate knowledge from a number of disciplines, taking as a central premise, well articulated by Beatrice Rogers (1988b) of Tufts University, that "the solution to the world hunger problem will be achieved only through the integration of knowledge from the whole range of relevant scientific disciplines." Thus, we have drawn on nutrition science, economics, demography, biology, chemistry, health science, geography, agronomy, history, anthropology, philosophy, and public policy analysis.

To a large extent, this book is data-driven. From the data on malnutrition in Part 1, to the numbers on elasticity and population in Part 2, to the future projections about production, consumption, and prices in the last chapter, the text is larded with illustrative tables and figures. It is our hope that the data themselves will, in large measure, back up our arguments.

But the larger lesson of the book is this: policy decisions need to be made not based on plausible-sounding arguments, but on careful fact-based analysis of the likely impacts of alternative policies. Throughout the book, examples of fact-based analysis are presented to illustrate the various points. The numerous text boxes serve not only to provide visual variety, but also to bring the reader's attention to interesting related material and examples.

The second edition of this text acknowledged our debt of gratitude to, among others, Leslie Whittington, who at one time taught the course from which the book was developed. We dedicate this fourth edition to the memory of Professor Whittington, who was murdered, along with her husband and children, on September 11, 2001.

1

Introduction

> Hunger. It was prevalent everywhere. Hunger was pushed out of the tall houses in the wretched clothing that hung upon poles and lines; Hunger was patched into them with straw and rag and wood and paper; Hunger was repeated in every fragment of the small modicum of firewood that the man sawed off; Hunger stared down from the smokeless chimneys, and started up from the filthy street that had no offal, among its refuse, of anything to eat. Hunger was the inscription on the baker's shelves, written in every small loaf of his scanty stock of bad bread; at the sausage-shop in every dear-dog preparation that was offered for sale. Hunger rattled its dry bones among the roasting chestnuts in the turned cylinder; Hunger was shred into atomies in every farthing porringer of husky chips of potato, fried with some reluctant drops of oil.
>
> —Charles Dickens, *A Tale of Two Cities,* Book 1, Chapter 5

For many, the word "starvation" brings to mind the image of an emaciated infant, the victim of an Ethiopian famine. For history buffs, places such as Bengal or Ukraine may spring to mind. Images in the media of families migrating in search of food, or of babies with bloated bellies and bodies too weak to sit up, have flooded our consciousness with the horror of hunger. Each decade seems to produce its own horror stories: Famine in North Korea reached such an acute stage in late 1997 that there were reports of people eating grass and tree bark. Southern Africa was the site of an international effort to avoid widespread death from famine in 2002. One estimate puts worldwide deaths from starvation during the 1990s at 100,000 to 200,000 per year (Kates 1996). Another study estimates that as many as 576,000 Iraqi children died in the five years following the 1991 Gulf War because of food shortages resulting from economic sanctions against the country (Zaidi and Fawzi 1995).

When a person dies of hunger, what happens? Describing famine-related death, an anonymous author writing for *Time* magazine put it eloquently and succinctly:

> The victim of starvation burns up his own body fats, muscles, and tissues for fuel. His body quite literally consumes itself and deteriorates rapidly. The kidneys, liver, and endocrine system often cease to function properly. A

> shortage of carbohydrates, which play a vital role in brain chemistry, affects the mind. Lassitude and confusion set in, so that starvation victims often seem unaware of their plight. The body's defenses drop; disease kills most famine victims before they have time to starve to death. An individual begins to starve when he has lost about a third of his normal body weight. Once this loss exceeds 40 percent, death is almost inevitable. (Anonymous 1974:68)

Although the drama of famine tends to capture our attention, most hunger-related deaths do not occur in famines. They happen daily—quietly and largely unchronicled—all around the world. Figures vary, but one conservative estimate (UNICEF 2006:1), using data provided by the World Health Organization (WHO) of the United Nations, says that some 6 million children die annually from hunger. This amounts to one death every five seconds.

In previous editions of this book, we asked readers to imagine the media coverage that would ensue if a 747 jet crashed, killing all of its 220 children passengers (the deaths attributable to hunger are equivalent to seventy-five of these jet crashes every day). After the terrorist attacks of September 11, 2001, in New York City and Washington, D.C., we no longer have to imagine the media coverage, or the grief, or the universal resolve never to let it happen again. The tragedy of undernutrition fails to stir such outrage or determination. The world hunger problem is too pervasive, too commonplace, too remote, too hopeless. The Food and Agriculture Organization (FAO) of the United Nations estimates that 850 million people—about 13 percent of the world's population—suffer from undernutrition (FAO 2008:8).

Of course, as the numbers above make clear, not all hunger is fatal. Consider the following common scenario, played out again and again in developing countries. Picture a loving but poorly educated, poverty-stricken mother with several children. Food is scarce. Her youngest child, an infant boy, has not grown for months because of undernourishment, and the baby's resistance to disease has fallen dangerously low. The family's supply of water is unsanitary. The older members of the family can handle the microorganisms in the water, but the baby develops diarrhea. He loses interest in eating. He seems more willing to take liquids, so the mother removes solids from his diet. Because liquids cannot provide enough nourishment to conquer his illness, the diarrhea continues. Finally, in a desperate but seemingly logical attempt to stop the diarrhea, his mother removes liquids from his diet as well. By now, the baby is feverish, and limiting liquids accelerates his loss of fluids. Severe dehydration follows, with death not far behind.

While adult males do die of hunger during times of famine, most hunger-related deaths, whether from famine or from chronic undernutrition, occur among preschoolers. Pregnant and lactating women are also at substantial risk, although less so than children. Malnutrition in children takes many forms. Undernourished children may be crippled by vitamin D deficiency, blinded by vitamin A deficiency, or stunted by protein deficiency. But the most common form of child undernutrition results simply from a lack of sufficient calories, with disease and death too often the result.

The purpose of this book is to provide a general introduction to the world food problem, its causes, and possible ways of addressing it. Our intention is to encourage the reader to be objective and analytical. We try to avoid advocating any particular point of view, or turning a blind eye to inconvenient facts. As the past few paragraphs suggest, the pages that follow are replete with statistics and with inferences drawn from those statistics. We invite the skeptical reader to contend with us, to point out other inferences that can be drawn or other factual information that leads in a different direction.

Part 1 of this book presents some factual background: What is undernutrition? How does being undernourished affect a person? How can we determine whether a person is malnourished? What do we know about the extent of undernutrition in different periods of time and in different geographical areas?

Factors Influencing Food Supply and Demand in the Future

Part 2 deals with factors that influence the extent of undernutrition. The framework we use to outline these factors is the framework of economics: supply and demand. As we look to the future, global quality of life will hinge on whether world food supply grows faster or slower than world food demand. If supply grows more rapidly than demand, average quality of life in the world will almost certainly improve—food prices will fall, making it easier for poor people to afford an adequate diet and freeing up income for the rich to spend on other goods and amenities. By the same token, if demand outpaces supply, quality of life is likely to deteriorate.

In analyzing future prospects for food supply and demand, there are four particularly critical factors, which we refer to as the "four Ps":

- Population
- Prosperity
- Pollution (or environmental quality)
- Productivity in agriculture

The impact of population growth on food demand is obvious: more people to feed means more demand for food. Widespread economic prosperity means that more people can afford adequate diets and that people are more likely to have access to healthcare, a sanitary water supply, and education. Income levels also affect food demand: as people attain higher income levels, they tend to buy more food and a wider variety of it, including meat and animal products. So 6 billion relatively affluent people require significantly more agricultural production than do 6 billion relatively poor people.

Pollution, environmental quality, and the availability of land and water resources are critical factors in analyzing the future of agricultural production. To what extent can a population expand the area it devotes to agricultural production? Will soil erosion or water pollution result in land that is less arable or less irrigable? How will global climate change affect agricultural production?

Agricultural productivity refers to the amount of food produced on a given area of agricultural land. Regardless of environmental quality and land and water resources, the food supply will continue to grow if productivity grows quickly enough. Productivity per acre may increase when farmers apply more fertilizer, or use more labor. Productivity can also increase because of new technology, such as new seed varieties.

Population, prosperity, pollution, and productivity interact with each other in complex ways. For example:

- As population grows, urban and industrial water users compete with agriculture for scarce water.
- Population growth slows as people become more prosperous.
- As agricultural productivity increases, economic prosperity improves for the entire economy.
- Increased use of agricultural chemicals may improve productivity while harming the environment.

Government policies can influence the long-term supply-and-demand balance of food. However, the complexity of these interactions illustrates how difficult it can be to decide among various policy alternatives. Appropriate policy changes are the subject of the last section of this book.

The Main Nutrition Policy Alternatives

Part 3 focuses on policy interventions that may help alleviate the world hunger problem. For the most part, the policy interventions we examine are aimed at the factors identified in Part 2. Undernutrition can be reduced by increasing hungry people's access to food.

The most important actions a government can take to alleviate malnutrition are to promote general economic growth and to promote agricultural research. These contribute significantly to a second tier of governmental objectives: reducing population growth and maintaining or improving natural resource quality. Governmental efforts to address the world food problem by regulating prices or by redistributing food between rich and poor countries have been attempted again and again, but have been largely unsuccessful.

Part 1

Malnutrition: What Are the Facts?

MALNUTRITION IS A LEADING KILLER. In high-income countries, one variant of malnutrition—overnutrition—is the main problem. In the third world, another variant—undernutrition—is the main problem. The problem of third world undernutrition is exacerbated by secondary malnutrition—malnutrition stemming from causes such as disease.

Before considering the causes of undernutrition, and policy alternatives to alleviate it, we must examine the facts and provide answers to these questions: What is malnutrition? What are its effects? How do we measure it? Who is malnourished? What are the trends?

2

Famines

Famines get the spotlight. The television specials and the historical controversies and the sad Irish songs are about famines. But famine is a fairly small part of the world food problem. If through some magical intervention we could end famines, we would still have an enormous problem of widespread, pervasive, and permanent undernutrition. Although most of this book is focused on this pervasive and permanent condition, this chapter discusses famine.

Famines are localized, temporary, and severe food shortages. They are almost always the result of a confluence of forces that include natural disaster and poor policy response. Of course, there is a connection between the permanent state of widespread undernutrition and the crisis of famine: in countries where undernutrition is a serious and common problem, it does not take much of a natural disaster to create a famine.

Brief descriptions of present and historical famines illustrate how natural disasters and policy responses have interacted to create or exacerbate famines. These examples also illustrate some of the ways that economists have studied famines and policy approaches to famine.

The Irish Potato Famine

The Irish potato famine of the late 1840s is fairly well known in the West because it spurred a wave of Irish emigration to the United States, transforming US culture in ways that continue to be seen, especially on St. Patrick's Day, and because the famine became emblematic of Britain's repression of Ireland.

Ireland of the 1840s was a country of deep and widespread rural poverty. Nearly two-thirds of the Irish people were illiterate (Abbot 2003), and over a third lived in mud houses with a single room (Abbot 2003; Donnelly 2001:2). Per capita income was only about 60 percent of the level in Britain (Mokyr 1985).

Poverty was especially prevalent in rural areas. About two-thirds of the Irish population depended on agriculture for their livelihoods (Kinealy 2002: 18), and 40 percent of these were landless laborers (Donnelly 2001:9). Much of the land in the Irish countryside was owned in large tracts by landlords. Landless laborers acquired plots of land from the landlord and in return either worked in the landlord's fields (primarily growing grain or linen for export, or producing butter for sale in urban areas) or paid a rent to the landlord. As a result of this social structure that trapped labor in the agricultural sector, Irish agricultural workers were about half as productive as their British counterparts (Donnelly 2001:9).

In this environment of poverty, a third of Irish households depended almost exclusively on potatoes for food (a farmer in pre-famine Ireland might have consumed 12 or more pounds of potatoes per day). Potatoes have a number of advantages as a low-cost food source in Ireland: they can be grown in relatively poor soil, they yield a high number of calories per acre, and they are rich in protein, carbohydrates, vitamins, and minerals. A diet of potatoes and buttermilk (a byproduct of producing butter) provides better nutrition than a diet consisting primarily of wheat or maize.

Because of the potato-based diet, and despite the widespread poverty, "the Irish poor were amongst the tallest, healthiest and most fertile population in Europe" (Kinealy 2002:32). The cheap and nutritious potato diet served as a foundation for low-wage agriculture; cheap food exported from Ireland in turn fueled the industrial revolution in Britain. In addition, the low-wage labor provided a cushion that protected some landlords from the consequences of their inefficient farming practices.

The potato blight—a fungus that causes potatoes to turn black and rotten as they grow in the ground—had appeared in small areas prior to 1845. But the blight hit about half the crop in 1845, and destroyed nearly the entire crop in 1846, 1848, and 1849 (the 1847 crop was partially successful). Estimates of famine-related deaths range from 290,000 to 1,250,000, compared to Ireland's pre-famine population of about 8 million (Abbot 2003).

Once the severity of the potato blight was understood, a tremendous amount of attention was devoted to the appropriate "policy response": What could or should the government do? Throughout the nineteenth century, Ireland was governed by Great Britain. The choices made by the British government, and the criticisms of these choices, illustrate a philosophical or ideological debate about the appropriate relationship between government action and private action. The policy decisions fall into three categories (identified here with today's nomenclature): technology policy—what the government should do to encourage better scientific understanding of the causes and consequences of the potato blight; trade policy—what the government should do to increase food imports or reduce food exports during a time of famine; and poverty alleviation policy—what the government should do to help the poor.

The British government recognized the possibility of ending the famine with a technological fix, but its efforts never came to fruition. The government instituted a board of scientific experts to draw conclusions about how to save potatoes that had been infected by the blight. The board's recommendations involved complex chemical procedures requiring materials and training unavailable to the starving Irish masses. Even if followed, the program promised little hope of success. The government also appealed to the private sector, by promising to purchase and donate to all farmers any treatment that would kill the blight. No successful antifungal treatment was discovered until years after the Irish famine.

Trade policy in the mid-nineteeth century was the subject of intense ideological debate. The individuals in power during much of the famine were ardent proponents of free trade, or laissez faire—a policy of minimal government intervention in markets. The Irish famine put pressures on both sides of the debate over free trade. On the one hand, the famine provided the impetus for repeal of the Corn Laws that restricted imports of food into Ireland. On the other hand, exports of food from Ireland continued. The rigidity of the position in favor of free trade is reflected in an exchange between Randolph Routh, an official in Ireland who administered food distribution, and Charles Trevelyn, the permanent head of the treasury for the British government:

> Routh: "I know there is great and serious objection to any interference with these [food] exports, yet it is a most serious evil."
>
> Trevelyn: "We beg of you not to countenance in any way the idea of prohibiting exportation. The discouragement and feeling of insecurity to the [grain] trade from such a proceeding would prevent its doing even any immediate good; and there cannot be a doubt that it would inflict a permanent injury on the country." (quoted in Donnelly 2001:69)

Some scholars point to evidence of substantial reductions in grain exports to conclude that "even if exports had been prohibited, Ireland lacked sufficient food . . . to stave off famine" (Gray 1982:46). Kinealy notes that exports of other food commodities remained high, and concludes: "The Irish poor did not starve because there was an inadequate supply of food within the country, they starved because political, commercial, and individual greed was given priority over the saving of lives" (2002:116).

If trade policy illustrates the role of ideology in policy, then poverty assistance or relief policy illustrates the law of unintended consequences. Policies to help the poor during the famine were under constant discussion and revision. The government policies included such aspects as:

- Importation of grain from the United States.
- "Work houses" where poor families could live.
- Public works programs to provide incomes to the jobless.
- Soup kitchens to distribute prepared food.

The cost of these programs was financed in large part through a tax on Irish landlords. The amount of the tax depended on how many poor households or tenants lived on the landlord's property. Landlords realized that they could reduce their tax burden by evicting tenants from their farms and destroying the tenant cottages. In this way, a policy intended to help the poor actually ended up separating many poor people from their shelter and from their means of growing food.

The evictions had the impact of consolidating land holdings into larger farms. Between 1841 and 1851, the number of small farms (5 acres or less) dropped from over 300,000 to less than 100,000. The number of large farms (30 acres or more) tripled (Abbot 2003). Many landlords, having lost their rent-paying tenants, went bankrupt. Over the ensuing decades, the landlord-tenant system died out, and it became commonplace for Irish farmers to own the land that they worked.

Some of the better-off tenants who lost their homes to eviction had sufficient resources to emigrate to the Americas. During the 1840s, an estimated 1.3 million Irish people emigrated. The conditions of their voyages were harsh: perhaps as many as 40 percent died during the passage to the Americas (Abbot 2003).

Famines Created by Government Policies

Two of the worst famines in the past century occurred in centrally planned economies: the Ukrainian famine of the 1930s and the Chinese "Great Leap Forward" famine of the 1950s. If the Irish potato famine illustrates that a famine can occur in a country governed by those who embrace a laissez-faire ideology, these two famines illustrate that state socialism is not immune to poor policy choices that cause or exacerbate famine conditions.

The Ukraine Famine, 1932–1933

By the early 1930s, the urban industrial regions of the Soviet Union had been transformed into a collectively owned, centrally planned system. In 1929, Joseph Stalin introduced a policy of compulsory collectivization of agriculture. Under the collectivization plan, all of the productive assets—land, machinery, cattle, and so forth—of 25 million farmers were to be aggregated into 250,000 collective and state farms.

Even if collectivization had been enthusiastically embraced by Soviet farmers, the process of reorganization would no doubt have been awkward, and aggregate agricultural production may have decreased. There were difficulties in obtaining agricultural machinery and managing the transportation of agricultural goods, as well as in finding experienced managers to oversee the new large farms.

In addition to these problems, the collectivization process was resisted by farmers. For example, farmers slaughtered their horses and cattle rather than surrender them to the collective. This resistance was especially strong in the Ukraine, where peasants had always cultivated their own land and therefore "had a much stronger sense of private ownership and deeper feeling of freedom and independence" compared to Russian peasants (Dolot 1985:xiv).

The objectives of the central Soviet government during the early 1930s were therefore to maintain ample food supplies for the urban industrial sector while completing the transformation of agriculture into a collective system. In the Ukraine, these objectives were pursued by giving farmers a quota of grain that had to be shipped. Farmers who resisted joining the collectives were forced to ship their entire crops:

> Stepan Schevchenko was a poor farmer . . . like the rest of us [but different] from us in only one way: he had categorically refused to join the collective farm. He paid off all his taxes for the year 1932, and apparently thought that the government would leave him alone. . . . But he was overly optimistic. One day he received a requisition order demanding him to deliver 500 kilograms of wheat to the state. He delivered it in full. But no sooner had he done so when he received another order. This time they demanded twice as much wheat, . . . [even though] he had none left. . . . The officials . . . threatened him with Siberia, . . . [and] he was forced to sell everything he had of value, including his cow, to buy the order of wheat. . . . He soon received the inexorable third order: 2,000 kilograms of wheat immediately! . . . The Bread Procurement Commission paid him a visit. . . . He and his family were ordered to leave their house. . . . All that belonged to the Shevchenkos was confiscated and [became] "socialist property." (Dolot 1985:146)

The seizure of all available stocks of food in the Ukraine caused widespread starvation among the very people who produced the food in the first place. The most extreme famine conditions were suffered in the Ukraine for several reasons: the more active resistance to collectivization in the Ukraine, a nationalistic or ethnic bias against Ukrainians on the part of the Russo-centric decisionmakers in Moscow, and a desire to hide evidence that the agricultural collectivization experiment was less than perfectly successful.

An estimated 6–8 million Ukrainians died during the famine (of a pre-famine population of about 34 million) (see Mace 1984:vi). This leads some historians to compare it to Nazi Germany's use of concentration camps during the Holocaust. James Mace draws these sobering conclusions: "The Great Famine of 1932–33 is unique in the annals of human history in that it was wrought neither by some natural calamity nor even by the unintentional devastation created by warring armies. It was an act of policy, carried out for political ends in peacetime. It was deliberately man-made" (1984:i). (See Box 2.1 for a story of famine in Russia in the 1600s.)

Box 2.1 Global Climate Change and Famine in the 1600s

In 1600, Huaynaputina—a volcano in Peru—erupted. The eruption lasted for two weeks, and spewed 27 cubic miles of ash. The ash and the sulfur dioxide emitted by the volcano caused severe weather thousands of miles away in Europe. Sweden had record snowfalls. Estonia had the coldest winters in 500 years of record keeping. But in Russia, the cold weather was just part of the climate disaster. The summer of 1601 had such heavy rain that crops failed.

Boris Godunov, who had become tsar in 1598 following the death of Ivan the Great's invalid son, attempted to deal with the famine. He permitted serfs to leave their masters and move to other farms or to cities, and he financed public works (notably the erection of the Bell Tower, dedicated to Ivan the Great) by paying workers with grain drawn from government stores.

But those efforts fell short. In 1602 and in 1603, crops failed again. The crop failures resulted not just from floods, but also from drought and freezing.

During the 1600–1603 period, one-third of the Russian population died of famine. A Dutch merchant, traveling in Russia during these years, wrote: "So great was the famine and poverty in Moscovia that even mothers ate their children."

Sources: Mataev 2001; Perkins 2008.

The Chinese "Great Leap Forward" Famine, 1959–1961

The most destructive famine in terms of human lives lost occurred in China during Mao Tse-tung's "Great Leap Forward." During this period, a large number of social and economic changes were being instigated by the central government. In the agricultural sector, collectivization began in 1952 and was successful in increasing agricultural output through 1958. Beginning in 1958, the government insisted that farmers undertake untested production methods based on unorthodox (and, as the Chinese experience was to prove, flawed) science. The government forced a reorganization of smaller group or cooperative farms into larger communes (see Box 2.2). In addition, a commitment to an ideology of regional self-sufficiency led to a program that forced farm workers to divert some of their working hours to industrial production, such as in small-scale steel plants. China was also seeking to establish its economic independence from the Soviet Union, so food exports were increased during 1959 and 1960 to repay debts.

Simultaneous with these changes in government policies, poor weather conditions occurred from 1959 to 1961; conditions were especially poor in the latter two years of that period, with 15–20 percent of agricultural land being hit by natural calamity. Agricultural production, which had risen by 28 percent from 1952 to 1958, fell back below 1952 levels. Houser and Sands (2000) examined regional data to see whether higher mortality rates were uniform throughout

Box 2.2 Incentives in Chinese Agricultural Communes

One of the changes that accompanied the Great Leap Forward campaign was a change in the way group farms were organized. Economist Justin Yifu Lin (1990) examined the details of farm organization and concluded that the changes in the rules of these organizations contributed to the famine of 1959–1961.

After the revolution of 1949, many Chinese farmers voluntarily formed cooperatives of different types. In "mutual aid teams," a handful of neighboring families would share tools and draft animals and would help on each other's plots when needed. Each farm household continued to own land, tools, and animals; each family made its own decisions about what crops to plant; each family received the output from its land for consumption or sale. In "elementary cooperatives," twenty to thirty households agreed to combine into a single farm. Here, each family continued to own land, tools, and animals, but the decisions were communal, and output was shared. The sharing of output depended on how much land, tools, and animals the household contributed to the cooperative, and on how much work each household contributed. In "advanced cooperatives," the cooperative itself owned the land, tools, and animals, and members shared in output based solely on their labor contribution.

Each of these three kinds of organization provided an incentive for people to work hard. Even in the most "communal" of these organizations—advanced cooperatives—each farm household received a bigger share of the commune's output if it contributed more labor. Moreover, as Lin pointed out, the fact that membership was voluntary created an additional reason for families to work. If one family became known as a lazy household, the other hardworking households could form a new cooperative the following year, leaving out the lazy household. The initial collectivization effort in China was quite successful; agricultural output increased 28 percent from 1952 to 1958.

In 1958, the central government disbanded existing agricultural cooperatives and forced all farmers to join large communes of about 5,000 households and 10,000 acres each. Not only were these communes larger than previous cooperatives, but the rules of organization were different in two important respects. First, peasants were paid "based mainly on subsistence needs and only partly on worked performed" (p. 1236). Second, membership in the communes was no longer voluntary, and peasants were forbidden from withdrawing from the commune. With this change, it became "impossible to use withdrawal [from the commune] either as a way to protect oneself or as a means to check the possibility of shirking by the other members. . . . Since supervision in agricultural production is extremely difficult, . . . incentives to work in a compulsorily formed . . . collective must be low. A peasant will not work as hard as on the household farm. Therefore the productivity level of a collective will be lower than the level reached on the individual household farm" (p. 1242).

Lin noted that the downturn in production coincided exactly with the adoption of these new rules for cooperatives. The rules were abandoned in 1961, again coinciding exactly with the rebound in agricultural production.

the country (as we would expect if policy caused the famine), or whether there were geographical differences (as we would expect if weather problems caused the famine). They concluded that about two-thirds of the famine-related deaths were attributable to policy mistakes, and about one-third to poor weather.

An estimated 30 million people died prematurely as a result of the Great Leap Forward famine. The enormity of the problem emboldened political leaders in the provinces to abandon the policies imposed by Mao's central government. The famine can be said to have had political as well as demographic consequences, as power devolved to the provinces until the Cultural Revolution in the late 1960s reasserted the primacy of the central government. Chen and Zhou (2007) find that the famine had long-term consequences, reducing the stature, health, labor supply, and earnings of those who were exposed to it during early childhood.

A recent column by Anne Applebaum focuses attention on a recently published exhaustive history of the Great Leap Forward famine. The book (*Tombstone* by Yang Jisheng, not yet published in English) "establishes beyond any doubt that [the famine] was caused by China's misguided charge toward industrialization. A combination of criminally bad policies . . . and official cruelty . . . created, between 1959 and 1961, one of the worst famines in recorded history" (Applebaum 2008:A13).

Recent Famines: North Korea and Southern Africa

The famines described above were in a sense "national" problems. In none of the cases was there a discussion about how the world community could or should respond. Recent experiences with famine illustrate an evolving internationalist perspective.

North Korea

Since the 1990s, North Korea (the Democratic People's Republic of Korea) has suffered from famine conditions of varying intensity. The roots of this famine are found in the Cold War. During the Cold War, North Korea was closely allied with the Soviet Union and China, often playing one of those superpowers against the other to obtain increased aid. With the breakup of the Soviet Union and the end of the Cold War, this aid dried up. Not only did food donations dwindle, but domestic agriculture in North Korea, which had been designed to utilize subsidized imports of energy and fertilizer, now required radical restructuring. Drought struck in 1995, and an estimated 2–3 million people (about 10 percent of the population) died from famine-related illness in the 1994–1998 period (see Box 2.3).

Box 2.3 A Diet of Grass, Trees, and Children

Starving people resort to extreme behavior. Reports from North Korea illustrate this in striking ways. In 1997, Oxfam, a leading hunger-related nongovernmental organization, reported: "people in North Korea are eating 'wild' foods like bark, leaves and grasses, in their desperation to survive the famine that is presently sweeping the country."

In 2003, in a macabre echo of Jonathan Swift's "Modest Proposal," reports emerged of a particularly horrible form of cannibalism in North Korea. "Aid agencies are alarmed by refugees' reports that children have been killed and corpses cut up by people desperate for food. . . . Anyone caught selling human meat faces execution, but . . . one refugee said: 'Pieces of special meat are displayed on straw mats for sale. People know where they came from, but they don't talk about it. . . . If a funeral takes place during the day and the burial is performed that evening, the grave may be dug open and the body stolen before morning.'" (Nicol 2003).

The response of the Korean government undoubtedly made things worse. Prior to the famine, nearly all grain in Korea was produced on state or communal farms. The workers on these farms were given a grain ration from the harvest that was nutritionally sufficient. As grain yields began to fall because of poor weather and insufficient inputs, the central government cut peasant worker rations by over a third. The intention was to preserve more of the grain harvest for shipment to hungry urban areas. The unintended impact was to reduce the grain harvest even further, for two reasons (Natsios 1999). First, farm workers secretly "preharvested," taking grain from the communal fields before it was ready for harvest and hiding it for their own consumption. Because the grain was harvested before it was fully ripe, the grain yield was lower than it would have been if harvested according to plan. Second, farmers diverted effort from communal fields to legal and illegal private plots. These private plots were often located on poor terrain and had lower yields compared to the collective farms, but the output did not have to be shared.

The international response to the famine was substantial. Shipments from all sources averaged over 1 million metric tons per year from 1995 to 1998 (Natsios 1999) (to put this in context, the World Food Programme [WFP] estimated that in 2001, North Korea would produce 3 million metric tons of food, but would require 4.8 million metric tons to feed its population [Struck 2001]). Reliance on food donations from abroad has created pressure on the North Koreans to bring their foreign affairs and military policies into conformance with demands by countries that make the food donations. For example, in 2002 the United States suspended food aid to North Korea (see Dao 2003). Observers suspected that the suspension was a reaction to North Korea's refusal to halt its nuclear weapons program, although US officials denied that this was the reason. Support for aid to North Korea was also undermined by reports of corruption in the aid distribution system and fears that the aid was not helping those most in need.

In the years that followed, North Korea refused to allow representatives of the World Food Programme and other international relief organizations to visit much of the country. WFP efforts were reaching only 1.2 million of the over 20 million North Koreans. Then, in the summer of 2008, the government of North Korea made some concessions regarding its nuclear program and agreed to allow unrestricted access; food aid shipments began again. One observer warned: "However, the North seems headed toward a major food crisis. Two consecutive years of bad harvests and rising grain prices are making it harder for the impoverished North to import food, and assistance from South Korea and China, traditionally its most generous aid providers, is dwindling" (Sang-Hun 2008).

Southern Africa

International response also played a major role in ameliorating the impacts of famine conditions in southern Africa in 2002 and 2003. By June 2002, it was obvious that poor weather conditions (a drought followed by heavy rains during the harvest season) would devastate crop production in a large part of southeastern Africa, affecting Zimbabwe, Zambia, Lesotho, Malawi, Mozambique, and Swaziland.

The situation in Zambia was one of unusually poor weather occurring in an extremely poor country where much of the population suffers from undernutrition during "normal" or nonfamine times. Production of Zambia's staple crop—maize—fell in 2002 to a level about half that of the average output over the previous four years. Between mid-2002 and early 2003, the World Food Programme directed shipments of 130,000 metric tons of food to Zambia, helping to feed 1.7 million people. The weather conditions improved considerably for the 2003 crop, with maize production double that of 2002. Zambia faces ongoing problems of poverty, AIDS, and undernutrition, but the acute crisis of the 2002 famine has ebbed.

The weather situation in 2001–2002 in Zimbabwe was similar to that in Zambia. But in Zimbabwe, food output was suppressed further by a government program to redistribute agricultural land. During Zimbabwe's period as a colony of Britain (when it was known as Southern Rhodesia), prime agricultural land along the railway line was given over to white commercial farmers. After Zimbabwe's independence in 1980, these farmers continued to farm the land, but the government came under increasing pressure from its supporters to seize the white-owned land and distribute it to indigenous supporters of the ruling party.

In the first eighteen years of independence, less than one-quarter of the land in commercial farms was acquired by the government and reassigned to black Zimbabwean families. In 1998 the land reform process was accelerated, and in 2000 a new law was passed authorizing the government to force white farmers to give up their farms. In addition, white farmers were forced off their farms by "farm invasions" that were not legally authorized. The area planted with maize on large commercial farms fell from 163,000 hectares in 1998 to 74,000 hectares in 1999, and to 61,000 hectares in 2000. Much of the land remained in cultivation in smaller parcels by the recipients of the land reform, but yields dropped precipitously. Zimbabwe's cereal production in 2002 was less than one-quarter of the peak production of 1996.

The Zimbabwean government purchased substantial amounts of food, but additional help was needed from the international community. The ruthlessness of the land reform activities and the impression that the Zimbabwean government was turning a blind eye to lawless farm invasions created a reluctance on the part of developed-country donors to provide assistance. Ultimately, the World Food Programme did make substantial amounts of food aid available (in 2002–2003 amounting to about 500,000 metric tons of food). However, as of the summer of 2003, nearly half of Zimbabwe's population continued to face the specter of famine.

Over the next five years, little changed. The actions of the Zimbabwean leadership were widely condemned (Lynch 2008). Rural violence and land seizures continued (Dixon 2008). Price controls kept farm prices low, discouraging food production (Lynch 2008). Food production per capita has since remained at the low levels of 2002–2003 (FAOSTAT various years). The World Food Programme reports: "The overall production of maize in 2008 is estimated at 575,000 metric tons—28% lower than last year's already low levels . . . [and] leaving a shortfall of around . . . 1 million tons of maize, which needs to be imported. . . . The number of people requiring food assistance possibly [will rise] to a peak of 5.1 million [out of a population of 11 million] in the first quarter of 2009" (WFP 2008).

* * *

These recent examples illustrate the difficulties in making a clear distinction between famine and the problem of permanent undernutrition. Famine is a shock or a disaster—it is a disturbance to the normal condition. There is both good news and bad news in this. The good news is that the needed policy response to famine is a temporary (though urgent) response. The bad news is that existing institutional frameworks (laws, power structure, social mores) often lack the flexibility to respond, and thus exacerbate the impact of the natural disaster. The failed responses of the governments of North Korea and Zimbabwe have drawn out famine conditions so long that it is hard to characterize them as "temporary."

Famine and Disaster Relief

In times of famine, domestic and international disaster relief agencies respond as best they can to get food into the hands of the starving. Sometimes, the best efforts achieve excellent results.

Singh (1975) described a successful effort to address potential famine in India in 1967. The district of Bihar suffered severe drought—the worst in a century. Crop production fell to half the normal amount. An effort by the national government created jobs for food, employing 18,000 people and paying them with food drawn from government stocks. As a result, there was no starvation.

More recently, there was a huge international outpouring of charity in response to the tsunami of 2004: aid donations from governments and private individuals and organizations amounted to $7,100 per affected person (Telford, Cosgrave, and Houghton 2006). And the World Food Programme can point to successes, as in the Zambian emergency described above. The emergence over the past decade of the WFP as an organizing entity for international assistance to famine-struck countries has resulted in better coordination of disaster assistance to fight famine.

But sometimes the relief can be a mixed blessing. The authors of a well-documented and detailed Oxfam report warn that poorly supervised or uncontrollable distribution of food aid can do more harm than good. To quote one example, a field worker helping out in a drought-relief food aid program (where the food handouts were supposed to be free of charge to the recipients) said:

> In Haiti we had . . . a problem of theft and mishandling. In [a] town . . . fairly near to us and very badly hit by drought, the magistrate [appointed mayor] was known to sell PL480 [international relief] food for $7.00 a 50 pound bag. At other times the CARE food distributors were so desperate that they would just throw bags of food off the truck and drive on, so that the food would go to the strong and the swift. (Jackson and Eade 1982:9)

When disaster creates the need for assistance, but thc local food supplies are adequate, supplying emergency food relief can be counterproductive. It depresses the local price of food, in turn depressing the income of the local farming community, and may lead to other socially undesirable results (see Box 2.4).

In areas where food aid has become a yearly occurrence, "dependency" can develop. Thielke (2006) describes the situation in Kenya:

> Only as a result of regular food deliveries by the United Nations World Food Program can the people in Kenya's harsh northern region avoid starvation. . . . According to [a German expatriate], the aid has become a threat to the entire region. "People just hang about, waiting for food deliveries from Nairobi," he says. "They are losing traditional survival strategies. The area is pretty much made up of only dependent recipients of charity." [A] Swiss biologist . . . who headed the ICIPE ecology research institute in Kenya for

Box 2.4 When Food Aid Is Not Needed

Tony Jackson and Deborah Eade

Imported food may not be necessary at all, despite a major disaster, and its arrival may do more harm than good. The classic example of this comes from Guatemala where the earthquake in 1976 killed an estimated 23,000 people, injured over three times as many and left a million and a quarter homeless. The earthquake occurred in the middle of a record harvest. Local grain was plentiful and the crops were not destroyed but left standing in the fields or buried under the rubble but easy to recover.

During the first few weeks, small consumer items—salt, sugar, soap, etc.—were in short supply and temporarily unavailable in the shops. Some of these small items, such as salt, were lost when the houses collapsed. People expressed a need for these food items in the short period before commercial supplies were resumed. However, during that year, about 25,400 tons of basic grains and blends were brought in as food aid from the US. A further 5,000 tons of US food aid already stored in Guatemala were released and supplies were also sent in from elsewhere in the region.

Catholic Relief Services (CRS) . . . field staff objected to the importing of food aid but they were overruled by their headquarters in New York. . . . The Co-ordinator of the National Emergency Committee of the Government of Guatemala asked voluntary agencies to stop imports of food aid. . . .

Finally, the Government of Guatemala invoked a presidential decree to prohibit imports of basic grains from May 1976 onwards. Yet after this decree, quantities of food aid were still imported in the form of blended foodstuffs. One article refers to these blends as "basic grains in disguise."

Field staff and local leaders identified three negative results. Firstly, they considered that food aid contributed to a drop in the price of local grain that occurred soon after the earthquake and continued throughout 1976. As to the need for basic grains, a peasant farmer explained: "There was no shortage. There was no need to bring food from outside. On the contrary, our problem was to sell what we had."

. . . An OXFAM–World Neighbors official reported: "Virtually everyone in the area is selling more grain this year than he does normally. Furthermore, emergency food shipments have drastically curtailed demand for grains. Thus the prices of the farmers' produce have plummeted."

Later, the then Director of CRS in Guatemala was to tell the New York Times: "The general effect was that we knocked the bottom out of the grain market in the country for nine to twelve months."

. . . The second negative effect of the continuing supply of free food was to encourage the survivors to queue for rations instead of engaging in reconstruction or normal agricultural work.

Thirdly, it brought about a change in the quality and motivation of local leadership. The OXFAM–World Neighbors official, quoted above, noted: "Immediately after the earthquake, we tended to see the same leaders whom we'd seen before the earthquake—people [with] a high degree of honesty and personal commitment to the villages. But gradually . . . I began seeing fellas who I

(continues)

Box 2.4 continued

knew were totally dishonest. They'd go into the different agencies and . . . say that theirs was the most affected village in the Highlands, and they'd get more food. So largely because of the give-aways, the villages started to turn more to leaders who could produce free things like this, whether they were honest or dishonest, rather than to the leaders they'd been putting their trust in for years. With larger and larger quantities of free food coming in, there are increased incentives to corruption. . . . Groups that had worked together previously became enemies over the question of recipients for free food."

Source: Extracted from Jackson and Eade 1982:9–11.

many years agrees. "Northern Kenya is hopelessly overpopulated," he says, "and this overpopulation is the result of years of western aid shipments to areas that really don't have the potential to feed so many people." . . . [A former health commissioner explains:] "There are more and more people and there is less and less water. . . . The desert is expanding. . . . In the past people slaughtered their animals for food during difficult times. . . . But ever since the World Food Program began feeding us, hardly anyone does this anymore. Everyone just waits for the next delivery."

General Studies of Famines

Reutlinger and colleagues (1986:27) list the groups most likely to fall victim to famine:

- Small-scale farmers or tenants whose crops have failed and who cannot find other employment in agriculture (the Wollo in Ethiopia in 1973).
- Landless agricultural workers who lose their jobs when agricultural production declines (Bangladesh 1974) or who face rapidly rising food prices and constant or declining wages (the great Bengal famine of 1943).
- Other rural people, including beggars who are affected by a decline in real income in the famine regions (almost all famines).
- Pastoralists who get most of their food by trading animals for food grains; their herds may be ravaged by the drought, or animal prices may collapse relative to food-grain prices (the Harerghe region of Ethiopia in 1974 and the drought-stricken Sahel in 1973).

Martin Ravallion (1997), in a thorough review of the economics literature regarding famines, cites examples of failed policy responses to famine: "The

British government's . . . non-intervention in food markets during [nineteenth-century] famines almost certainly made matters worse. . . . At the other extreme, . . . food procurement policies implemented [by] . . . the Soviet Union . . . resulted in severe famine in the Ukraine in the 1930s" (p. 1225). Ravallion draws the following lessons for policies in response to famine:

- *Better governance.* Greater democratization and freer flow of information in a society make it more difficult for a government to ignore famines.
- *Early warning and rapid response.* Policy interventions are likely to be more effective if they take place before famine conditions are firmly entrenched.
- *Increased aggregate food availability.* Policies to increase the total amount of food available in famine areas include food aid, discouragement of hoarding in private or public storage, and encouragement of domestic food production.
- *Distribution policies.* "Although the case is often strong for increasing aggregate food availability during a famine, food handouts need not be the best form of intervention from the point of view of minimizing mortality. Cash or coupon payments to potential famine victims can provide more effective relief than the usual policy of importing and distributing food" (p. 1230).
- *Stabilization policies.* "An effective but affordable . . . stabilization policy in famine-prone economies . . . will probably combine buffer stocks and . . . a relatively open external trade regime" (p. 1233). Buffer stocks are programs in which the government purchases food in periods when it is plentiful and sells food from its stocks when shortages occur.
- *Other policies.* Ravallion argues that there are potential synergies between policies to address famines and other policies to spur economic development, including credit programs, improved infrastructure, and assignment of property rights.

For students of history, a review of famines in Europe and Asia can be found in Mabbs-Zeno 1987.

3

Malnutrition Defined

One common definition of *malnutrition* is "overconsumption or underconsumption of any essential nutrient." This chapter is devoted to exploring this definition.

Four Types of Malnutrition

Internationally famous nutritionist Jean Mayer (1976) identifies four types of malnutrition: (1) overnutrition, (2) secondary malnutrition, (3) dietary deficiency or micronutrient malnutrition, and (4) protein-calorie malnutrition.

Overnutrition

When a person consumes too many calories, the resulting condition is called *overnutrition.* Overnutrition is the most common nutritional problem in high-income countries such as the United States, although high-income people in low-income countries also suffer from this type of malnutrition. The diet of the world's high-income people is usually overladen with calories, saturated fats, salt, and sugar. Their diet-related illnesses include obesity, diabetes, hypertension, and atherosclerosis. Overnutrition is a serious problem. For example, Martorell (2001) reports research showing that in Latin America, 35 percent of women age 15–49 are either overweight or obese. Another report shows the percentage of the Chinese population who are obese rose from 16.4 percent in 1992 to 22.6 percent in 2002 (Luo, Mu, and Zhang 2006). The problem of overnutrition in the developing world is being addressed through an international effort coordinated by the World Health Organization (WHO 2002).

Secondary Malnutrition

When a person has a condition or illness that prevents proper digestion or absorption of food, that person suffers what is called *secondary malnutrition.* It is called "secondary" because it does not result directly from the nature of the diet, as do the other types of malnutrition, which are termed "primary." Common causes of secondary malnutrition are diarrhea, respiratory illnesses, measles, and intestinal parasites. The following mechanisms cause secondary malnutrition:

- *Loss of appetite (anorexia).*
- *Alteration of the normal metabolism.* For example, when the body shifts some of its attention to fighting infection, among other things, production of disease-fighting white blood cells may be increased and body temperature raised.
- *Prevention of nutrient absorption.* For instance, diarrheal infections irritate the lining of the gastrointestinal tract, creating difficulty in absorbing nutrients and at the same time causing it to shed contents before full digestion has had time to occur.
- *Diversion of nutrients to parasitic agents.* Parasites such as hookworms, tapeworms, and schistosome worms rob the body of nutrients it would otherwise retain (Briscoe 1979; Martorell 1980; Brooker, Hotez, and Bundy 2008).

Public health measures such as providing sanitary human-waste disposal and clean water are especially important in reducing secondary malnutrition. A worldwide effort has increased the number of people with access to safe drinking water by 1.6 billion since 1990. In 2006, 84 percent of the population of all developing countries had access to an improved drinking water source (United Nations 2008:42).

Low-income people in developing countries are at risk for undernutrition (insufficient calories), which is commonly exacerbated by secondary malnutrition (especially parasites and diarrhea caused by unsanitary drinking water). Because of the strong link between the two, undernutrition and secondary malnutrition are commonly grouped together and called, simply, *undernutrition.*

Dietary Deficiency or Micronutrient Malnutrition

A diet lacking sufficient amounts of one or more essential micronutrients, such as a vitamin or a mineral, results in dietary deficiency. Although a deficiency of any micronutrient can become a serious problem, most nutritionists are primarily concerned about deficiencies in vitamin A, iodine, and iron. In recent years, this category of malnutrition is usually referred to as *micronutrient malnutrition* or *micronutrient deficiency.* The three most significant micronutrient deficiencies are for vitamin A, iodine, and iron:

- *Vitamin A.* Deficiency of vitamin A can cause "xerophthalmia," or night blindness. It is also associated with increased mortality from respiratory and gastrointestinal disease. One study suggests (Gopalan 1986) that vitamin A supplements could reduce deaths of children age 6 months to 5 years by 23 percent. Recent research suggests that vitamin A plays a role in maintaining the immune system and in fighting cancer.

- *Iodine.* Iodine deficiency causes goiter and leads to a reduction in mental abilities. Babies born to iodine-deficient mothers can suffer from "cretinism," which can result in learning disabilities in children. One study indicates that even mild iodine deficiency can reduce intelligence quotient (IQ) by 10–15 points (Ma, Wang, and Chen 1994). Iodine deficiency is the greatest single cause of preventable brain damage and mental retardation. The WHO estimates that one-third of the world's people live in iodine-deficient environments (WHO 2007).
- *Iron.* Iron deficiency, or anemia, causes reduced capacity to work, diminished ability to learn, increased susceptibility to infection, and greater risk of death during pregnancy and childbirth. More than 40 percent of people in developing countries are estimated to suffer from iron deficiency (WHO 2001).
- *Other micronutrients.* In recent years, nutritionists have become concerned about zinc deficiencies. Zinc now appears to be effective in increasing the growth of very young children; it also reduces the incidence of diarrhea and assists in absorption of other micronutrients. Other diseases caused by micronutrient deficiencies include rickets (soft bones), caused by vitamin D deficiency; scurvy, caused by vitamin C deficiency; and beri-beri and pellagra, caused by deficiencies in B-vitamins. Some research (Tang et al. 1993) indicates that the risk of contracting AIDS is substantially lower among those who consume very high levels of niacin (a B-vitamin), vitamin A, and vitamin C. Recent research has emphasized the importance of folic acid in reducing the incidence of spina bifida (Erickson 2002) and the importance of vitamin D in fighting cancer, diabetes, and heart disease (Holick 2004).

When compared with underconsumption of proteins or calories, the problem of underconsumption of micronutrients appears relatively easy to solve. The missing elements are inexpensive, and programs to provide them are relatively easy to initiate. In the United States, iodized salt (salt to which iodine has been added) protects us from iodine deficiency, specially fortified milk provides vitamin A (and vitamin D), and iron pills (or vitamin supplements) are a common source of iron. There are many examples of successful micronutrient interventions in the developing world:

- In Guatemala, dietary anemia was greatly reduced in a rural community after the inhabitants were persuaded to substitute iron cooking pots for aluminum pots.

- Also in Guatemala, fortification of sugar with vitamin A has been effective, and experiments are now under way to fortify sugar with iron.
- In Brazil, a school's drinking water was fortified with iron, creating a noticeable improvement in students' iron levels at a cost of about 15 cents per student per year.
- In China, iodine deficiency was treated by dripping potassium iodate solution into the water of an irrigation canal. Iodine intake by people in the area increased significantly.

Other interventions to combat micronutrient malnutrition include making vitamin pills available to the population and cultivating plants that contain one or more micronutrients. The World Bank estimated in 1994 that it would cost about $3 per year to meet a person's entire needs for vitamin A, iron, and iodine. (Other micronutrients are not so inexpensive: an antioxidant formula of vitamin E, beta-carotene, and vitamin C may cost $60 per person per year.)

Despite the apparently easy solution to these vitamin and mineral deficiencies, the problems have remained surprisingly persistent, though progress is being made. For iodine: "Out of the 130 countries . . . there are only 47 countries where [deficiency] still remains as a public health problem [in 2006], compared to 54 in 2004 and 126 in 1993" (WHO 2007:14). Under the multinational "Micronutrient Initiative," in 2002, 300 million children under the age of 5 were given a high-dose vitamin A supplement. The United Nations Children's Fund (UNICEF 2007:9) reports that the proportion of children receiving the full (two-doses per year) treatment for vitamin A deficiency increased from 16 percent in 1999 to 72 percent in 2005. According to the World Health Organization's most recent estimates, anemia affects about 25 percent of the world's population, compared to 37 percent of the world's population in the early 1990s, although it still affects almost 50 percent of preschool-aged children and 40 percent of pregnant women (WHO 2001, 2008).

Policies to reduce micronutrient malnutrition will continue to be important because of the huge impact on public health that relatively small expenditures can have. As we shall see in the next chapter, the health impacts of micronutrient deficiencies are significant. However, the solutions to these deficiencies are more likely to come from fortification programs such as those described above, and less likely to come from additional food consumption and production.

Protein-Calorie Malnutrition

The underconsumption of calories or protein—known as *protein-calorie malnutrition* (PCM) or *protein-energy malnutrition* (PEM)—is a problem that can only be solved by increasing the amount of food that an individual eats. A person suffering from PCM is not obtaining enough of the protein or calories needed for normal growth, health, and activity. PCM hardly ever occurs in families with enough income to satisfy their basic needs for food, shelter, clothing, and heat; it is found predominantly in low-income countries where poverty is widespread.

In extreme forms, PCM manifests itself as the potentially fatal nutritional disorders known as *kwashiorkor* and *marasmus* (see Boxes 3.1 and 3.2). Kwashiorkor is most likely to be encountered among populations where the diet is

Box 3.1 Kwashiorkor

Eleanor Whitney and Eva Hamilton

The word *kwashiorkor* originally meant "the evil spirit which infects the first child when the second child is born." It is easy to see how this superstitious belief arose among those Ghanaians who named the disease. When a mother who has been nursing her first child bears a second child, she weans the first and puts the second on the breast. The first child soon begins to sicken and die, just as if an evil spirit had accompanied the new baby into the world and set out to destroy the older child. What actually happens, of course, is that protein deficiency follows soon after weaning, for while breast milk provides these children with sufficient protein, they are generally weaned to a protein-poor gruel.

. . . By the time children with kwashiorkor are four, their growth is stunted; they are no taller than they were at two. Their hair has lost its color; their skin is patchy and scaly, sometimes with ulcers or open sores which fail to heal. Their bellies are swollen with edema; they sicken easily, and are weak, fretful, and apathetic.

The swollen belly of the kwashiorkor child is due to edema; blood protein is so low that fluid leaks out into the body. Since the child is too weak to stand much of the time, the fluid seeks the lowest available space—in this case the belly. The picture of such a child is one of skinny arms and legs and a greatly swollen belly. On first glance you might think the child is fat, but if the fluid could be drawn off, his true condition would be revealed: he is actually a wasted skeleton, just skin and bones.

The body follows a priority system when there is not enough protein supplied to meet all its needs. It abandons its less vital systems first. When it cannot obtain amino acids enough from dietary sources, the body switches to a metabolism of wasting, and begins to digest its own protein tissues in order to supply the amino acids needed to build the most vital internal proteins and keep itself alive. Hair and skin pigments (which are made from amino acids) are dispensable and are not manufactured. The skin needs less integrity in a life-and-death situation than the heart, so its maintenance ceases and skin sores fail to heal. Many of the antibodies are also degraded in order that their amino acids may be used as building blocks for heart and lung and brain tissue. Children with a lowered supply of antibodies cannot resist infection and readily contract dysentery, a disease of the digestive tract. Dysentery causes diarrhea, leading to rapid loss of those nutrients—including amino acids—which these children may be receiving in food. Thus dysentery worsens the protein deficiency, and the protein deficiency in turn increases the likelihood of a second or third or tenth attack of dysentery.

Source: Reprinted from *Understanding Nutrition* by Whitney, Hamilton, and Rolfes, copyright 1990 by West Publishing Company. Used by permission of Wadsworth Publishing Company.

Box 3.2 Marasmus
Eleanor Whitney and Eva Hamilton

When children are almost totally deprived of food, they cannot obtain the energy necessary to maintain their body systems, much less that necessary for growth. Marasmus, a wasting disease, results. Invariably, protein deficiency occurs with this condition, as available protein is used not to build body protein but to supply energy (which takes priority). As a result, the marasmic child has many, though not all, of the same symptoms as the child with kwashiorkor.

Marasmic children are wizened little old people in appearance, just skin and bones. They are often sick because their resistance to disease is low. Their hearts are weak, and all their muscles are wasted. Their metabolism is slow. They have little or no fat under their skin to insulate against cold. Their body temperatures may be subnormal. The experience of hospital workers with victims of this disease is that their primary need is to be wrapped up and kept warm. They need love, since they have often been deprived of maternal attention as well as food.

Unlike the kwashiorkor child, who is fed milk until weaning, the marasmic child may have been neglected from early infancy. The disease occurs most commonly in children from six to eighteen months of age in all the overpopulated city slums of the world. Since the brain normally grows to almost its full adult size within the first two years of life, marasmus impairs brain development and so may have a permanent effect on learning ability.

Marasmus also occurs in adults in countries where calorie deficiency is prevalent.

Source: Reprinted by permission from *Understanding Nutrition* by Whitney, Hamilton, and Rolfes, copyright 1990 by West Publishing Company. Used by permission of Wadsworth Publishing Company.

heavily based on cassava (as in West Africa) or on plantains (as in parts of Latin America and southern Uganda). These particular plant foods are almost completely devoid of available protein, and children who are weaned on them are at high risk for severe protein deficiency. Marasmus is most likely to occur under conditions of extreme poverty in which children are weaned onto a gruel that contains modest amounts of protein, but where available food is nutritionally inadequate. Marasmus is thus more common among the poorest populations of the world, such as those of Ethiopia, Nepal, and Bangladesh. Without warmth, loving care, and expert medical attention, children with kwashiorkor or marasmus can die quickly.

Calories and Protein

Because protein-calorie malnutrition is a major source of nutrition-related disease, and because reducing it requires increasing food consumption, much of this book will focus on this form of malnutrition. These two elements—calories and protein—are both derived from food and are both necessary for growth, health, activity, and survival. But their nutritional roles are different. More important, calories and proteins are derived from foods in different ways, and a diet must be carefully planned if it is to be adequate in both.

Nutritional Role of Calories and Proteins

Calories are a measure of the energy contained in food (see Box 3.3). The body obtains energy from carbohydrates (e.g., sugar and starch) and fats (e.g., oil and butter). Calories are used by the body to provide energy for various needs:

- Involuntary functions such as breathing, blood circulation, digestion, and maintaining muscle tone and body temperature.
- Physical activity.
- Mental activity.
- Fighting disease.
- Growth.

The human body makes the millions of different proteins that it needs from some twenty amino acids, which are the building blocks of the body's proteins. Proteins function in ways other than providing a source of calories:

- They are necessary for building the cells that make up muscles, membranes, cartilage, and hair.
- They carry oxygen throughout the body.
- They carry nutrients into and out of cells and help assimilate food.
- They contribute to the development of antibodies that fight disease.
- They work as enzymes that speed up the digestive process.

Simple organisms such as yeasts and algae can synthesize almost all of the amino acids they need. But humans cannot synthesize or make sufficient

Box 3.3 Calories and Kilocalories

In physics, chemistry, and engineering, a calorie is the amount of heat energy required at sea level to raise the temperature of 1 gram of water by 1 degree centigrade. A kilocalorie (abbreviated Kcal) is the energy it takes at sea level to raise the temperature of 1,000 grams of water (a kilogram—also a liter) by 1 degree centigrade.

Nutritionists always quote their data in kilocalories, but unfortunately they commonly shorten the word to "Calories"—capitalized to indicate kilocalories as distinguished from calories. But the convention is not always observed. Regardless, in nutrition literature the lowercased word *calorie* usually means kilocalorie.

quantities of some amino acids. To live, therefore, we must consume enough *essential amino acids* (those that the body cannot produce in sufficient quantities). Of the approximately twenty amino acids needed, nine are essential. Because the body cannot manufacture them, we must consume them as part of our diets. All of these amino acids are found in eggs, milk, and meat—proteins from these dietary sources that contain all the essential amino acids are referred to as *complete proteins.*

Proteins from vegetable sources tend to be deficient in at least one of the essential amino acids and are therefore "incomplete." To understand the importance of this for diet, we need to understand the way that the human body takes the amino acids from protein in food and constructs protein for the body's use. Protein molecules in food are composed of fixed proportions of amino acids. A balanced protein is one in which the essential amino acids appear in the same ratio to each other as the body needs them for constructing its protein molecules. (Information about dietary reference intakes for essential amino acids can be found in National Academy of Science 2002:10-65–10-66.)

If you consume more amino acids than you need, your body cannot use them for making proteins; instead, it burns the amino acids for energy. If you consume less of an amino acid than you need, a portion of other amino acids goes to waste for want of the "matching" part needed to manufacture protein molecules. Whitney and Hamilton provide a delightful analogy:

> Suppose that a signmaker plans to make 100 identical signs, each saying LEFT-TURN ONLY. He needs 200 L's, 200 N's, 200 T's and 100 each of the other letters. If he has only 20 L's, he can make only 10 signs, even if all the other letters are available in unlimited quantities. The L's limit the number of signs that can be made. The quality of dietary protein depends first on whether or not the protein supplies all the essential amino acids, and second on the extent to which it supplies them in the relative proportions needed. (1977:92)

Complete proteins can be obtained from animal products such as meat, milk, eggs, and cheese, but not from most grain and vegetable products. However, complete proteins can be obtained by combining different *types* of grains and vegetable products. For example, a person living on a diet of beans might be getting plenty of the amino acid lysine, but might be deficient in the amino acids methionin and cystine. A person living on a diet of wheat might be getting enough methionin and cystine, but not enough lysine. A diet that combines wheat and beans would give the appropriate balance of these amino acids. (See Scrimshaw and Young 1976 for more detail.)

The Chemical Process of Producing Dietary Calories and Protein

The energy that the human body uses comes from the sun, but the body cannot absorb solar energy directly and use it for physical growth and activity. The chemical process by which plants transform solar energy into a form of energy (in plants) that can be absorbed by humans or animals is known as *photosynthesis:* carbon dioxide plus water plus radiant energy from the sun yields (in the presence of chlorophyll) a carbohydrate plus oxygen plus water. Using enzymes, a plant can rearrange the carbon, hydrogen, and oxygen atoms of the carbohydrate, or sugar, to form starch or fat.

The conversion of sugar into starch is energy-efficient: practically all the energy in the original sugar can be released in burning the starch made from it. The conversion of sugar to fat, however, is about 77 percent energy-efficient, so we can say that most of the energy in the original sugar can be released in burning the fat that is made from it.

Making amino acids is not so easy. Like carbohydrates and fat, amino acids contain carbon, hydrogen, and oxygen, but they also contain nitrogen; and nitrogen in the form that can be used to build amino acids is scarce. In the process of photosynthesis, plants can use carbon dioxide (CO_2) straight from the atmosphere. But in making amino acids, nitrogen cannot be used as N_2, the form in which it is found in the atmosphere. It has to be converted to more complex forms, such as ammonia (NH_3), before the plant can use it.

Converting atmospheric nitrogen to usable nitrogen is called *nitrogen fixation.* It can be done in a commercial fertilizer plant, where energy is combined with some raw organic stock such as naptha. Or it can be done by nature, which provides three other ways of fixing nitrogen: lightning storms, nitrogen-fixing bacteria, and blue-green algae. For the most part, the nitrogen fixed by blue-green algae is not available to plants useful to man. Leguminous plants, such as beans, peas, and alfalfa, provide a suitable environment on their roots for nitrogen-fixing bacteria and thus have an extra boost of nitrogen available.

The fixed nitrogen taken up by plants becomes available in the food chain to make amino acids and, ultimately, proteins. Plants and animals must spend energy (use up sugar) to synthesize amino acids. They spend still more energy to recombine these amino acids into enormous molecules of protein. As a result, the energy available from burning a protein is substantially less than the energy used to produce that protein. By contrast, the amount of energy released in the burning of a carbohydrate or fat is closer to the amount of energy it took to make it in the first place.

Which Is the Bigger Problem: Protein Deficiency or Calorie Deficiency?

The chemistry involved in producing calories and proteins helps explain why proteins are scarce relative to carbohydrates and fats. The relative scarcity implies that putting adequate protein into our diets will be more expensive than consuming adequate calories. And, generally speaking, we do pay a premium for protein-rich foods. For instance, hamburger, which is a richer source of protein than rice, costs more per pound than rice (and in contrast, rice provides about twice as many calories per pound compared to hamburger).

Because protein is more expensive than carbohydrates, the question arises: Should we show more concern about the protein intake of poor people than we do about their calorie intake? Until the 1970s, nutritionists believed that protein was the central concern.

No doubt, protein deficiency can and does occur. However, the experience of the past thirty years demonstrates that calorie deficiency is a more widespread problem. For example, a study (Gopalan 1970) of 15,000 Indian preschool children in the 1960s found that 35 percent showed evidence of both calorie and protein deficiency, 57 percent showed evidence of calorie deficiency but not protein deficiency, but virtually none showed evidence of protein deficiency without an accompanying calorie deficiency.

Recent data on food availability also indicate that calorie deficiency is likely to present a bigger problem than protein deficiency. The Food and Agriculture Organization (2004b) showed that in developing countries of Africa, food available per capita would provide 2,407 calories and 60 grams of protein per person per day in 2003. If we compare this to the average requirements listed in Table 3.1 for adults, we see that available protein is substantially higher than the requirements for all groups, but that available calories are lower than the requirements for most adult men and for pregnant and lactating women.

Nevin Scrimshaw, a renowned advocate of the importance of protein in the diet, puts it like this: "It is true that adult protein needs are met by most traditional developing country diets when they are consumed in sufficient quantity to meet normal energy needs" (1988). A review of studies of protein intake in developing countries reached the same conclusion: "At the habitual levels of intake, only 75 to 80 percent of the . . . requirement for energy is satisfied, while more than the current safe level for dietary protein is supplied" (Rand, Uauy, and Scrimshaw 1984: sec. 5.1).

How Much of a Nutrient Is Enough?

Before describing what can happen when a person does not get enough of a nutrient, we need to explore what "enough" means. Of course, no single standard applies to everyone. Each person is different. Growing children have different nutritional needs than mature adults. Men have different nutritional needs than women. Active people have different nutritional needs than sedentary people. Box 3.4 describes how energy requirements can be different for two people of the same sex and age. And some people are just different for unexplainable reasons. For example, we all know someone who can eat and eat yet never gain weight; that person has a higher metabolism, or a higher daily need for calories, than the rest of us.

Table 3.1 Dietary Reference Intakes for Calories and Protein

		Calories	Protein (grams per day)	
Category	Age (years) or Condition	Estimated Average Requirement (EAR) (Kcal per day)	Estimated Average Requirement (EAR)	Recommended Daily Allowance (RDA)
Males	0–1/2	570	9.1[b]	9.1[b]
	1/2–1	743	9.9	13.5
	1–3	1,046	10.6	13.0
	4–8	1,742	15.2	19.0
	9–13	2,279	27.4	34.0
	14–18	3,152	44.5	52.0
	19–30	3,012[a]	46.2	56.0
	31–50	2,852[a]	46.2	56.0
	51+	2,607[a]	46.2	56.0
Females	0–1/2	520	9.1[b]	9.1[b]
	1/2–1	676	9.9	13.5
	1–3	992	10.6	13.0
	4–8	1,642	15.2	19.0
	9–13	2,071	28.1	34.0
	14–18	2,368	38.3	46.0
	19–30	2,345[a]	37.6	46.0
	31–50	2,253[a]	37.6	46.0
	51+	2,081[a]	37.6	46.0
Pregnant	1st trimester	+0	+12	+25
	2nd trimester	+300	+12	+25
	3rd trimester	+300	+12	+25
Lactating	1st 6 months	+500	+22	+25
	2nd 6 months	+500	+22	+25

Sources: Protein RDA: National Academy of Science 2002. Protein EAR: National Academy of Science 2003: summary tab. 1. Calories EAR: Thomson-Wadsworth 2003 (developed from National Academy of Science 2002).

Notes: a. For 19-year-olds: 3,067 calories for males, 2,403 calories for females. For each year above age 19, subtract 10 calories for males and 7 calories for females. Calories shown are for individuals 24.5 years old, 39.5 years old, and 65 years old for the three age groups.

b. For infants under 6 months, adequate intake of protein is reported.

Nutritionists have several different ways of describing the nutritional needs of a group. The recommended daily allowance (RDA) shows the nutrient level at which 97–98 percent of the group will be adequately nourished. RDAs are useful as targets for an individual: an individual who gets his or her RDA of a nutrient can be quite confident of obtaining an adequate level of that nutrient. However, "RDAs are not useful in estimating the prevalence of inadequate intakes for groups" (National Academy of Science 2002:13-7). Nor are they appropriate as targets for the *average* intake of a group. For these purposes, a better measure is the estimated average requirement (EAR) (for calories, the EAR is often referred to as the estimated energy requirement). The EAR is the daily intake amount that will be adequate for half the individuals in the group.

Box 3.4 How Many Calories Do I Need Each Day?

Table 3.1 shows the dietary reference intakes for people in different groups. But not everyone in each group is the same. To obtain an absolutely accurate calculation of how many calories you burn each day, you would need to undergo a complicated and expensive clinic assessment. But you can make a good estimate by using the following formulas (applicable to people over age 19 who are not excessively over- or underweight).

For men:

$$661.8 - 9.53 \times \text{AGE (yrs)} + \text{PAC} \times [(15.92 \times \text{WEIGHT (kg)}) + (539.6 \times \text{HEIGHT (m)})]$$

For women:

$$354.1 - 6.91 \times \text{AGE (yrs)} + \text{PAC} \times [(9.36 \times \text{WEIGHT (kg)}) + (726 \times \text{HEIGHT (m)})]$$

Plug in your age, weight (1 pound = 0.4536 kg), and height (1 inch = 0.0254 m).

The physical activity coefficient (PAC) takes on one of four values, depending on how active you are:

Daily Exercise Level	PAC	Example of Daily Exercise
Sedentary	1.00	None
Mildly active	1.12	30 minutes of moderate walking
Active	1.27	30 minutes of moderate walking, 25 minutes of moderate bicycling, and 40 minutes of tennis
Very active	1.45	45 minutes of moderate cycling, 25 minutes of jogging, and 60 minutes of tennis

Depending on height and weight, calorie requirements can vary widely for different people in the same PAC category. Consider two 21-year-old men. One is 5 feet 3 inches tall, weighs 104 pounds, and has a sedentary lifestyle. His daily calorie requirement is about 2,000. The other is 6 feet 3 inches tall, weighs 199 pounds, and has an active lifestyle. His daily calorie requirement is about 4,000.

Thus the EAR is always lower than the RDA (compare the EAR and RDA values for protein shown in Table 3.1). The RDA for calories is no longer reported, since it might encourage people to consume too many calories (Trumbo et al. 2002).

The difference between the EAR and the RDA is shown in Figure 3.1. This bell-shaped curve shows the distribution of people in a certain group—say, men age 19–22—according to how much of a certain nutrient—say, calories—each

Figure 3.1 Distribution of Nutrient Requirements in a Typical Population of Healthy Individuals

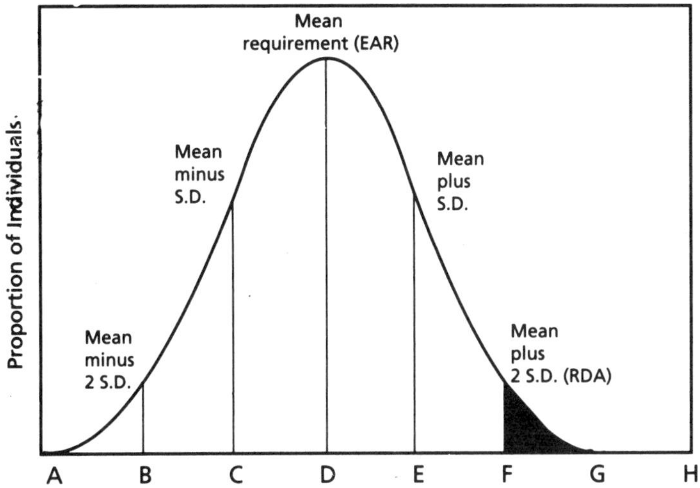

Source: From "The Requirements of Human Nutrition," by Nevin S. Scrimshaw and Vernon R. Young. Copyright © 1976 by Scientific American, Inc. All rights reserved.

Note: The curve is bell-shaped. S.D. = standard deviation.

person needs. Some men in this age group need few calories (near point A); some need a lot of calories (near point G). On the curve, point D is the mean (or average) requirement. This is the EAR. The degree to which the bell curve is spread out around the mean is measured by a statistic called the *standard deviation* (S.D.)—a large standard deviation means that the curve is a relatively flat, wide bell; a small standard deviation means a tall, skinny bell. An intake level at point F is calculated by taking the standard deviation, multiplying it by two, and adding the result to the mean. Statisticians have shown that point F, calculated in this way, will show the following characteristics: nearly all (97.5 percent) of the people in the group will have a requirement that is less than F; only 2.5 percent of the population will have a requirement greater than F. The RDA for a nutrient in a given age-sex group is set at point F by adding two standard deviations to the mean requirement for the group.

RDAs and EARs for micronutrients (vitamins and minerals) can be found in National Academy of Science 2003. A general description of the process of determining calorie and protein requirements is contained in WHO 1985.

4

Measuring Undernutrition

In the previous chapter, we identified underconsumption of calories, micronutrients, and protein as the most serious types of malnutrition. We now turn to the question of how we can measure the extent of undernutrition in a group. In this country, or city, or village, is undernutrition rare or is it commonplace? Because resources for coping with undernutrition are scarce, they must be spent wisely, and this requires that we accurately identify where the problem is most serious.

Measuring the Nutritional Status of the Individual

Measurements of undernutrition in a group will be based on determinations of nutritional status, individual by individual. The common methods of direct assessment of the nutritional status of an individual are: clinical, biochemical, dietary, and anthropometric. Each method has shortcomings and each may result in a somewhat different assessment of the nature and extent of nutritional disorders.

Clinical Assessment

Clinical assessment of nutritional status relies on the examination of physical signs on the body that are symptomatic of nutritional disorders (Jelliffe 1966). Kwashiorkor, for instance is accompanied by loss of pigment in the hair (it often turns reddish) and by edema (swelling) of the ankles. Examples of clinical assessment can be found in a national Philippine nutrition survey (Philippines Naional Science and Technology Authority 1984) that examined schoolchildren for goiter (a generalized swelling at the base of the neck above the collarbone) and found that 3.1 percent of children over the age of 9 suffered from iodine deficiency. More recently, researchers examined the extent of vitamin A deficiency among preschool children in Mali by giving eye tests and diagnosing xerophthalmia (Schemann et al. 2003).

Correct identification of nutritional disorders using physical signs depends not only on the training and skill of the clinician but also on how significantly the signs are manifested in the particular individual. Clinical signs are difficult to quantify and are usually obvious only in the advanced stage of the disease; for this reason, clinical assessment can be used in only the most severe and specific types of nutritional disorders.

Biochemical Assessment

Biochemical assessment requires examination of bodily fluids such as blood or urine for the complex metabolic changes that accompany nutritional disorders. In the Philippine nutrition survey, over 14,000 blood samples were drawn and analyzed. From these samples and based on standards established by the World Health Organization, 27 percent of the population were found to be anemic. The highest rates of anemia occurred in those younger than 1 year (51 percent) and in pregnant women (49 percent).

Biochemical tests provide an accurate indication of short-term nutritional problems (especially micronutrient deficiencies), but their complexity and expense are impediments to widespread use in field surveys. Imagine the problems associated with persuading a sample of over 14,000 individuals living throughout the Philippines to submit to having blood drawn, and then transporting these samples to appropriate laboratory apparatus for analysis before they deteriorated in the heat.

Dietary Assessment

Dietary surveys are often employed to assess nutritional status. Two approaches are used: (1) dietary recall, in which the subject is asked to remember what he or she ate, such as during the past twenty-four hours or the past seven days; and (2) dietary record, in which someone records the amount of food consumed at mealtimes, often by weighing it. Both methods have their advantages and drawbacks.

Dietary recall (see Box 4.1) is advantageous in that researchers can interview subjects when they are not expecting to be surveyed; they are less likely to adjust their consumption because of the survey. Yet it is often difficult for people to remember exactly what they or members of their family ate during the past twenty-four hours, much less during the past week. And estimates of quantity consumed are particularly prone to error in recall surveys.

When keeping food records, especially if every portion of food must be weighed, the cook tends often to simplify the diet to make record keeping easier (Quandt 1987). Subjects in food record-keeping surveys are especially likely to adjust their diets so that things will "look better" to the surveyor, especially if the surveyor is in the household for the sole purpose of making and recording the measurements.

In both types of surveys, measuring the quantity of food consumed by breast-fed babies presents difficulties. And in both cases, seasonal variation in

Box 4.1 Use of a Dietary Recall Survey to Evaluate Nutrition in Indonesia

A study of the prevalence of undernutrition among pregnant women in Indonesia (Hartini et al. 2003) used a dietary recall survey. Four hundred fifty women in their second trimester of pregnancy were interviewed repeatedly (up to six times) and asked to recall their food intake in the twenty-four hours preceding the interview. The results were averaged to estimate each woman's typical dietary intake. From this, daily nutrient intake (how many calories, how many grams of protein, etc.) was estimated using coefficients about the nutritional composition of the food items. The daily nutrient intake for each woman was compared to the estimated average requirement for the nutrient in Indonesia. The study ran from 1996 to 1998, spanning the Indonesian financial crisis that began in 1997. Therefore the study was able to analyze the impact of the crisis on nutrition. Urban consumers suffered the most from the crisis: protein intake declined from 52 grams per person per day before the crisis to 44 grams during the crisis; iron intake dropped from 17 grams to 12 grams. In rural areas, nutrient intakes held steady, or even increased, during the crisis.

consumption may confound the data unless appropriate adjustments are made. For instance, the Philippine survey was conducted from February through May, a time when access to the countryside is easier because of the relative absence of monsoon rains and typhoons. The retail price of tomatoes, for example, is typically 250 percent higher in November than in April. Similarly, the price of rice tends to be low from February to May, while the price of corn tends to be high (Philippines Ministry of Agriculture 1981a, 1981b). The unadjusted survey data thus probably overestimated annual consumption of tomatoes and rice, which were in abundant supply during the survey period, and probably underestimated the consumption of corn.

In either case (dietary recall or dietary record), results of the survey can be used to determine amounts of various nutrients consumed; these amounts can then be compared to a dietary standard appropriate to the particular country being observed, to determine nutritional status.

Dietary assessment is useful in studies relating consumption and income, or in determining food allocation patterns within the family. But care must be taken in interpreting the results of such surveys. Food intake is not always a good index of nutritional status. For instance, secondary malnutrition can substantially degrade the nutritional status of an otherwise appropriately fed individual.

Anthropometric Assessment

Anthropometry is the science of measuring the human body and its parts. It serves as the most commonly used measure of nutritional status. To understand how anthropometric assessment works, we must first understand how human physical growth and development are responsive to nutritional status.

Impact of Undernutrition on Physical Growth and Development

Because calories and protein are necessary for the growth of the human body, the most obvious physical effects of undernutrition manifest in the individual's size. In considering this impact, we should discriminate between *acute undernutrition* and *chronic undernutrition.* Acute undernutrition is short-term, severely inadequate food intake, such as one might see during famine or war. The human body can recover from a relatively short bout of acute undernutrition; people who lose weight in a famine can gain it back when the famine ends. In some cases, children whose growth has slowed during a famine will regain their normal size when the famine ends. Chronic undernutrition refers to long-term inadequacy of protein or calories or both, and causes physical effects even when the inadequacy is moderate. The physical effects of undernutrition manifest in various ways.

Low Height-for-Age, or Stunting

An individual whose height is low for his or her age is said to be *stunted.* Such a person may have suffered from chronic undernutrition at some time during their growth years. Low height-for-age is a symptom of past undernutrition; the person may or may not be undernourished today. The link between nutrition (and indirectly, income) and height is the focus of a small group of economic historians (see Fogel 2004; Steckel 1995; Komlos 1989; for a nontechnical overview, see Bilger 2004; for an example of how anthropometric history is practiced, see Box 4.2).

Evidence of the link between nutrition and height is found in a large number of sources. Average height in nations increases over time as nutritional status improves (see Tanner 1977:349; Steckel 1995). At a local level, in 1981 in the Indian village of Bagbana, studied by coauthor Phillips Foster and colleagues (unpublished), 61 percent of adult sons were taller than their fathers by an average of 3.1 centimeters (1.25 inches). John Strauss and Duncan Thomas (1998) present data on height increases over time for the United States, Côte d'Ivoire, Brazil, and Vietnam. In the United States, the average man born in 1930 was 4 centimeters taller than the average man born in 1910, reflecting a marked improvement in nutrition; but the increase from 1930 to 1950 was only 1 centimeter, perhaps because there was less room for improvement in nutritional status.

Other data show that in periods of war or economic depression, when nutritional status declines, average height stagnates or declines. In Vietnam, average

Box 4.2 An Anthropometric Mystery

In 1850, the average American man was 3 inches taller than the average Dutchman. A century later, the average Dutchman was 1 inch taller than the average American (Steckel 1995:1919). Since the 1950s, Europeans have grown an extra three-quarters of an inch per decade, and some Asian populations several times more, compared to Americans, who haven't grown taller in the past half century. Even the Japanese—once the shortest industrialized people on the planet—have nearly caught up with Americans, and Northern Europeans are now three inches taller than Americans (Bilger 2004).

Economic historians attribute the early American advantage in height to abundant sources of protein from farm animals and wild game. But why hasn't the United States maintained its "lead"? Per capita incomes have grown in Europe, but they've grown in the United States too. Food consumption has increased in Europe (calories per capita per day grew by 14 percent in the Netherlands between 1961 and 2003), but it has increased faster in the United States (by 30 percent over the same time period).

Could the explanation have something to do with immigration? Perhaps the US height average is brought down by immigration of shorter adults from Central America and Asia. But analysts have checked, looking only at heights of native-born men of European ancestry, and the height disparity still exists (Bilger 2004).

Students of anthropometric history are now considering other explanations. Perhaps growing income inequality in the United States is creating a new, shorter economic underclass. John Komlos of the University of Munich is examining heights of different demographic groups (college-educated whites, for example) to see if that provides any evidence about this possibility. Richard Steckel of Ohio State University notes that, relative to Europeans, Americans grow more slowly in infancy and adolescence; perhaps, then, inferior pre- and postnatal care, or too much junk food in teenage diets, is stunting growth in the United States (Bilger 2004).

height increased with birth year from 1920 to the late 1950s, when the Vietnam War started. From the late 1950s through the 1970s there was little change in height, reflecting the impact of war on nutritional improvement. In what amounts to a kind of naturally occurring controlled experiment, Alderman, Hoddinott, and Kinsey (2003) examined the impact of the 1982–1984 drought in Zimbabwe. In comparing the height-for-age of children before and after the drought, they found an average loss of stature of 2.3 centimeters.

Low Weight-for-Height, or Wasting

People who are currently undernourished are thin. In scientific jargon, they exhibit low weight-for-height, or *wasting*. This fact is known to anyone who has tried to lose or gain weight by changing their food intake. Low weight-for-height is a symptom of current undernutrition.

Low Weight-for-Age, or Underweight

Low weight-for-age is a symptom of either past or present undernutrition. Individuals with low weight-for-age are referred to as *underweight.*

Fat Composition of the Body

The human body stores excess calories as fat. In periods when calorie intake exceeds requirements, fat is added, and in periods when calorie intake is deficient, fat is depleted. No single part of the body represents the fat content of the whole body. Even a careful calculation of total body density (by weighing people and then comparing their weights to the weight of the water they displace when submerged in a tank) may not estimate accurately the percentage of the entire body that is fat. Nevertheless, the mid-upper arm is considered to be fairly representative of the body as a whole and is used as an indicator of nutritional status.

Nature vs. Nurture, or Heredity vs. Environment

Of course, nutrition is not the only determinant of body size. We observe differences from country to country, or ethnic group to ethnic group. For example, a 1967 study of body weights in the United States and seven Latin American countries showed that boys in all eight countries started life with similar average weights, but that there was considerable variation among countries in average adult weights, with US adults being heavier than adults in the Latin American countries by as much as 20 kilograms (44 pounds) (US White House 1967). Foster's unpublished 1981 study of the Indian village of Bagbana shows that median adult height in that village is about the same as the height of a person in the shortest 5 percent of Americans. Even more striking, the average weight of a 70-year-old woman in Bagbana is about half the average weight of her counterpart in the United States. Are Indian women underweight? Are US women overweight? Are these weight and height differences attributable to nutrition?

Some ethnic groups have reputations as being unusually small or unusually large. Weiner (1977:419–420) reports a number of studies indicating that protein composition of a group's diet may explain size. The meat-eating Sikhs of northern India are noticeably larger than the vegetarian Madrassi of southern India. In Kenya, the farming Kikuyu tribe live on cereals, tubers, and legumes. The Masai tribe—also in Kenya—are nomadic cowherds. Their diet includes meat and milk; in addition, Masai get high-protein nutrition from blood drained from living cattle. On average, Masai men are 7.5 centimeters (3 inches) taller and 10 kilograms (22 pounds) heavier than Kikuyu men.

Although studies of twins show that "heritability of height" is about 0.9 (out of a maximum of 1.0), Bowles and Gintis (2002) help put this number into perspective:

> It is sometimes mistakenly supposed that if the heritability of a trait is substantial, then the trait cannot be affected much by changing the environment. The fallacy of this view is dramatized by the case of stature. The heritability of height estimated from U.S. twin samples is substantial (about 0.90 . . .). Moreover there are significant height differences among the peoples of the world: Dinka men in the Sudan average 5 feet and 11 inches—a bit taller than Norwegian and U.S. military servicemen and a whopping 8 inches taller than the Hadza hunter-gatherers in Southern Africa. . . . But the fact that Norwegian recruits in 1761 were *shorter* than today's Hadza shows that even quite heritable traits are sensitive to environments. What *can* be concluded from a finding that a small fraction of the variance of a trait is due to environmental variance is that policies to alter the trait through changed environments will require non-standard environments that differ from the environmental variance on which the estimates are based. (2002:13)

Stephenson, Latham, and Jansen compared growth data from US children, privileged African children, and underprivileged African children, and concluded that ethnic differences were less important than other factors as determinants of growth: "Poverty, poor food intakes, infectious and parasitic diseases, and other environmental factors combine to prevent children from realizing their growth potential" (1983:53). There are, of course, genetic influences that lead to differences of body size, and especially of stature, but it seems that in prepubertal children, heredity is a much less significant cause of below-average growth than are other factors.

The contributions of nature and nurture can be understood most easily by considering a person's height. Nature (or genetics, or heredity) determines the person's maximum potential height; nurture (or environment, and to a considerable extent nutrition) determines the degree to which the person attains the maximum potential height. So if we had two groups—one that had been well nourished during childhood and one that had been poorly nourished during childhood—we very likely could find some individual in the poorly nourished group who was taller than some individual in the well-nourished group. The taller person in this example had a greater genetic height potential, but didn't fully reach it, while the shorter person had a lesser genetic height potential, and did fully reach it. But if we were to measure *average* heights in the two groups, the average for the well-nourished group would be higher: though the two groups, on average, may have had about the same height potential, the well-nourished group, on average, would have achieved more of that potential. The fact that environmental influences are major determinants of the average body-size characteristics of a population provides the basis for anthropometric measurement of malnutrition.

Anthropometric Assessment of Nutritional Status

As discussed in the previous section, human physical growth and development are responsive to variations in calorie and protein intakes. Because of this, measurements of the human body, when compared to a reference standard, provide clues as to protein and calorie nutrition. This is especially true during youth, when growth is so rapid, but to some extent is also true during adulthood, when various dimensions change gradually with aging. Thus anthropometry can be used to suggest the protein and calorie nutritional status of adults as well as children.

An anthropometric assessment of an individual's nutrition status has three steps: (1) take measurements of the individual; (2) compare that individual to a "reference group"—the measurements of other individuals of the same age, sex, and ethnic group, for example; and (3) make a determination of nutritional status based on that comparison.

How Is Anthropometry Used?

As noted above, nutrition is not the only influence on bodily growth and development. Growth and development can also be affected by such variables as genetic disposition, health, hormonal abnormalities, or deficiencies in micronutrients (e.g., zinc deficiency is associated with poor growth). Because anthropometry does not provide particularly useful clues as to dietary deficiencies (micronutrient shortages), it is not normally used to measure them. Nevertheless, because nutritional status is by far the most common variable influencing bodily growth and development, and because deficiencies in calories and protein are by far the most important causes of subnormal growth and development in the third world, anthropometry is generally assumed to provide strong clues as to protein and calorie nutrition.

Despite substantial genetic variation in height and weight within any population, the growth trajectories of infants and very young children are much more uniform than those of older children. Therefore, anthropometric measurements can be particularly accurate in identifying nutritional problems at very early ages.

In developed countries, anthropometry is used most often as a measure of overnutrition, which is the most common nutritional problem among high-income people. Adults who are trying to keep their weight down are using the anthropometric measure weight-for-height as a reference standard, although they seldom think of their activities in such technical terms.

Since 1966, when the World Health Organization published a monograph by D. B. Jelliffe titled *The Assessment of the Nutritional Status of the Community,* which provided a set of standardized measurements, anthropometry has become the most widely used tool in nutritional assessment. The most common measures used for anthropometric assessment of nutritional status are height, weight, and a combination of arm circumference and skin-fold thickness. They are usually used in conjunction with sex and age (as in weight-for-age or height-for-age) and sometimes in combination with each other (as in weight-for-height).

Low cost and ease of data collection combine to make this method so popular. Training time for field surveyors is minimal. The measurements are relatively easy to take, and the tools required are simple to use: scales, tape measures, measuring boards, and skin-fold calipers. Intelligent amateurs can be trained in a matter of days to take accurate measurements. By contrast, training technicians to recognize and diagnose undernutrition from clinical symptoms can take weeks or months.

From the point of view of accuracy, age data may be the biggest problem in using anthropometry. Adults in developing countries often have only a vague idea of how old they are, and illiterate mothers sometimes have difficulty telling the age of their children.

Capital requirements for anthropometry are minimal. Compared to the laboratory apparatus involved in biochemical analysis, anthropometric tools are inexpensive. And anthropometry measures the results of nutrition (the size and shape of the body) rather than nutritional inputs (as in dietary intake). Because of its quantitative nature, anthropometry can be used to judge varying degrees of undernutrition—not just its presence or absence. And by using a mix of anthropometric indexes, researchers can make judgments about a nutritional disorder (protein or caloric) as well as the time dimension of the disorder (past or present).

Reference Groups

A person's size can only be evaluated by comparing it to the sizes of other people (the *reference population*). These comparisons lead to statements such as "Tommy is shorter than average," meaning that Tommy is shorter than the average person in the reference population. Or we might say, "Adele is in the 95th percentile of height," meaning that 95 percent of people in the reference population are shorter than Adele.

The purpose of a reference population is to provide a standard against which the growth and maturation of particular individuals can be judged. By observing how much one person deviates from others of a similar status (age and sex, for example), we can draw inferences about the subject's nutritional status.

Similarly, whenever you try to guess someone's age, you use the person's body size and configuration (among other clues). That is, your life experience teaches you what to expect in the way of variations in height and weight as people grow and develop, and from this (and other clues, such as amount of gray hair and wrinkled skin) you deduce age.

In setting up a reference population, it is important to control for as many variables as possible that might influence the observed variables. For example, if an individual is much shorter than the reference group, we don't learn much if the individual is 4 years old and the reference group is composed of teenagers. Therefore, reference groups are chosen so that we can compare persons of approximately the same age and of the same sex.

The curves in Figure 4.1 show how the reference population can be described. The top part of the figure shows the distribution of height (length) among girls age 3 years and younger. The bottom part of the figure shows the

Figure 4.1 Height-for-Age and Weight-for-Age Percentiles for Girls Age 0–36 Months, United States, 2000

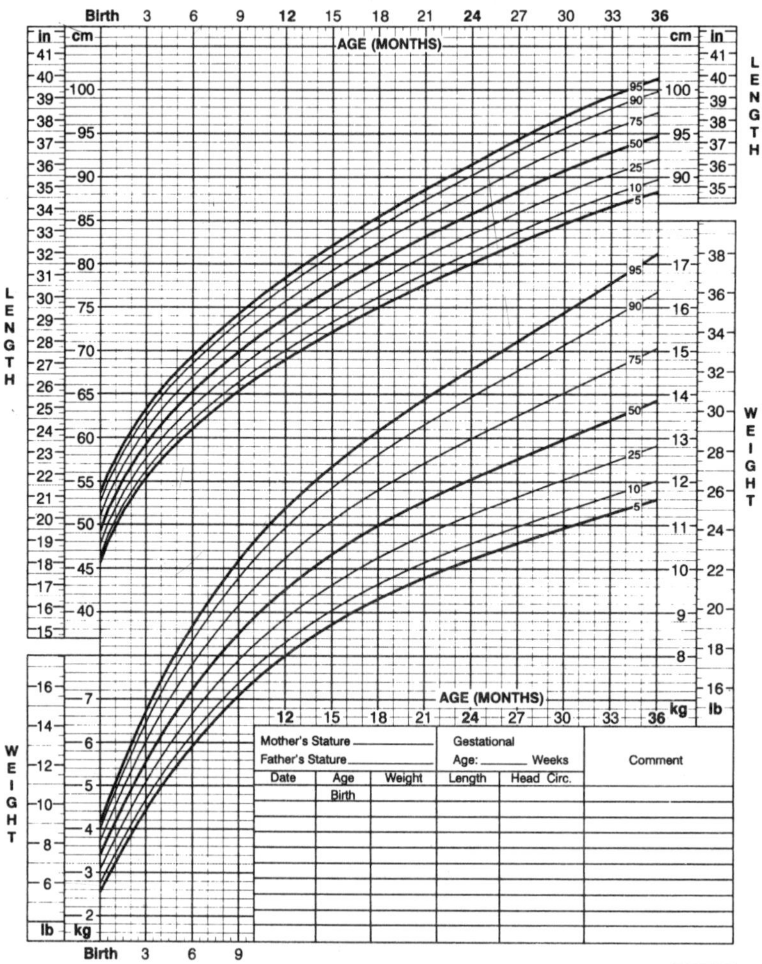

Source: Centers for Disease Control, http://www.cdc.gov/nchs/data/nhanes/growthcharts/set1 clinical/cj411018.pdf.

weight-for-age of the same reference group. A 24-month-old girl who weighs 12 kilograms is at the 50 percentile (in other words, 12 kilograms is the "median weight" for girls this age). A 24-month-old girl who weighs 10.2 kilograms is at the 5th percentile, meaning that about 95 percent of girls in the reference population weigh more than 10.2 kilograms. A similar set of curves can plot weight against height and serves as a description of the reference population for the measure weight-for-height.

As noted above, there may be systematic genetic differences among people of different ethnic backgrounds. Where those differences are substantial, reference populations for different ethnicities or countries-of-origin may be necessary. Table 4.1 compares Philippine and US weight-for-age reference standards. Average weights of Philippine and US children are the same for the first two years of life, but by the sixth year, Philippine children are 10 percent lighter than US children.

Classification System for Identifying Undernutrition

Choosing a classification system for undernutrition involves selecting an appropriate anthropometric measure or measures in conjunction with an appropriate set of criteria (deviations from the reference standard) for identifying undernutrition, or various levels of it. If we have measured the body-size characteristics of the individual, and know the body-size characteristics of the reference population, what can we conclude about nutrition by comparing the individual to the reference group? When is the individual's height or weight so low that it indicates undernutrition? No matter which system is chosen, the aim is to identify those most in need of help, either because they are undernourished or because they are in danger of becoming so.

Table 4.2 shows five commonly used classification systems. The first three can be used with any measure (height-for-age, weight-for-age, or weight-for-height). The whole point of using a reference population is to decide, roughly, whether an individual's height or weight is abnormally low for that group. If an individual's height or weight is below a certain amount (referred to as a *cutoff*

Table 4.1 Philippine and US Weight-for-Age Reference Standards for Preschool Children

	US Weight-for-Age (median kilograms)				Philippine Weight-for-Age as Percentage of
Age (years)	Boys	Girls	Average	Philippine Weight-for-Age (median kilograms)	Philippine Weight-for-Age as Percentage of US Weight-for-Age
1	11.7	10.7	11.2	11.2	100
2	13.5	12.7	13.1	13.1	100
3	15.4	14.7	15.0	14.7	98
4	17.6	16.7	17.2	16.1	94
5	19.4	19.0	19.2	17.8	92
6	22.0	21.3	21.6	19.4	90

Sources: US Department of Health and Human Services 1987; Philippines National Science and Technology Authority 1984:218.

Table 4.2 Anthropometric Classification Systems for Comparing Individuals to Reference Groups

Classification System	Range	Malnutrition Category
World Health Organization	Z-score of –1 to –2	Mild
	Z-score of –2 to –3	Moderate
	Z-score lower than –3	Severe
"Road to Health" (RTH)	Greater than 80% of median	Normal
	60–80% of median	Mild-to-moderate
	Less than 60% of median	Severe
Gómez	Greater than 90% of median	Normal
	75–90% of median	Mild
	60–75% of median	Moderate
	Less than 60% of median	Severe
Body mass index (BMI)	Greater than 18.5	Normal
	17–18.5	Grade I
	16–17	Grade II
	Less than 16	Grade III
Mid-upper arm circumference (MUAC)	Less than 18.5 cm	Moderate
	Less than 16 cm	Severe

Source: Cogill 2003:42, 72–73.

point), then he or she may be presumed to have a nutritional problem. The three most commonly proposed candidates for cutoff points are each based on deviation from the *median score*. "Median" means that half the reference population is larger and half is smaller. Thus, if the median weight of a reference population is 90 pounds, then half the people in that group weigh more than 90 pounds, and half weigh less. The three candidates for cutoff points are: (1) percentile, (2) percentage of the median, and (3) standard deviation unit. For the purposes of explanation, we use weight as the variable being measured and compared to the reference population. The terms can also be applied to any other variable, such as height.

Percentile indicates the percentage of the reference population who weigh less than the individual. Thus, if the individual's weight is at the 30th percentile, this indicates that 30 percent of the reference population weighs less than the individual. The 50th percentile is the same as the median. In discussing the reference-groups chart in Figure 4.1, we noted that a 24-month-old girl who weighs 10.2 kilograms is at the 5th percentile.

We calculate percentage of the median by dividing the individual's weight by the median weight for the reference population. The 10.2-kilogram girl in our example is at 85 percent of the median weight, since median weight for girls this age is 12 kilograms.

Figure 4.2 illustrates how one classification system—the Gómez classification—can be used to describe the nutritional status of different age groups in an Indian village. Examination of this figure yields some interesting speculations. For example, middle-aged to older Indian villagers are mostly moderately to severely undernourished. Because middle-aged to older people in the United States are overweight, optimal nutrition must lie somewhere in between. Figure 4.2 also illustrates a phenomenon commonly found in low-income third world populations: severe undernutrition tends to decline during the first six years of life. This is partly because, as children grow, they more easily command their share of the family's food resources (children who are being weaned are notoriously difficult to feed) and partly because the weakest individuals have already succumbed to malnutrition and disease. With the early deaths of weaker children, the remaining population of older children "looks better."

Standard deviation unit, or "Z-score," is a statistical measure of dispersion away from the mean. For example, a weight measurement that is two standard deviation units below the mean will be approximately at the 2nd percentile, meaning that only 2 percent of the reference population will weigh less than this amount. For measures of height and weight, experience shows that two standard deviations below the mean is usually fairly close to 75 percent of the median. Experience also shows that the average weight (and height) of population groups is usually quite close to the median (Krick 1988:326–328). Our reference chart (Figure 4.1) does not tell us the standard deviation of the sample, but an estimate allows us to calculate a Z-score of –1.65 for the 10.2-kilogram girl we have been using as an example. This means that a 24-month-old girl

Figure 4.2 Gómez Classification by Age, Bagbana Village, India, 1981

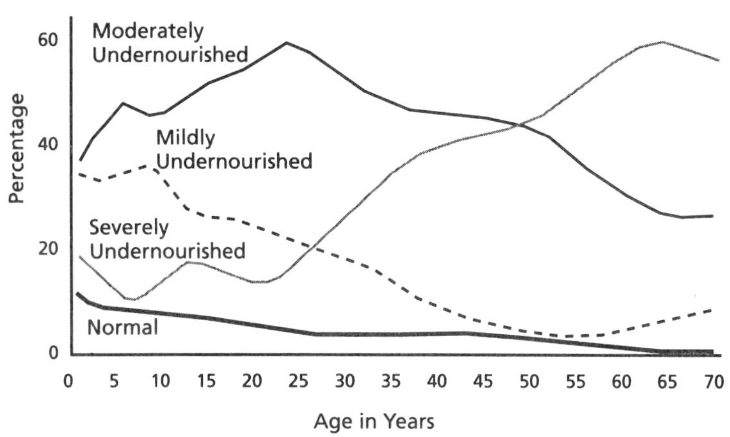

Source: Adapted from Dever 1983:88.

weighing 10.2 kilograms is 1.65 standard deviations below the mean weight for girls of this age.

As mentioned above, these three categorizations can be used for any anthropometric measure. Waterlow and colleagues (1977) suggest that a more complete picture of a person's nutritional status can be obtained by combining weight-for-height (which identifies present undernutrition) and height-for-age (which identifies past undernutrition). The beauty of this system is that it is likely to correctly identify individuals who are presently at risk for undernutrition, and to eliminate from consideration those who merely have been undernourished in the past.

The remaining two categorizations shown in Table 4.2 are body mass index and mid-upper arm circumference.

Body mass index. As a measure of undernutrition (or overnutrition) among adults, the FAO and WHO now regard body mass index (BMI) as the most suitable. BMI is calculated as weight in kilograms divided by the square of height in meters. According to the FAO (1996c), a BMI below 18.5 is regarded as lower than normal and thus indicative of undernutrition. In China, 12.5 percent of adults have a BMI below this critical level, as do 48.6 percent of adults in India. A person who is 6 feet tall and has a BMI of 18.5 would weigh about 135 pounds. In the United States, the National Institutes of Health (NIH) have set the ideal ranges for BMI at 21 to 23 for women and 22 to 24 for men.

Mid-upper arm circumference. One of the more interesting features of child growth and development is that, while the body of a well-nourished child is gaining in length and weight from one to five or six years of age, his or her mid-upper arm circumference (MUAC) remains essentially the same. Using this observation, Shakir (1975) developed a simple screening device to identify severely undernourished preschoolers. It consists of a tape that is wrapped around the child's mid-upper arm, lightly enough to not compress the skin but firmly enough to fit exactly. The reference standard is 16.5 centimeters. A circumference of greater than 14.0 centimeters (85 percent of standard) is considered normal. A circumference of 12.5–14.0 centimeters (76–85 percent of standard) is classed as undernutrition. A circumference of less than 12.5 centimeters (less than 76 percent of standard) is classed as severe undernutrition. (Table 4.2 shows MUAC nutritional classifications for adults.) The tape is inexpensive and can be made by hand, if necessary. If the tape is coded with culturally appropriate colors, even illiterate health workers can do the classification.

The system is highly accurate for severe cases of undernutrition and moderately accurate for mild cases. Nevertheless, because of its simplicity and because it works on preschoolers without age data, it has become an important screening tool. As Cogill says:

> Mid-upper arm circumference (MUAC) is relatively easy to measure and a good predictor of immediate risk of death. It is used for rapid screening of acute malnutrition from the 6–59 month age range (MUAC overestimates rates of malnutrition in the 6–12 month age group). MUAC can be used for screening in emergency situations but is not typically used for evaluation purposes. MUAC is recommended for assessing acute adult undernutrition and for estimating prevalence of undernutrition at the population level. (2003:12)

Measuring Nutritional Status of Large Groups

In addition to determining the nutritional status of individuals, policymakers also need to learn about the nutritional status of large groups. They need to know the extent of undernutrition on a continent or subcontinent, in a country, in a state or region of a country, or among demographic or ethnic groups.

Drawing Inferences from a Sample

If we can measure the nutritional status of a number of representative individuals in a country or in a group, we can draw inferences about the extent of malnutrition in the whole group. The science of statistics deals with the problem of how to determine the characteristics of an entire population (in this case, the extent of malnutrition in a country or among a group) by observing the characteristics of a sample from that group. So if we have anthropometric evidence categorizing, say, 11,000 children under the age of 5 in Nigeria, which tells us that 34 percent of these children are undernourished, this might allow us to deduce that 34 percent of all children in Nigeria are undernourished.

Mortality or Disease Rates

A second way of drawing inferences about the incidence of malnutrition in a country or other large group is to examine aggregate data about effects of malnutrition and to draw inferences from those data. For instance, low birth weights or high infant mortality rates in a country or region are assumed to indicate high rates of undernutrition. Aggregate data such as these can provide a good initial approximation of where substantial numbers of undernourished people are likely to be found. Because undernutrition and poor health often occur together, aggregate data on morbidity can also provide indirect measures of the incidence of undernutrition in a country or region.

Among the developing countries, the Philippines conducts among the most careful nutrition surveillance of its people. According to its 1982 national nutrition survey (Philippines National Science and Technology Authority 1984), 52 percent of the country's population was afflicted with roundworms (*ascariasis*).

Hookworm infection was noted among 19 percent of the male population aged 13 to 59. Of the 14,785 subjects examined, 69 percent tested positive for some kind of parasite. With morbidity data such as these, one could expect to find substantial rates of undernutrition in the Philippines.

As we will see in the next chapter, high infant and child mortality rates, or high morbidity rates, are suggestive of high rates of undernutrition in a population, but these variables do not measure nutritional status directly. Rather, undernutrition is inferred from the nonnutritional aggregate data.

Food Balance Sheets or Food Availability Measures

Another indirect approach to identifying regions or countries with nutritional problems is to look at aggregate nutrient intake or average per person nutrient intake. These are available on a country-by-country basis in food balance sheets published annually by the FAO. The food balance sheet shows sources and uses of over a hundred separate food items on an annual basis. Listed sources of food include beginning stocks, production, and imports; uses include ending stocks, exports, animal feed, and human consumption. The term "balance sheet" refers to the fact that total supply of each food item equals the total use: sources and uses are in balance. A condensed balance sheet for India in 2001 is shown in Table 4.3. This version aggregates many individual food items into groups, and also aggregates several sources and uses.

Once human consumption is estimated for every food commodity in a country, the food consumed can be converted into calorie and nutrient measurements, and per capita consumption figures can then be derived. If a country's per capita consumption turns out to be below amounts recommended by nutritionists, we have good cause to assume that a substantial block of its population is undernourished.

Measures derived from food balance sheets—in particular, calories per capita per day—have become perhaps the most widely used measures of malnutrition. The big advantage of such measures is that they are readily available from the FAO for almost every country for a number of years or time periods (see FAO 2004b).

However, the measures are not free of problems. We have emphasized that calories are not the sole nutrient of concern in identifying malnutrition. In addition, as noted in Chapter 3, there is no single level of nutrient requirement that applies to all people. And if we are looking at national data (for a population that includes infants, children, and adults), we see an especially wide variation in nutritional requirements. In addition, the information on food availability shows us only the average nutrient intake. Some people in the country or region consume less than the average, and some consume more. So even if average caloric intake is greater than average caloric requirement, a country can still suffer significant undernutrition. Therefore, comparing caloric intake to caloric requirement does

Table 4.3 Food Balance Sheet for India, 2001

	Production	Imports-Exports	Stock Changes	Total Supply	Feed, Seed, Nonfood Uses	Food	Per Capita Supply, Kilograms per Year	Calories per Capita per Day	Grams of Protein per Capita per Day	Grams of Fat per Capita per Day
	Million Metric Tons									
Cereals	196,843.40	–5,333.13	–6,160.47	185,349.90	19,016.02	166,345.80	162.27	1,486.82	35.14	6.66
Starchy roots	30,242.70	–23.43	0	30,219.27	6,306.38	23,916.59	23.33	46.56	0.79	0.07
Sugar and sweeteners	327,832.30	–1,524.63	–863.91	32,5443.70	286,158.70	39,290.07	38.32	255.03	0.25	0.07
Pulses, nuts, oil crops, and oils	48,551.90	6,237.61	109.68	54,899.18	26,445.92	28,481.10	27.78	375.06	7.67	30.38
Fruits, vegetables, and miscellaneous	135,192.80	–1,262.73	0	133,930.00	14,495.36	119,441.90	116.51	128.77	4.09	1.26
Meat and meat products	8,646.58	–248.51	0	8,398.06	179.00	8,219.97	8.02	76.25	2.17	7.56
Milk and eggs	86,025.75	–374.93	0	85,650.81	17,277.02	68,718.34	67.03	110.54	6.91	5.27
Fish	5,352.30	–335.27	0.66	5,016.89	473.01	4,544.68	4.44	8.19	1.35	0.26
Vegetal products total								2,292.23	47.93	38.45
Animal products total								194.98	10.43	13.10
Grand total								2,487.21	58.36	51.55

Source: FAOSTAT 2004b.
Note: For a full version of this food balance sheet, see the FAOSTAT website.

not always give an accurate view of the extent of undernutrition. For example, the FAO reports that during the 1990–1992 period, India and Senegal had virtually the same calories available per person per day (2,310 in India, 2,320 in Senegal). However, anthropometric data for the two countries in the late 1980s found that 63.9 percent of children in India were underweight, as were 21.6 percent in Senegal.

Aggregate food intake measures can also be used to estimate the percentage of a population that is undernourished. The FAO's *Sixth World Food Survey* (1996c) uses the following method: average caloric intake per capita is calculated from food balance sheets, and an estimate of the statistical variance of caloric intake per capita is derived from studies of individuals. These two statistics (average and variance) give an estimate of the distribution of food intake—for example, what percentage of the population has an intake below 2,000 calories per day, what percentage has an intake below 2,500 calories, and so on. Thus, once a minimum food requirement has been identified for the population, we know what percentage of population is undernourished. Minimum food requirements are identified for different age and sex classes (e.g., children under 10, males over 10). Minimum requirements are calculated in two ways. First, we can determine the number of calories that would give a person of average height a minimally healthy weight. A minimally healthy weight is determined as a BMI of 18.5 (see prior discussion of body mass index). A second method is to determine the *basal metabolic rate* (BMR) for the population. The BMR shows the number of calories needed for survival when the body is at rest. The minimal food requirement is calculated by multiplying the BMR by a constant (e.g., 1.56 was the constant used in the *Sixth World Food Survey*).

5

Impacts of Undernutrition

Throughout this book, we use the term "undernutrition" to refer to a physical condition, and we reserve the term "hunger" to refer to the subjective feeling that comes from not having enough food. But undernutrition does not just make a person feel hungry; it stunts their growth (as we saw in the previous chapter), makes them more susceptible to disease, reduces their capacity to do physical work, and interferes with their intellectual growth and achievements. In all these ways, it reduces the person's ability to be economically productive, and in the aggregate—when undernutrition is widespread in a country—it can reduce the pace of economic growth.

Undernutrition and Mortality Risk Among Adults

The connection between nutrition and health is widely and intuitively understood. Hundreds of Internet websites include "BMI calculators" that can tell you whether your weight is outside the "ideal range" for your height (see Chapter 4 for discussion of body mass index). Economic historians (see Fogel 2004 for a comprehensive review) have confirmed the relationship. Costa and Steckel (1997) used medical records of US Civil War veterans to investigate their relative risks of dying during the twenty-five years following their initial evaluations (veterans varied in age from 47 to 64 at their initial evaluations). Their results track out a U-shaped curve. The veterans with BMIs in the mid-20s had the lowest risk of dying (bottom of the curve). Those with BMIs less than 20 (underweight) and those with BMIs greater than 30 (obese) were about 50 percent more likely to die than the mid-20 group.

The nutrition-health link has been studied in more detail by Norwegian epidemiologist Hans Waaler. He mapped the heights and weights of soldiers in the Norwegian army and discovered clear relationships between body shape and risk of dying. The general shape of a Waaler surface is a "basin," as shown in Figure 5.1. The low point (the darkest spot) shows the height-weight combinations at which mortality risk is the lowest. As we move to individuals of lower heights and weights (reflecting a history of undernutrition, on the sides

Figure 5.1 Waaler Surface and Cross Sections

a. Waaler Surface

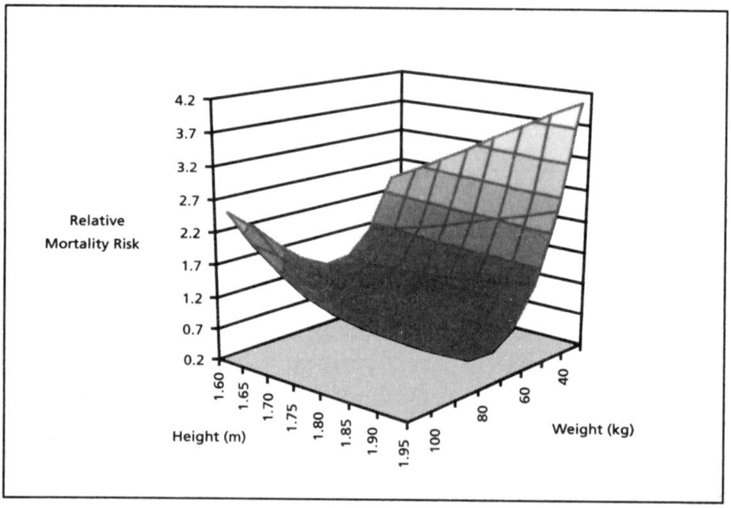

b. Mortality Risks of Men Standing 1.7 m Tall

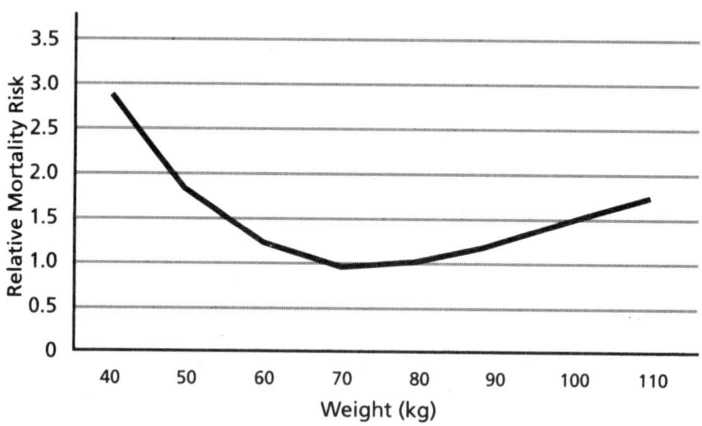

(continues)

Figure 5.1 continued

c. Mortality Risks of Men Weighing 70 kg

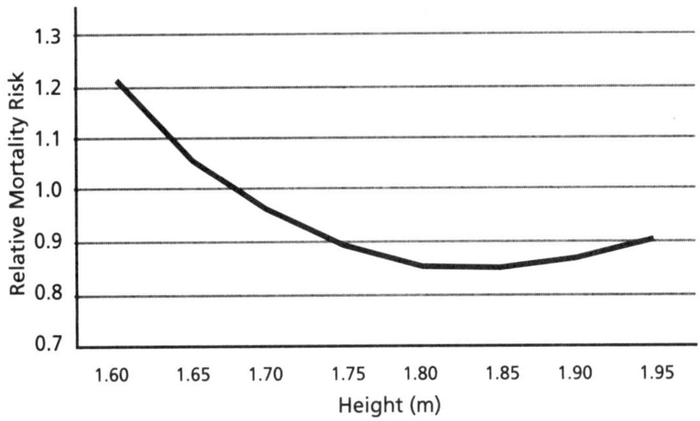

Source: Derived from data in Fogel 2004.

of the basin), the mortality risk increases. (Relative mortality risk measures the risk of dying for men of a particular height and weight compared to the average for the entire population. For purposes of illustration, suppose that 15 percent of all men in a group die between the ages of 50 and 64. Men who are 1.85 meters tall and weigh 80 kilograms have a relative mortality risk of 0.69; this means that their probability of dying between ages 50 and 64 is 15 × 0.69 = 10.35 percent. Men who are 1.6 meters tall and weigh 50 kilograms have a relative mortality risk of 1.62; their probability of dying between ages 50 and 64 is 15 × 1.62 = 24.30 percent.)

If we look at a cross section of the Waaler surface for individuals of a particular height—say, all men standing 1.7 meters tall—we see a picture that looks very much like the Costa-Steckel result (see Figure 5.1b; since everybody in this sample is the same height, those with low weight also have low BMIs, and those with high weight have high BMIs). This kind of result is the basis for a doctor advising a patient to modify their diet in order to achieve a "more healthy" weight.

A more surprising result from the Waaler surface is that mortality risk also varies in a predictable way with height. If we look at a cross section of the Waaler surface for individuals of a particular weight—say, all men weighing 70 kilograms—we see generally that mortality risk declines as height increases (see Figure 5.1c). This indicates that mortality risk is influenced not only by current nutritional status (weight as an adult), but also by nutritional status during the developmental years of childhood, which is an important determinant of adult height. (At the top heights in Figure 5.1c, we find individuals who are currently undernourished and therefore have low weight considering how tall they are; that accounts for the upturn in the mortality risk among the tallest people in this weight category.)

The link between adult height and mortality risk indicates that childhood undernutrition has repercussions throughout life. Such repercussions are being discovered in an analysis of medical histories of children born during the "Hunger Winter" in the Netherlands during World War II (see Box 5.1).

Undernutrition and Child Health

Undernutrition is an especially serious problem for infants and children because their immune systems are less fully developed than those of adults. They are therefore more susceptible to diseases than are adults. Undernourished children are especially susceptible, because their bodies are already weakened.

Mothers' Nutrition, Breast-Feeding, and the Health of the Baby

Mothers who are undernourished during pregnancy are likely to give birth to babies with low birth weights. These babies start life malnourished and are significantly more likely to die in the first year of life; in fact, during the period shortly after birth, low-birth-weight babies die at a rate forty times that of normal babies (Samuels 1986; Overpeck, Hoffman, and Prager 1992). Low birth weight for an infant indicates that the infant was malnourished in the womb or that the mother was malnourished during her own infancy, childhood, adolescence, or pregnancy. The malnourishment is typically due to underconsumption of calories and protein; however, it can also result from underconsumption of micronutrients such as iron (Levinger 1995).

Even poor nutrition early in a girl's life can affect the health of the babies she bears as an adult. One study in Guatemala (Stein et al. 2003) followed the lives of girls who were given nutritional supplements during their early childhoods in the 1960s and 1970s. Some girls were given a high-protein, moderate-energy supplement, and others were given a low-energy supplement with no protein. The girls who received the more nutritious supplement grew up to be larger women, on average, and when they gave birth, their babies were larger on average.

If the baby is breast-fed, the child's health is significantly improved. Breast milk contains all the nutrients a child needs during the first months of life. Breast

Box 5.1 The "Hunger Winter"

Scientists like controlled experiments. If a scientist wants to measure the impact of a new drug, he or she will split a group of patients into two subgroups and give members of one subgroup the new drug and members of the other subgroup a placebo. But nutritionists who want to learn about the impacts of starvation cannot deliberately starve one subgroup to compare to the normally fed subgroup.

The starvation conditions that existed in the Netherlands during World War II provided scientists with a kind of controlled experiment. At the end of the war, the Nazi-occupied Netherlands experienced severe food shortages (the *honger-winter,* or "hunger winter"); average food intake was about 1,000 calories per person per day, with some people surviving on as few as 400–800 calories per day. The famine, which began in late 1944, was short-lived, ending abruptly with the Allied victories in May 1945. This situation, and the existence of precise birth records, created an unplanned "controlled experiment": children born, or who were *in utero,* during the famine, could be compared to children born before and after the famine.

The Dutch Famine Birth Cohort has generated a number of studies that identify the impacts of maternal undernutrition on unborn children. For example, children who were *in utero* during the hunger winter had a higher rate of infant mortality, but also had higher rates of cardiovascular disease and late-onset diabetes during adulthood.

(More information about the Dutch Famine Birth Cohort is available at http://www.dutchfamine.nl/index_files/study.htm, and in the book *The Hunger Winter* by Henri van der Zee [1998]. A description of the research that has grown out of the hunger winter cohort can be found at http://ihome.ust.hk/~lbcaplan/dutchfamine.html, and in the book *Famine and Human Development* by Zena Stein [1975].)

milk also helps the baby fight infection. In developing countries, breast-feeding contributes to infant health in two indirect ways. First, breast-feeding provides the baby with a guaranteed food supply; the infant does not have to compete with other family members for scarce food. Second, breast-feeding ensures that the baby will have a clean food supply; babies who do not breast-feed are exposed to diseases caused by unsanitary food and water. According to one estimate, 1.5 million children in the developing world die because they are not breast-fed (UNICEF 2001:map 1). Even women who are suffering from mild to moderate undernutrition are able to produce sufficient milk to feed their infants, although severe undernutrition, such as that resulting from a famine, does compromise a woman's ability to produce breast milk (Prentice, Goldberg, and Prentice 1994).

In the past decade, AIDS has created a new problem for breast-feeding: breast milk can pass the AIDS virus from an infected mother to her child (Coutsoudis and Rollins 2003). But nutrition can also influence the mother-child

AIDS transmission. A study in Malawi indicated that women with vitamin A deficiencies were more likely to pass on the AIDS virus to their infants. Among pregnant women infected with the virus that causes AIDS—the human immunodeficiency virus (HIV), the transmission rate for those with the highest vitamin A concentrations was 7 percent; for women with the lowest vitamin A concentrations, the transmission rate was 32 percent (Semba et al. 1994).

High Infant and Child Mortality Rates

Death is the most dramatic of undernutrition's adverse effects. As mentioned in Chapter 1, deaths from acute undernutrition—such as those that occur in famines—are not nearly as big a worldwide tragedy as are deaths in which chronic undernutrition weakens the ability of people to resist diseases. This form of fatal undernutrition is an especially large problem among children.

Childhood mortality is a significant problem, as Figure 5.2 illustrates. More than 20 percent of the 52 million people who died in 1995 were under 5 years old. Yet this age category accounted for only about 11 percent of the population. The only category with more deaths was the 75+ age group. The death rate is especially high for third world children. The World Health Organization reported: "Defined as the probability of dying by the age of five years, the global average in 1995 was 81.7 per 1,000 live births; 8.5 in the industrialized world, 90.6 in the developing world, and 155.5 in the least developed nations" (WHO 1996b).

Figure 5.2 Death and Population in the World by Age, 1995

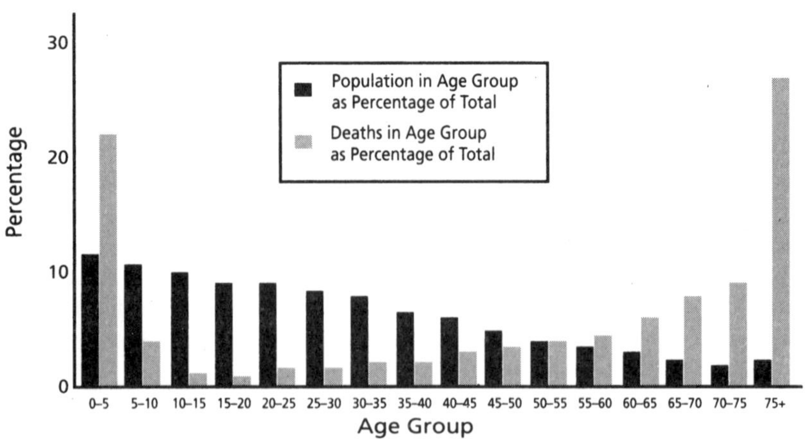

Source: UN, WHO, http://www.who.ch/whr/1997/fig13.gif.

Childhood diseases, to which undernutrition contributes by weakening the immune system, are major killers in developing countries (Dever 1983). Intestinal disorders that lead to diarrhea are the leading child killers in these countries, but other diseases associated with malnutrition, such as pneumonia, influenza, bronchitis, whooping cough, and measles, are also deadly.

The WHO reports that "malnutrition has been found to underlie more than half of deaths among children in developing countries" (1997). One study of fifty-three developing countries estimated that 56 percent of child deaths are caused by the "potentiating effects of malnutrition in infectious disease." This is such an important statistic, it bears repeating: 56 percent of child deaths are caused by weakening of the immune system resulting from undernutrition. Of these, 83 percent are caused by mild-to-moderate malnutrition; only 17 percent are caused by severe malnutrition (Pelletier, Frongillo, and Habicht 1995). When discussing the most cost-effective ways of reducing the impacts of undernutrition worldwide, we should keep in mind the fact that many deaths can be avoided by relatively low-cost interventions that move children from the mild-to-moderate undernutrition category to the adequately nourished category.

Keilmann and McCord (1978) showed that infant mortality doubles with each 10 percent decline below 80 percent of the median weight-for-age. Figure 5.3 is a graphic representation of their results.

The risk of death from nutrition-related disease in the third world decreases dramatically after the second year of life. This is evident from Table 5.1, which shows that mortality in the first year is always lower than mortality in the next

Figure 5.3 Mortality in Children Age 1–36 Months by Nutritional Status, Punjab, India

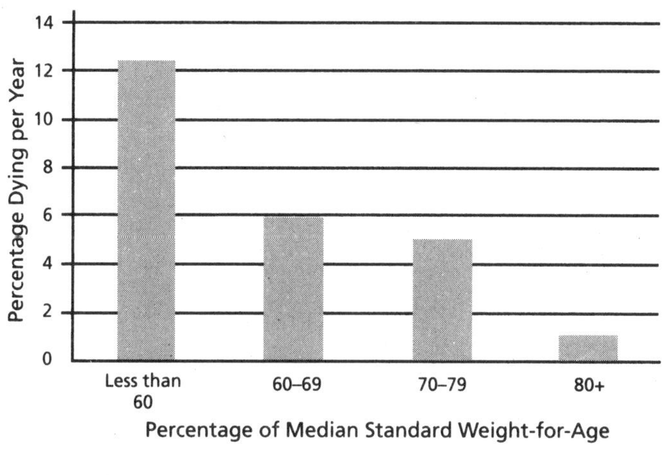

Source: Adapted from Galway et al. 1987:31. Data from Keilmann and McCord 1978.

Table 5.1 Neonatal, Infant, and Child Mortality, Selected Countries

	Percentage Who Die in 1st Month of Life, 2000	Percentage Who Die Before 1st Birthday, 2006	Percentage Who Die Before 5th Birthday, 2006
Sierra Leone	5.6	15.9	27.0
Angola	5.4	15.4	26.0
Afghanistan	6.0	16.5	25.7
Nigeria	5.3	9.9	25.3
Liberia	6.6	15.7	23.5
Mali	5.5	11.9	21.7
India	4.3	5.7	7.6
China	2.1	2.0	2.4
Brazil	1.5	1.9	2.0
United States	0.5	0.6	0.8
Japan	0.2	0.3	0.4
Sweden	0.2	0.3	0.3

Source: UNICEF 2008: statistical tab. 1.

four years. One study (de Zoysa et al. 1985) in the mid-1980s measured deaths per 1,000 per year for the developing world as a whole and found about 22 deaths per 1,000 in the first year of life, about 20 deaths per 1,000 in the second year, and about 6 deaths per 1,000 in each of the third through fifth years.

When we compare developed and developing countries in terms of child mortality, the differences are startling. For example, in Table 5.1, compare the data for the developing countries with the highest rates of infant mortality (top six rows) to the rates for representative developed countries (bottom three rows). Again, these differences are strongly related to differences in nutrition.

The emphasis in this section has been on deaths caused by disease related to undernutrition. Disease related to undernutrition is not always fatal, or course. Individuals who suffer from nutrition-related diseases, but who survive, are weakened during the (sometimes lengthy) duration of the disease. Analysts of health policy have introduced a concept called disability-adjusted life years (DALYs), which measures total amount of healthy life lost, whether from premature mortality or from temporary or permanent disability. The World Health Organization estimates that poor nutrition is responsible for 46 percent of DALYs lost to children (WHO 1996a).

Micronutrient deficiencies also cause health problems. The WHO's 1995 *World Health Report* states: "As a result of iodine deficiency—a public health problem in 118 countries—at least 30,000 babies are stillborn each year and over 120,000 are born mentally retarded, physically stunted, deaf-mute or paralyzed. A quarter of all children under age 5 in developing countries are at risk of vitamin A deficiency" (WHO 1995c). The interaction between undernutrition and health is discussed in Chapter 15.

Undernutrition, Menstruation, and Breast-Feeding

One physical effect of undernutrition has important implications for population growth. Undernourished women are less fecund, because they begin ovulation later in life, and because the resumption of ovulation after childbirth is postponed, relative to well-nourished women. The resumption of ovulation after childbirth can be further postponed by breast-feeding, which puts additional nutritional demands on the nursing mother.

The female sex hormone, estrogen, is produced from cholesterol, a fat (Pike and Brown 1984:42). The fatter a woman is, the more estrogen she is likely to produce. An undernourished girl is likely to produce less estrogen than a well-nourished girl of the same age; therefore, the well-nourished girl is likely to begin menstruation at a younger age. Delayed age of menarche (age of first menstruation) is an indicator of low levels of calorie intake. Girls in the United States reach menarche earlier today than they did a century ago because they now produce a higher percentage of fat (and estrogen). Because vigorous exercise reduces body fat, well-fed young girls who are also athletes often reach menarche later in life than do their more sedentary counterparts.

Table 5.2 shows a geographic breakdown of menarche. Notice that the age of menarche in most industrialized countries is about 13 years. Developing countries show a greater range, with average age of menarche being as high as 16 in some of the poorest (most undernourished) parts of the world.

We have already briefly discussed the importance of breast-feeding to the health of infants. An additional benefit of breast-feeding is that it postpones the recurrence of menstruation after childbirth. The same hormone (prolactin) that stimulates the production of breast milk also suppresses ovulation. In addition, the production of breast milk tends to use up the body's supply of fat, reducing estrogen production and impeding ovulation. The use of breast-feeding to postpone ovulatory cycles after childbirth is referred to as the lactational amenorrhea method of birth control. The method is at least 98 percent effective if three conditions are met: (1) the mother has not experienced the return of her menstrual periods, (2) the mother is fully or nearly fully breast-feeding, and (3) the baby is less than six months old. Studies in developing countries confirm the effectiveness of breast-feeding as a birth control method. In Chile, for example, according to one study, only 1 of 422 breast-feeding women became pregnant during the six months after childbirth; in Pakistan, there was 1 pregnancy among 391 women; in the Philippines, there were 2 pregnancies among 485. The experience in Pakistan and the Philippines showed 98–99 percent birth control effectiveness for a full year after childbirth among women meeting the first two of the above three conditions (Family Health International 1997). Another study estimates that breast-feeding is by far the most widely used contraceptive method in India—six times greater use than the birth control pill, four times greater use

Table 5.2 Average Age of Menarche, Selected Countries

	Age at Menarche
Congo-Brazzaville	12.0
Italy	12.2
China	12.4
Japan	12.5
Dominican Republic	12.6
Colombia	12.8
United States	12.8
Australia	13.0
Indonesia	13.0
Switzerland	13.0
France	13.1
Jamaica	13.1
Israel	13.3
Britain	13.3
Philippines	13.6
Taiwan	13.6
Zambia	13.7
Guatemala	13.8
South Korea	13.9
Ghana	13.9
Nicaragua	14.0
India (Punjab)	14.3
Kenya	14.4
Cameroon	14.6
Somalia	14.8
Nigeria	15.0
Tanzania	15.2
Haiti	15.4
Bangladesh	15.8
Senegal	16.1
Nepal	16.2

Source: Thomas et al. 2001.

than sterilizations and intrauterine devices, and nearly two times greater use than condoms (Gupta and Rohide 1993).

Impacts of Undernutrition on Intellectual Development and Educational Attainment

Impact on Intellectual Development

Ancel Keys, working with conscientious objectors during World War II, found that male adults subjected to diets that led to measurable undernutrition experienced intellectual problems as the first symptom. Later, as their undernutrition continued, the men suffered problems with physical dexterity (Keys et al. 1950). Since Keys's work, nutritionists have found, in a wide variety of settings, that poor nutrition in the early years of life can have a negative impact on intellectual development (Grantham-McGregor, Fernald, and Sethuraman 1999a, 1999b).

More recent research has tended to focus on the potential effect of childhood undernutrition on later intellectual development and achievement. If a mother suffers from undernutrition during pregnancy, her baby can suffer from reduced intellectual capacity and cognitive functioning. When malnourished women are given protein supplements during pregnancy, improvements in their children's cognitive functioning can be observed through age 6 or 7 (Hicks, Langham, and Takenaka 1992). If pregnant women are given adequate calories and protein, the effects on their offspring can be sustained into adolescence and even young adulthood (Pollitt et al. 1993).

Chronic malnutrition in children—especially during the first two or three years of life—can impair mental development directly (since brain development is negatively affected) and indirectly (because undernourished children are less active, and therefore their brains are less stimulated) (World Bank 1997).

In a study conducted in Hyderabad, India, children who had previously suffered from kwashiorkor, a form of protein-energy malnutrition, scored an average of 35 points below their matched controls on IQ tests administered up to six years after their recovery. However, the authors note that it was difficult to determine "to what extent this is a result of the episode of kwashiorkor and to what extent it is due to other factors" (Champakam, Srikantia, and Gopalan 1968). Galler (1986), in a longitudinal study in Barbados, matched 183 children who had a history of kwashiorkor, with 129 classmates who had no such history but who were of similar age and sex, and who were from the same socioeconomic group. Both sets of children were followed from age 5 to 18 years. By sexual maturation, the previously undernourished children had essentially caught up with the matched group in terms of physical growth, but demonstrated small deficits in IQ throughout the growth period. UNICEF, in its 1998 *State of the World's Children* report, found evidence to support the conclusion that poor nutrition reduces intelligence: iron deficiency during infancy and early childhood reduced IQ by 9 points on average; 2-year-old children who were severely stunted (an anthropometric indicator of poor nutrition) had IQs 5–11 points below those of 2-year-olds of normal height; and low-birth-weight babies (born to mothers who had been poorly nourished during pregnancy) had IQs 5 points below those of normal children.

Iodine deficiency is the most significant avoidable cause of mental retardation worldwide. An overview of eighteen studies shows that iodine-deficient groups have IQs that are 13.5 points below those of non-iodine-deficient groups (Bleichrodt and Born 1994). To understand how important a difference of this magnitude can be, consider that a person with an IQ of 100 is at the median score for intelligence (half the population scores less than 100, half scores greater than 100); a person with an IQ of 86.5 (100 – 13.5) scores higher than only about 20 percent of the population, and lower than 80 percent. Deficiencies in other micronutrients, such as iron and vitamin A, are also associated with impaired intellectual abilities (Levinger 1994).

Using the Bayley scales for cognitive skills, Gretl Pelto (1987), professor of nutritional science at the University of Connecticut, in a seven-year longitudinal study of nutrition and cognitive development among seventy-eight Mexican preschool children, tested short-term memory, responsiveness to stimulus, attention and distractability, abstract categorization skills, and sedentary passivity. She found that children with little animal food in their diets were short in stature and delayed in cognitive development, and that delays in intellectual development resulted from nutritionally induced growth stunting.

Chavez and Martinez (1982) studied child development among poor Mexican peasant families. They set up a controlled experiment in which one set of families was given supplemental food for the child through 3 years of age, and for the mother while she was pregnant and lactating. The control group was a set of families matched to the treated group to have similar genetic and socioeconomic characteristics, but was given no food supplements. Of the many tests for neurological maturation and mental performance given to each set of children, in virtually every instance—walking, control of bladder, and language development—the control children lagged behind the treated children. The better-nourished children were found to be more precocious in constructing three-word sentences.

Chavez and Martinez stressed that it is not possible to determine the ultimate significance of the gap between undernourished and better-nourished children; undernourished children may catch up later in life. There is some evidence that intellectual impairment is at least partially reversible if nutrition is improved. Winick, Meyer, and Harris (1973) tracked severely undernourished Korean orphans and found no signs of mental impairment years after their adoption by US families. A study that followed Filipino children through the first years of life also found that IQ deficiencies resulting from infant malnutrition persisted in children up to the age of 12 (UNICEF 1998: panel 3). Lynn and Vanhanen conclude: "Rises in intelligence that occurred in Western populations during the twentieth century are largely attributable to improvements in nutrition" (2002:185).

Impact on Educational Attainment

Not surprisingly, given its impact on intellectual development, undernutrition can also impact the educational achievements of children.

In the United States, it was found that children who had low birth weight had problems succeeding in school. They were more likely to need special education services and more likely to repeat a grade (Levinger 1995).

The Barbados study mentioned above (Galler 1986) found that the most striking difference between the undernourished and well-nourished groups was a fourfold increase in the frequency of attention deficit disorder among the previously undernourished. This syndrome is characterized by decreased attention span, impaired memory, high distractability, restlessness, and disobedience, and was found to reduce educational progress among the previously undernourished children to a far greater extent than the slight deficit in IQ they experienced.

Recent studies in the Philippines (see Box 5.2) and Kenya found a significant relationship between nutritional status and educational attainment as measured by test scores (Glewwe, Jacoby, and King 1996; Bhargava 1996).

Other research shows that if children miss breakfast—if they fast for sixteen hours or more—their school performance suffers. In particular, students suffer from poorer memory and reduced ability to pay attention. World Bank nutrition specialist Alan Berg (1973) points out that education also suffers from missed days of school due to nutrition-related illnesses. He cites the case of four Latin American countries where illness caused children to miss more than fifty days of school a year.

A study of children born during the 1982–1984 drought in Zimbabwe found that those children started school an average of 3.7 months later than children born in a nondrought period; the children of the drought also finished fewer grades of schooling (Alderman, Hoddinott, and Kinsey 2003).

Recall our discussion of anthropometric results showing that nutrition is positively related to height-for-age (see Chapter 4). If nutrition also affects educational attainment, does this mean that we can find a correlation between height of an individual and years of schooling? Strauss and Thomas (1998) show that, in fact, taller individuals finish more years of schooling, both in the United States and in Brazil.

Box 5.2 The Cebu Longitudinal Study

Another example of nutritional research is the Cebu Longitudinal Study, led by Linda Adair of the University of North Carolina. It began as a study of about 3,000 pregnant women in Cebu, Philippines, who gave birth during the year spanning May 1983 to April 1984. The original plan was to study the infant-feeding practices of these women (how long they breast-fed, when they introduced different kinds of solid food, etc.). From 1991 to 1992, 2,400 of the families were located and revisited, and both mothers and children were interviewed and evaluated anthropometrically. The younger siblings of the 1983–1984 babies were also measured and evaluated. In 1994 and in 1999, the families were again revisited for interviews and tests. Thus the impact of nutritional practices in the first years of life could be traced through the children's first sixteen to seventeen years of life. (Studies of the data collected include Glewwe, Jacoby, and King 1996. For a more complete description of the study and analyses of its data, see http://www.cpc.unc.edu/projects/cebu.)

Effects of Undernutrition on Labor Productivity

In this chapter and the last, we have seen evidence that undernutrition has negative effects on physical growth and development, health, intellectual capacity, and educational attainment. For all of these reasons, we might expect that undernutrition can reduce labor productivity. A number of studies have analyzed this effect (see Strauss and Thomas 1998 for a good review of the literature):

- A study of Chinese cotton-mill workers (Li et al. 1994) found that the women were able to do 14 percent more work for each 1-gram increase in their hemoglobin; these increases were attained by giving the workers iron supplements.
- A study of agricultural workers in Colombia and the United States (US Department of State 1976) found a high correlation between nutritional status and physical work capacity. Among undernourished Colombian sugar cane cutters, physical work capacity was reduced by 50 percent.
- A survey of smallholder farmers in Sierra Leone (Strauss 1968) discovered that a 50 percent increase in calories per capita was associated with a 16.5 percent increase in farm output. When nutrition levels were extremely low, the impact of improved nutrition was higher: for families whose average intake was less than 1,500 calories per person per day, a 50 percent increase in calories led to a 25 percent increase in farm output.

Again, relying on the anthropometric measure height-for-age as an indicator of poor nutrition, studies have found that smaller adults are less productive workers in many jobs—they cannot lift and carry as much heavy weight, for example, as larger adults. Strauss and Thomas (1998) show that in a test of physical capacity, taller people are more likely to be able to carry a heavy load. The World Bank, in its 1995 *World Development Report,* estimates that this stunting causes an economic loss of $8.7 billion per year worldwide.

In that same 1995 report, the World Bank finds that an increase in a person's height by 1 percent is associated with an increase in that person's wages by 1.38 percent. Strauss and Thomas (1998) show that positive correlations exist between wages and height and between wages and body mass index in Brazil, where undernutrition is fairly widespread, but not in the United States. This is consistent with the explanation that lower wages are tied to undernutrition. Strauss and Thomas also show that lower wages are not just a result of less education; when the authors limit the comparison to Brazilians with no education, they still find that wages are higher for taller people or for people with higher BMIs.

Paul Schultz (1999) examined the impact of nutrition on wages in Ghana and Côte d'Ivoire. In Ghana, he found a positive correlation: greater body heights and higher BMIs were associated with higher wages. However, in Côte d'Ivoire, the nutritional indicators were not found to have a statistically significant impact on wages. Schultz concludes that the reason for this disparity is that nutrition was much worse in Ghana than in Côte d'Ivoire during the period studied; the implication is that undernutrition ceases to be a cause of low wages once the problem has been reduced to a certain level. This conclusion is buttressed by the findings in the previous paragraph about the lack of correlation between wages and nutrition in the United States.

Finally, Strauss and Thomas (1998) find that poorly nourished people are more likely to be unemployed. Among urban males in Brazil who were at least 165 centimeters, the average unemployment rate was less than 5 percent. For individuals who were 155 centimeters tall, the average unemployment rate was 10 percent. Likewise, for people who had BMIs of at least 24, the average unemployment rate was 4 percent, compared to 10 percent for individuals who had BMIs of 18.

In the previously cited study of the impact of drought on Zimbabweans born between 1982 and 1984, Alderman, Hoddinott, and Kinsey (2003) conclude: "We present calculations that suggest that this loss of stature, schooling and potential work experience results in a loss of lifetime earnings of 7–12 percent and that such estimates are likely to be *lower* bounds of the true losses."

Economic historians have also examined the link between nutrition and productivity. Nobel Prize–winning economist Robert Fogel concluded that food shortages were so severe in Europe in the eighteenth and early nineteenth centuries that "the bottom 20 percent subsisted on such poor diets that they were effectively excluded from the labor force," being too weakened from hunger to work. Fogel estimates that improved nutrition accounts for 30 percent of the growth in income per capita in Britain between 1790 and 1980 (see Fogel 1994, which also reviews other studies linking nutrition to economic development in Europe and the United States).

6

Undernutrition: Who, When, Where?

In Chapter 4, we explored a number of ways to measure undernutrition. Given that some measures are based on somewhat arbitrary cutoff points, we should not be surprised to find that the extent of undernutrition can only be approximated.

Nevertheless, we know enough about undernutrition to put the severity of the problem into perspective. In this chapter we consider the following questions: How widespread is undernutrition? How has the extent of undernutrition changed over time? What are the demographic and geographic characteristics of the problem? What time of year is worst for undernutrition?

Global Trends in Long-Term Perspective

Table 6.1 presents estimates of the number of people worldwide who are affected by undernutrition-associated conditions. For technically advanced readers, descriptions and criticisms of the method used by the Food and Agriculture Organization to measure the extent of undernutrition can be found in Svedberg 1999 and Gabbert and Weikard 2001.

Over time, the number of undernourished people has declined. This is illustrated by Table 6.2. For the developing world as a whole, the number of undernourished people dropped from 960 million in the early 1970s to 923 million in the early 1980s, to 830 million at the turn of the century. Expressed as a percentage of the population, the drop has been more dramatic; over one-third of the developing world's population was undernourished in the early 1970s; by the turn of the century, the proportion had dropped to 17 percent. However, these declines have not been spread evenly throughout the developing world. Progress in Asia has been remarkable; the number of undernourished people has dropped by nearly 250 million since the early 1970s, and the percentage of undernourished people has dropped from 41 percent to 16 percent. These declines offset worsening conditions in sub-Saharan Africa, where the number of undernourished people has increased by over 120 million, and the percentage of population that is undernourished has declined only slightly, from 36 percent to 33 percent.

Past editions of this book have noted the trends toward improvement and perhaps have invited the reader to infer that this progress would continue. The experience of recent years, however, shows a stagnation, with the percentage of the world population that is malnourished declining from 18 percent in the mid-1990s to only 17 percent in the mid-2000s—a relatively slow rate of decline by past standards. The recent food price inflation of 2007 and 2008 will be discussed later in this chapter, but it seems likely that when comparable

Table 6.1 World Population Affected by Different Types of Malnutrition

	Prevalence	
	Millions of People	Percentage of World Population
Iodine deficiency (1994–2006)	1,900	31
Vitamin A deficiency (1980–1995)[a]	254	42
Iron deficiency (among women and children, 1993–2005)	818	35
Inadequate caloric intake (2002)	854	14[b]
Stunting among children from inadequate protein and energy (under age 5, 2008)	62	31

Sources: WHO 1995a, 2007, 2008; FAO 2008; UNICEF 2008.
Notes: a. Based on surveys of 73 out of 191 countries.
b. Based on world population of 6.23 billion.

Table 6.2 Trends in Undernutrition by Continental Area

	Number of People Who Are Undernourished (millions)				
	1969–1971	1979–1981	1990–1992	1995–1997	2002–2004
Sub-Saharan Africa	92.8	127.0	169.0	196.6	213.4
Near East and North Africa	42.8	20.8	25.0	35.0	37.3
Asia and the Pacific	769.9	730.0	569.7	510.8	527.2
Latin America and Caribbean	55.1	46.0	59.4	54.5	52.1
All developing countries	960.7	923.8	823.1	796.9	830.0
	Proportion of the Population That Is Undernourished (%)				
Sub-Saharan Africa	36	37	35	36	33
Near East and North Africa	24	9	8	10	9
Asia and the Pacific	41	32	20	17	16
Latin America and Caribbean	20	13	13	11	10
All developing countries	37	28	20	18	17

Source: FAOSTAT, Food Security Statistics, http://www.fao.org/faostat/foodsecurity/index_en.htm.

data become available for this period, they will show increases in both numbers and percentages of undernourished people.

Additional evidence of a long-term trend toward an improving food situation is shown in Figures 6.1 and 6.2. The former shows that worldwide food production per capita increased steadily from 1960 to 2004 (with a very slight dip between 2004 and 2005), and the latter shows that world food prices fell throughout most of this period. Still, food prices were more volatile than food production. They increased dramatically between 1972 and 1974 (see Johnson 1975 for a discussion of the reasons), and then fell sharply. The upward trend in real food prices since 2001 is consistent with our earlier observation that worldwide progress in reducing the prevalence of undernutrition has stagnated in recent years. As mentioned, the much sharper upturn in food prices during 2007 and 2008 will be discussed later in this chapter. Box 6.1 discusses a situation thousands of years ago when nutrition did not improve.

The Seasonality of Undernutrition

People at risk for undernutrition are not usually at risk all the time; it tends to come in fits and starts. We are all aware of the periodicity of famine—the word connotes a time of extreme food scarcity as contrasted with normal times when food is less scarce. But we are less aware that third world hunger usually follows the rhythm of the seasons. In the third world, a strong seasonality is usual in the production, price, and availability of food, as well as in the availability of employment (Sahn 1989). All these factors can influence the nutritional status of a family at risk for undernutrition.

Figure 6.1 Index of Worldwide Food Production per Capita, 1960–2005

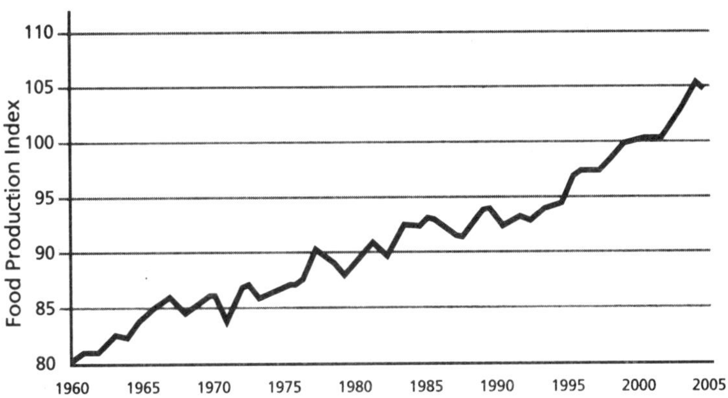

Source: FAOSTAT various years.

Figure 6.2 Index of Real Food Prices, 1961–2008 (inflation-adjusted, 2005 = 100)

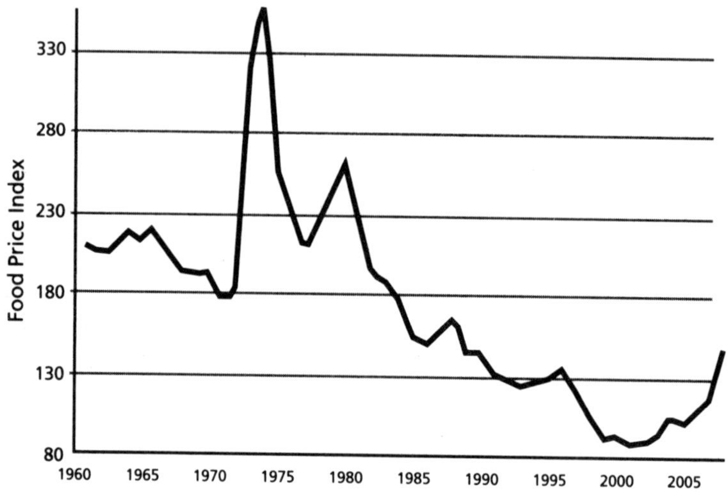

Source: Food price index in current dollars from IMF, deflated by US GDP deflator.

The seasonality of undernutrition is often linked to the agricultural year, which in the tropics is usually heavily dependent on rainfall patterns. In monsoon Asia, for instance, rice is planted at the beginning of the wet season, which typically starts in July. Harvest begins about four months after planting. With irrigation, farmers can harvest more than one crop per year, but most of the rice is still grown during the wet season. Likewise, in the Sahel region of Africa (including parts of Chad, Niger, and Mauritania), crops mature in September and October, but the food supply begins to run out in May, and the hungry season can run for four months, from May through August (FAO 1997b).

Consequently, the price of rice is lowest just after harvest and rises gradually as supplies dwindle. During the growing season, supplies may become short and prices may rise more sharply. The pattern of seasonal variation in food price and food consumption in the Philippines is shown in Table 6.3. Consumption is at its lowest in September, when the price has been high for three months, and picks up during the following months, when the rice is harvested and price falls below the annual average. Similar seasonal price patterns occur throughout the developing world. For example, millet prices in Niger rose 80 percent between July and August of 1996 (FAO 1997b).

A detailed study of dietary intake in Mozambique (Rose et al. 1999) vividly illustrates the impact of the hungry season. As shown in Table 6.4, average calories per person per day in the village of Montpuez drop from 91 percent of requirements (about 2,000 per person per day) in the harvest season, to 63 percent (about 1,430 colories) during the hungry season. During the hungry season, the

Box 6.1 Nutrition, the Natural Environment, and the Development of Agriculture

It is tempting to presume that the improvements in worldwide nutrition seen over the past half century are simply an extension of a long history of gradual improvement in human nutrition. But some anthropological evidence indicates that precivilized hunter-gatherers obtained good nutrition with relatively little effort.

Angel (1975) found that, 30,000 years ago—before the birth of agriculture—the average adult male was 177 centimeters (5 feet 11 inches) tall, but that, 20,000 years later, average height was only 165 centimeters (5 feet 6 inches). Other information from Angel (1984) indicates that residents of the eastern Mediterranean at the beginning of the age of agriculture were taller than modern Greeks. Harris notes that it is "hard to reconcile [a view of starving hunter-gatherers] with the enormous quantities of animal bones accumulated at various Paleolithic kill sites. . . . The skeletal remains of the hunters themselves bear witness to the fact that they were unusually well-nourished" (1977:10). Cohen (1984) notes that skeletal evidence indicates low rates of anemia and infection among hunter-gatherers.

Other evidence that hunter-gatherers had good nutrition comes from studies of modern hunter-gatherer societies. Lee (1968a, 1968b, 1969, 1972) studied the !Kung bushmen of Africa and found that they consumed 93 grams of protein per day, mostly from meat and nuts. The composition of the diet of many hunter-gatherers is varied and more likely to contain sufficient quantities of a broad array of nutrients.

An even more remarkable finding from Lee's study of the !Kung is that they achieved this nutritious diet with relatively little effort—less than three hours per day per adult, despite the fact that the !Kung lived on the edge of the Kalahari desert. The rest of the time was spent resting, visiting with others, doing embroidery, playing games, and the like. Similarly, Sahlins (1968) found that the aborigines of Australia worked for two days and took the third as a holiday. These observations are consistent with studies of great apes, which spend about half their time grooming, playing, and napping, and the other half foraging for food. "As collectors of food . . . [Paleolithic populations] were certainly no less effective than chimpanzees" (Harris 1977:10).

Why—if hunting and gathering was so nutritionally efficient—did agriculture ever develop? After all, Diamond (1987) refers to the development of agriculture as "the worst mistake in the history of the human race." According to Harris, the answer is global warming: "As long as . . . exploitation of [naturally occurring animal and plant resources] is kept relatively low, hunter-collectors can enjoy both leisure and high-quality diets. . . . Then, about 13,000 years ago, a global warming trend [began, and] . . . forests invaded the grassy plains which nourished the great herds [of large animals, known as 'megafauna']. . . . The collapse of the big-game hunting cultures . . . was followed by . . . [a system of] preying on smaller species . . . called . . . 'broad spectrum' hunting and collecting. . . . [H]unters then intensified predation of [smaller species], and these too soon became extinct. . . . As they fought their long and futile delaying action against the consequences of the depletion of animal species, [they] . . . shifted their primary subsistence effort away from animals toward plants. . . . It seems clear that the extinction of the Pleistocene megafauna triggered the shift to an agricultural mode of production" (1977:10–25).

Table 6.3 Seasonal Prices and Food Consumption in the Philippines

	Index of Food Price	Food Consumption (annual kilograms per capita)
March	97	108
June	100	110
September	110	98
December	92	104

Source: Adapted from Philippines Ministry of Agriculture 1981b, 1983.

Table 6.4 Comparative Nutrient Intakes for the Harvest Season and the Hungry Season for Smallholders in Montpuez, Mozambique, 1995–1996

	Percentage of Requirements	
	Harvest	Hungry
Calories	91	63
Protein	148	79
Vitamin A	23	88
Niacin	103	64
Calcium	42	42
Iron	92	64

Source: Rose et al. 1999.

calorie intakes of 30–40 percent of the villagers are insufficient to replace the calories they expend in normal daily activity. Maize—the source of 60 percent of calories in the average diet during the harvest season—disappears from the diets of many during the hungry season; it is replaced by manioc, peanuts, and sorghum—less desirable substitutes that are not eaten as much when maize is available.

One nutritional bright spot for the hungry season diet described in Table 6.4 is the abundance of vitamin A, due to an increase in consumption of fruits and vegetables such as pumpkin squash. Such increases in consumption of wild fruits and vegetables during the hungry season appear to be a widespread phenomenon. Falconer (1990), in a review of the literature, finds a number of studies that stress the importance of wild plants and animals as sources of food during the hungry season in different parts of the developing world. In one Zambian village, wild foods make up 42 percent of the diet during the hungry season, compared to 7 percent during the rest of the year. In rural Bangladesh, people eat almost no wild plants—an average 1 gram per person per day—during the ten months following the rice harvest; then, during the May–June hungry season, this increases to 191 grams per person per day. (See Box 6.2 for descriptions of the hungry season.)

Box 6.2 The Hungry Season in Zambia and Malawi

Data such as those presented in the accompanying text give us some idea of the quantitative aspects of the hungry season. But a more complete picture emerges from some vivid descriptions.

A *Washington Post* reporter visiting Zambia during the hungry season found this: "When the people by the lake began to starve, they fell back on the knowledge of their ancestors. They picked poisonous fruits from the bush and boiled them for three days to eliminate the toxin concocting a barely palatable dish. But sometimes hungry children would sneak a taste early . . . and the poison would make them ill. Kebby Kamota, father of 11, could take it no longer. 'Three days! Three days!' He shouted explaining how long his children would sometimes go without food" (Gillis 2003:A1).

A Peace Corp volunteer in Zambia describes the onset of the hungry season like this: "I can tell the change of season by signs rather than months. When there are mangoes, rain, disease, funerals and kids crying at night, it's the hungry season" (Kluender 2003).

A report by the Centro Internacional de Mejoramiento de Maize y Trigo describes how Agness Pungulani, a single mother in Malawi, copes: "During the . . . hungry season, markets were devoid of grain [and] local traders were selling maize at [two-and-a-half times the normal price]. When the food stocks disappeared, Pungulani . . . foraged, drank tea from wild occra leaves in place of evening meals, and pounded banana tree roots into a crude flour approximating their preferred maize staple. . . . A neighbor . . . says, 'This flour tastes sour, but we eat it because we have no choice.' The hungry season normally arrives during January-February in this part of Malawi, but lately families have run out of grain as early as September and must survive until the March harvests" (CIMMYT 2003)

A *New York Times* reporter describes the joyful end of the hungry season in Malawi like this: "Late one afternoon, during the long melancholia of the hungry months, there was a burst of joyous delirium in Mkulumimba. Children began shouting the word 'ngumbi,' announcing that winged termites were fluttering through the fields. These were not the bigger species of the insect, which can be fried in oil and sold as a delicacy for a good price. Instead, these were the smaller ones, far more wing than torso, which are eaten right away. Suddenly, most everyone was giddily chasing about; villagers were catching ngumbi with their fingers and tossing them onto their tongues, grateful for the unexpected gift of food afloat in the air. . . . Most every year, Malawi suffers a food shortage during the so-called hungry months, December through March. . . . Families often endure this hungry period on a single meal a day, sometimes nothing more than a foraged handful of greens" (Bearak 2003:33).

Who Is Undernourished?

In which countries is the problem of undernutrition the worst? Tables 6.5, 6.6, and 6.7 show that the answer to this question depends on the way in which the extent of undernutrition is measured.

Table 6.5 Countries That Rank High in Undernutrition Based on Dietary Requirements

Average Calories per Capita per Day as Percentage of Average Daily Requirement, 2001[a]		Percentage of Population Suffering from Undernutrition, 2002–2004	
Congo, Dem. Rep.	69.8	Eritrea	75
Burundi	72.5	Congo, Dem. Rep.	74
Tajikistan	73.4	Burundi	66
Eritrea	75.7	Comoros	60
Armenia	83.2	Tajikistan	56
Mongolia	83.6	Sierra Leone	51
Zambia	84.5	Liberia	50
Cambodia	86.2	Zimbabwe	47
Sierra Leone	86.3	Ethiopia	46
Cent. African Rep.	86.6	Haiti	46
Angola	87.0	Zambia	46
Liberia	87.4	Cent. African Rep.	44
Mozambique	88.4	Mozambique	44
Tanzania	89.1	Tanzania	44
Madagascar	89.9	Guinea-Bissau	39
Kenya	90.4	Madagascar	38
Haiti	90.8	Yemen	38

Sources: FAOSTAT 2008b.

Note: Afghanistan and Somalia are excluded because comparable data are not available for all of Tables 6.5–6.7.

a. Calculated using calorie requirements for age and sex catagories listed in Table 3.1 and corresponding population in each category from US Census Bureau for 2000.

Many countries rank high in terms of both calorie deficiency and prevalence of undernutrition: the Democratic Republic of Congo, Eritrea, Tajikistan, and Burundi all appear in the top five of both lists in Table 6.5. Nearly all the countries in Table 6.5 are sub-Saharan African countries. In the ranking of "worst case" countries using anthropometric indicators of child nutrition in Table 6.6, we see many countries that do not appear in Table 6.5. Moreover, there is less consistency among the three different measures of child undernutrition in Table 6.6. Eight countries appear in all three lists, and four countries appear in two of the three lists, but twenty-eight countries appear in only one of the three lists, but not the other two. Furthermore, the rankings in Table 6.6 according anthropometric indicators of childhood undernutrition are quite different from the rankings in Table 6.7 according to under-5 mortality and low birth weight.

Again, most of the worst-off countries in all these lists are sub-Saharan African countries. However, the lists of low-birth-weight babies and anthropometrically small children contain a number of Asian countries near the top. Ramalingaswami, Jonsson, and Rohde (1996) suggest that this phenomenon may be explained by a combination of factors:

Table 6.6 Countries That Rank High in Undernutrition Based on Anthropometric Measurements of Children Under Age 5

Percentage Who Are Stunted (low height-for-age)		Percentage Who Are Underweight (low weight-for-age)		Percentage Who Are Wasted (low weight-for-height)	
Yemen	53.1	Bangladesh	48	Burkina Faso	23.1
Burundi	52.5	Timor-Leste	46	Djibouti	20.7
Niger	50.0	Yemen	46	India	19.8
Zambia	50.0	Niger	44	Sudan, N.	15.7
Timor-Leste	49.4	India	43	UAE	15.2
Guatemala	49.3	Madagascar	42	Laos	15.0
Nepal	49.3	Sudan, N.	41	Togo	14.3
India	48.0	Eritrea	40	Sri Lanka	14.0
Madagascar	47.7	Laos	40	Mauritius	13.7
Ethiopia	46.5	Burundi	39	Chad	13.5
Malawi	45.9	Nepal	39	Congo, Dem. Rep.	13.4
Rwanda	45.3	Ethiopia	38	Maldives	13.2
Angola	45.2	Pakistan	38	Pakistan	13.2
Comoros	44.0	Burkina Faso	37	Bangladesh	12.8
Sudan	43.3	Chad	37	Madagascar	12.8
Bangladesh	43.0	Cambodia	36	Mauritania	12.8
Laos	42.4	Mali	33	Eritrea	12.6
Mozambique	41.0	Mauritania	32	Nepal	12.6
Chad	40.9	Myanmar	32	Timor-Leste	12.4
Guinea Bissau	40.9	Angola	31	Yemen	12.4

Source: UNICEF, http://www.childinfo.org/undernutrition.html, 2008.
Notes: Data for various years, mostly in the 2002–2006 range. Afghanistan and Somalia are excluded because comparable data are not available for all of Tables 6.5–6.7.

- "Girls and women in South Asia are less well regarded and less well cared for than in sub-Saharan Africa." Poor nutrition among pregnant women causes low-birth-weight babies.
- "Differences in standards of hygiene between the two regions [South Asia and sub–Saharan Africa] are very pronounced. . . . This all-round poor hygiene increases the burden of illness, and [causes] significantly higher levels of malnutrition among South Asia's children."
- Quality of childcare is higher in sub-Saharan Africa than in South Asia.

It is interesting that the small size of babies and children in South Asia does not translate into high rates of child mortality.

Some readers may wonder if we have forgotten to consider undernutrition in the developed world. But the incidence of undernutrition in the developed world is generally so close to zero that it does not contribute much to worldwide numbers. A comparison of Tables 6.1 and 6.2 reveals that, of the 854 million undernourished people worldwide, 830 million are in developing countries. The situation in the United States is described in Box 6.3.

Table 6.7 Countries That Rank High in Undernutrition Based on Infant and Child Health

Under-5 Child Mortality (deaths per 1,000)		Low Birth Weight (percentage of births)	
Sierra Leone	270	Yemen	32
Angola	260	Sudan	31
Niger	253	India	30
Liberia	235	Comoros	25
Mali	217	Haiti	25
Chad	209	Guinea Bissau	24
Equatorial Guinea	206	Sierra Leone	24
Congo, Dem. Rep.	205	Mali	23
Burkina Faso	204	Sri Lanka	22
Guinea-Bissau	200	Maldives	22
Nigeria	191	Chad	22
Zambia	182	Bangladesh	22
Burundi	181	Nepal	21
Cent. African Rep.	175	Ethiopia	20
Swaziland	164	Philippines	20
Guinea	161	Gambia	20
Rwanda	160	Pakistan	19
Cameroon	149	Senegal	19
Benin	148	Trinidad-Tobago	19
Mozambique	138	Micronesia	18

Source: UNICEF, http://www.childinfo.org/undernutrition.html.
Notes: Data for various years, mostly in the 2002–2006 range. Afghanistan and Somalia are excluded because comparable data are not available for all of Tables 6.5–6.7.

What Groups of People Are Undernourished?

Of course, hunger knows no boundaries: it afflicts the young, the old, the healthy, the sick, the working, the unemployed. But there are groups of people among whom undernutrition is more common.

Within any given country, malnutrition is likely to be more prevalent in rural areas. The World Bank, in its 1990 *World Development Report,* calculated that, among the forty-two countries it classified as low-income in 1988, rural people represented 65 percent of the population. Because rural incomes are usually considerably lower than urban incomes, and because undernutrition is so closely associated with low-income populations, a strong argument can be made that the majority of the world's hungry are rural. Because of this, and because so many policies affecting the price and availability of food to both rural and urban consumers impinge on the rural sector of the economy, we place heavy emphasis in this book on policies affecting the rural sector. The tension that exists between the rural and urban sectors of developing countries is discussed in more detail in Chapter 21. As the world moves further into the twenty-first century, developing economies are expected to become increasingly

Box 6.3 Undernutrition in the Developed World

All of our attention here has been focused on poor countries. What about rich countries such as the United States? Perhaps you have heard claims that thousands of children in this American city or that American state go to bed hungry every night. The source of those claims, and of the most in-depth study of hunger in the United States, is a food security report by the US Department of Agriculture (Nord, Andrews, and Carlson 2003).

That report concludes that food insecurity afflicts 12.5 percent of people and 18.1 percent of children in the United States (pp. 16–17). But what is food insecurity? The report based its definition on a ten-question survey (plus an additional eight questions for households with children). Some of the questions are rather subjective, such as: "In the past 12 months, did you worry that your food would run out before you got money to buy more?" Other questions are more concrete, such as: "Did you ever not eat for a whole day because there wasn't enough money for food?"

A household that answered yes to more than two of these questions was labeled "food-insecure." A household with children could be categorized as "food-insecure" if they "couldn't afford balanced meals" and "relied on a few kinds of low-cost food." If the household answered yes to six or more of the ten questions (or eight or more of the eighteen questions for households with children), the household was labeled "food-insecure with hunger." This more severe form of food insecurity afflicts 3.4 percent of people and 0.8 percent of children in the United States.

Of the households in the "food-insecure with hunger" category, only a tiny number (about 0.1 percent of all households) reported that their children went without eating for a whole day at any time during the year, and only 1 percent reported that adults went without eating for the whole day at any time in the year; 0.2 percent of households said that adults went without eating for at least one day in almost every month.

For poor families in the United States, running out of food and having no money to buy more can be difficult. But compare the descriptions in this box to the descriptions of the "hungry season" in Box 6.2. By international standards, the undernutrition problem in the United States is almost undetectable. The incidence of hunger is low because incomes in the United States—even incomes of the poor—are high by world standards, and because of public programs (such as food stamps) and private efforts (such as church-run food pantries) to help poor families.

urban, but even then, rural-oriented policies will remain especially significant among those aimed at undernutrition.

Children as a group are by far the most vulnerable to undernutrition, especially at weaning time—that transitional period during which an infant's diet is changed from 100 percent breast milk to 100 percent other foods. Though this transition can be abrupt, in the third world it often takes place during an eighteen-month time span, such as between 6 months and 2 years of age. While infants are being moved from breast milk toward other foods, their requirements for a calorie- and protein-dense diet are still very high, and in cultures where the diet is dominated by grains, providing an appropriate diet for weaning children takes a special effort.

Pregnant women and lactating mothers are the next most vulnerable to undernutrition, perhaps followed by elderly women. During times of extreme food shortages or famines, these groups are almost always the most at risk, but during such times a broader segment of the population, including large numbers of able-bodied men, is likely to be affected also.

We hear occasional reports of food deprivation based on gender. Roger Winter (1988), director of the US Committee for Refugees, writes about the Dinka people in Sudan, who have been plagued by both drought and the scorched-earth strategy of the Sudanese army, which has sought to subdue the Dinka's rebellious tendencies. Winter reports that many refugee groups consist chiefly of physically weakened young men and boys: "Women and children often are left behind, displaced and without access to international assistance or protection. They are dying in shockingly large numbers. In some areas, virtually all children under 3 are dead. Young girls are rare: In a society beset by war, with an economy based on cattle herding, girls are allowed to starve so that resources can be devoted to their brothers" (p. A25).

The Punjab in northwest India has the highest ratio of males to females in India. A study of the area (Das Gupta 1988) reported that the youngest daughters in families with many children are often selectively deprived of both medicine and the more nutritious foods in order that their brothers may be better cared for. Although coauthor Phillips Foster and colleagues looked, they did not find evidence of this practice during their unpublished 1981 study of Bagbana village in India.

A bias against girls is by no means universal. WHO data on underweight children break down the prevalence by sex of child. In seventy-six countries, the percentage of boys who are underweight is higher than the percentage of girls who are underweight. In forty-three countries, the percentage of girls who are underweight is higher. In three countries, the percentages are the same for boys and girls (see http://www.childinfo.org, a website administered by UNICEF).

When examining the at-risk groups, the emphasis on women and children does not mean that undernutrition never affects adult working men. Although clinical signs of undernutrition are rare in this group, significant numbers appear to suffer an energy depletion that limits work capacities and productivity.

So far, we have examined the number of people suffering from undernutrition, as well as the trends, geographic locations, and seasonality of hunger, and the people most vulnerable to it. But undernutrition occurs in individuals on a case-by-case basis. If we can identify the characteristics of a family that predispose them to undernutrition, we may, at the same time, uncover clues about appropriate policies for reducing its prevalence.

Arnold and colleagues (1981) examined data collected in 1979 by the National Nutrition Council of the Philippines. Their sample contained 722 families from seven provinces. To be included in the sample, a family had to have at least one preschool child. The purpose of the study was to see if family data could be used to predict the presence of an undernourished child. So for each family, the preschool child with the lowest level of nutrition was chosen as the subject. This child, therefore, became the dependent variable, and Arnold's group tried to predict the percentage of standard weight-for-age of this child. See Table 6.8 for some of the main findings of the study. Variables with a positive correlation coefficient have a positive influence on percentage of weight-for-age. That is, the higher the value of the variable, the more likely the subject child is to be well nourished. Some of the relationships are fairly obvious. The more education the father and mother have, and the more income the family has, the better nourished the subject child is likely to be.

The importance of weaning age in this case becomes clear when we realize that rice, which is low in both protein and fat, is the staple food in the Philippines. Therefore, the longer a child is breast-fed, the longer it receives an appropriately protein-rich, calorie-dense diet.

Variables with a negative correlation coefficient (those at the bottom of Table 6.8) have a negative influence on percentage of weight-for-age. Thus, mothers who bottle-fed their babies almost exclusively were more likely to have undernourished children than those who breast-fed exclusively. Children of mothers who mixed bottle- and breast-feeding tended to be better-off than those who were bottle-fed only, and worse-off than those who were breast-fed only. Children in large families were more likely to be undernourished than

Table 6.8 Correlation of Socioeconomic Variables with Percentage of Standard Weight-for-Age, Philippine Preschoolers (Under 6 Years Old), 1979

Socioeconomic Variable	Correlation Coefficient	Number of Families Sampled
Number of years of formal education of the mother	.27	721
Number of years of formal education of the father	.26	716
Income, farming families	.12	213
Income, nonfarming families	.35	499
Age of weaning if subject child is weaned	.34	545
Type of infant feeding (1 = breast alone, 2 = mixed, 3 = bottle alone)	–.37	718
Total number of household members	–.25	722
Birth order of subject child	–.21	722

Source: Arnold et al. 1981.

Note: All variables were significant at the .01 level except income, which was significant at the .07 level.

children in small families. And within the same family, children born later were more likely to be undernourished than their older siblings. Speculation has it that mothers sometimes give up, if slightly, on the youngest child after they have already borne three or four children (Scrimshaw 1978:389; 1984).

Another study has shown that the age of the mother at the time of giving birth may influence nutritional status. Children whose mothers are younger than 20 or older than 30 at the time of birth are more likely to be undernourished than children born of mothers between the ages of 20 and 30 (Rustein 1984).

Sometimes special circumstances exacerbate an already unfortunate situation and increase still further the possibility that a child will be undernourished. In 1980, coauthor Phillips Foster and colleagues visited a number of Filipino families who had at least one severely undernourished child. In most cases, in addition to the usual problems of low income, low education of the parents, and large number of children in the household, some special situation or stressor existed that might have significantly contributed to the undernutrition—for example, the mother suffered from tuberculosis and was always tired; the mother liked to gamble and left her 1-year-old in the care of her 4-year-old; the mother had borne twins and thought they carried a curse; the father was living with another woman; or the father was working in the city and came home only every other Sunday.

The Food Price Crisis of 2007–2008

Earlier in this chapter we referred to declining real food prices as an indicator that is consistent with, and that partially explains, the long-term progress in reducing the prevalence of undernutrition worldwide. But Figure 6.2 also shows a substantial upturn in prices, especially in 2007 and 2008. In fact, food prices rose so quickly and so drastically that weekly news magazines raised the alarm with cover blurbs such as "The End of Cheap Food?" (*The Economist,* December 7, 2007) and "The Silent Tsunami: The Food Crisis and How to Solve It" (*The Economist,* April 8, 2008).

Our general concern in this book is long-term trends and prospects, not short-term blips in prices. Figure 6.2 shows that even a big blip in price, such as that in the 1972–1975 period (or the smaller blip between 1978 and 1980), can be a temporary phenomenon that does not interfere with the long-term trend toward improvement in the world food problem.

But it would be a mistake simply to assume that every price increase is a temporary price blip. This is especially true in the current situation: the rapid price rise in 2007–2008 is not a sudden turnaround, but follows five years of persistent upward-edging prices, and perhaps as much as a decade of lack of progress in reducing the prevalence of undernutrition. Therefore, we should consider the particular facts of the 2007–2008 food price increase:

- According to the World Bank's Commodity Price Index for Food, food prices rose 47 percent between 2000 and 2006—about 6.4 percent per year, and higher than the average price increase for all goods in the United States of 2.5 percent per year over the same period.
- But the explosion in food prices began after 2006. Food prices rose more than 25 percent between 2006 and 2007 and then rose nearly 60 percent between 2007 and June 2008; food prices in June 2008 were nearly three times the level of food prices in 2000. The food price index, which has a value of 100 for the year 2000, rose to 293 for June 2008.
- Grain prices rose even more dramatically—over 75 percent between 2007 and June 2008, and peaked in that month at 3.34 times the level of prices in 2000.
- Between 2007 and the peak price in summer 2008, wheat rose from \$255 to \$348 per metric ton (a 36 percent increase), maize (corn) rose from \$164 to \$287, soybeans rose from \$384 to \$651 (a 75 percent increase), and rice rose from \$326 to \$902.
- However, as sharply as food prices rose, they came down almost as fast and far. The food price index had fallen from the June peak of 293 to 183 for November 2008 (back below 200; less than twice the price level in 2000); soybean, maize, and wheat prices had returned nearly to their 2007 levels; and rice price had fallen to about \$600 per metric ton from its peak of over \$900.

What caused this rapid rise in food prices? We postpone this question until the end of the next chapter, which will introduce the concepts of supply and demand and provide a framework for understanding how prices are determined in a competitive market.

Part 2

Causes of Undernutrition

THE MAIN CAUSES OF UNDERNUTRITION can be traced to economic, demographic, agricultural, and environmental variables. Part 2 explores these factors and how they influence the extent of undernutrition worldwide.

7

Economics: Supply and Demand

To this point, we have focused on the effects, measurement, and prevalence of undernutrition. We now begin to describe the factors that influence the occurrence and extent of undernutrition. Economics serves as the organizing framework for this discussion. This chapter is intended to be a very brief introduction to economic principles.

An Elementary Example of Supply and Demand

Alfred Marshall, a nineteenth-century British economist, is the father of modern economics. He used a simple example of a boy gathering raspberries to illustrate the concepts of supply and demand. Imagine a boy walking through the woods who spots a patch of raspberry bushes. The boy reaches out and plucks a raspberry and pops it into his mouth. He gets a burst of pleasure: he was hot, thirsty, and hungry; the raspberry refreshes him. As he continues to eat raspberries, two things occur. First, the pleasure from eating additional raspberries subsides; the pleasure he gets from eating the twentieth raspberry is less than the pleasure he got from that first raspberry. This may be because he is no longer as hungry, hot, and thirsty, or because the novelty of eating raspberries has worn off, or because the twentieth raspberry is not as ripe as the first one. Second, the boy discovers it is harder and harder to find raspberries to pick. He has to kneel on the ground, or reach into the brambles, to find additional raspberries.

If we were to graph this experience, it would look like Figure 7.1. The downward-sloping curve shows the pleasure the boy gets from each successive raspberry: he gets a lot of pleasure from the first berry, slightly less additional pleasure from the second, still less from the third, and so forth. The upward-sloping curve shows the amount of effort needed to obtain each successive berry: the first requires very little effort, the second requires a little more, the third even more, and so forth.

We can also use the graph in Figure 7.1 to track the boy's reasoning. Should I eat the first raspberry? Yes: the pleasure from that is very high and the cost

Figure 7.1 The Boy in the Berry Patch

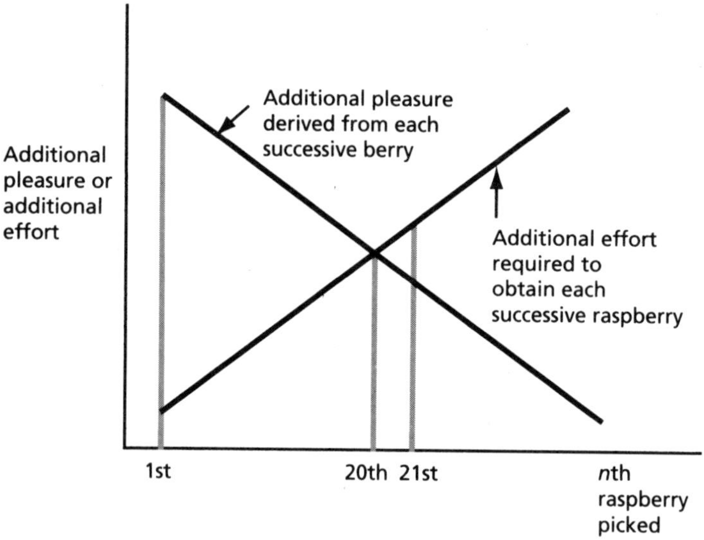

of obtaining it is very low. Likewise for the second, and the third. When he gets to the twentieth berry, the pleasure from that berry has dropped, and the cost of obtaining it has increased, but it is still (just barely) "worth it"—the pleasure derived exceeds the effort needed to obtain it. When he considers the twenty-first berry, he realizes that this berry is no longer worth the effort. In terms of the graph, the intersection of the "additional pleasure" curve and the "additional effort" curve shows the point at which the boy stops.

We can characterize the boy's decision as an "optimal" decision in the following sense. The boy's decision is one that maximizes his "surplus pleasure," defined as the pleasure from consuming berries, minus the discomfort of gathering them. This illustrates a fundamental principle of economics: an optimal decision is one in which the "marginal benefit"—in our example, the additional pleasure gained from eating one more berry—equals the "marginal cost"—the additional effort needed to pick that berry. Note that in the raspberry example, there is no money—the boy does not put a dollar value on his pleasure or on his effort. Yet there is an implicit comparability: the boy can say, "The pleasure from the berry is worth (or not worth) the additional effort."

A Producer's Supply Curve and a Consumer's Demand Curve

The fundamental concepts of economic behavior are contained in the story of the boy and the raspberries. The only difference when it comes to modern economics is that pleasure is calculated explicitly in monetary terms: Figure 7.2 is exactly like Figure 7.1, except that the Y-axis is now labeled "price" to indicate cost per unit of the good.

The downward-sloping curve is an individual consumer's *demand* curve. It shows how much of a good the person will want to consume at all the different possible levels of price. An alternative interpretation is that the demand curve shows the maximum amount per unit the person is willing to pay for all the different possible levels of consumption. The downward slope of the demand curve means that as the price of a good falls, a consumer will want to consume more of that good. (See Box 7.1 for an example of a downward-sloping demand curve.)

The upward-sloping curve in Figure 7.2 is an individual producer's *supply* curve. It shows how much of a good the producer will produce at all the different possible levels of price. Alternatively, it shows the minimum amount per unit that the person will require to produce at each of the different possible levels of production. The upward slope of the supply curve means that as the price of a good increases, a producer will produce more of that good. The concepts of costs of production and the technical relationships between input and output are discussed in Box 7.2.

Aggregate Supply, Aggregate Demand, and Markets

Of course, in real-world markets, the number of consumers does not equal the number of producers. A single potato farmer grows enough potatoes to feed

Figure 7.2 Supply and Demand Curves

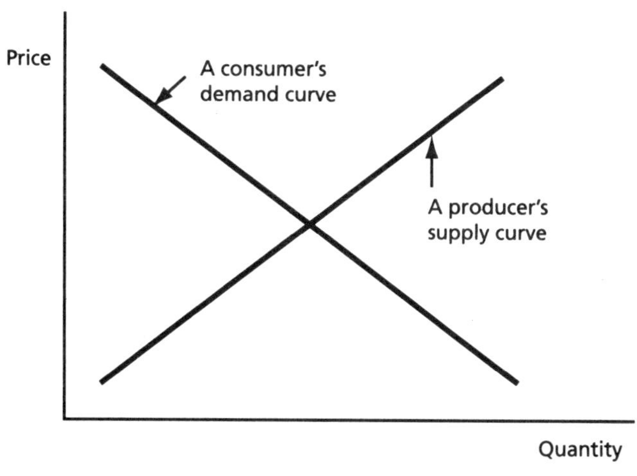

Box 7.1 Demand Curve Slopes Down

The city of London, England, has a problem: too many people driving too many cars. Economists have a solution: increase the costs associated with driving in the city. In February 2003, a congestion fee of 5 pounds (about $8) was imposed on all cars driving during weekdays (Monday to Friday) in a particular zone of central London. Cameras recorded the license plate numbers of cars that entered the zone to ensure that drivers paid the required fee. The hoped-for result was that automobile traffic would be reduced by 15 percent by 2010. Economists estimated that the fee would reduce traffic by 12–17 percent. But by early April 2003, traffic loads appeared to have stabilized at 20 percent lower than previous levels. The demand curve for driving in central London was flatter (economists would say, "the elasticity of demand was higher, in absolute value") than economists predicted. But the economists were right in the direction of their prediction: if you want people to consume less of a good, raise the price of the good.

Sources: Blow, Leicester, and Smith 2003; Transport for London 2003.

hundreds of people, and a single consumer buys different kinds of food produced by many different farmers. But the market serves as a place where many consumers and many producers can interact in a process that sets an equilibrium price at the point where the *aggregate supply curve* intersects the *aggregate demand curve,* illustrated in Figure 7.3. Recall that the individual producer's supply curve shows how much of a good the producer will produce at different price levels. If we add up (aggregate) those quantities for all producers, we have a picture showing the total quantities that will be produced at different price levels. Likewise, we add up quantities demanded by individual consumers to obtain the aggregate demand curve.

The point at which the aggregate supply and demand curves intersect shows the equilibrium price and aggregate quantity. If the price is above this equilibrium price, producers want to sell a greater quantity than consumers want to buy. Producers who cannot find a buyer for their output will try to attract buyers by offering a lower price. If the price falls below the equilibrium price, some buyers will be unable to buy the quantities they want, and the price will increase.

In a complex economy with many consumers, many producers, and many goods, the prices determined by the interaction of aggregate supply and aggregate demand for each good serve to organize or direct consumption and production patterns. Under certain restrictive assumptions, the consumption and production patterns dictated by the price mechanism will be the best choices in the following sense:

- Each consumer tries to obtain the most pleasure possible out of the money he or she has to spend. The consumer does this by trying to obtain the same amount of pleasure from the last dollar spent on each good. To understand

Box 7.2 Why the Supply Curve Slopes Upward

There is a physical basis and also a cost basis for the relationship between price and the amount that a farmer will try to produce. Let us start with the underlying physical relationship, using as our example the relationship between the amount of seed planted in one field (say, a hectare of land) and a crop yield. The raw data we assume are given in Table A and plotted in Figure A. The top curve in Figure A represents the total yield from varying amounts of fertilizer (total physical product or production function). The bottom curve represents the yield added by each successive increment of ten units of seed (marginal physical product).

Table A: Hypothetical Yield Response to Varying Amounts of Seed

Seed	Yield	Marginal Physical Product
0	0	39
10	39	13
20	52	9
30	61	5
40	66	0
50	66	–2
60	64	

Figure A

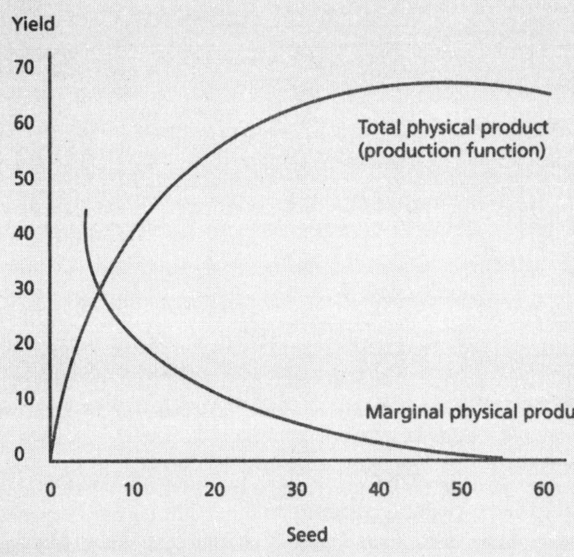

What we have in Figure A is a graphic representation of the fact that, as we increase the amount of an input used, holding other inputs constant, we experience diminishing returns to successive inputs (thus the downward-sloping marginal

(continues)

Box 7.2 continued

physical product curve—which is often called diminishing marginal returns). These functional relationships are based on observations that have been made in the real world.

Now let us combine the variable inputs commonly used to increase production on our hypothetical hectare of land. As we increase production, we add not only more seed, but also more fertilizer, more labor, and maybe more of other inputs such as irrigation water and pesticides. These things cost money. As we increase production, we could, at various amounts produced, add up the costs of the things we are using to increase production and plot these sums to obtain a cost curve.

The cost curve would show very much the same thing that Figure A shows, except that it would measure costs, instead of seed, along the horizontal input. (The relation between cost and production is based on the underlying physical relationships between inputs and production.) We diagram our cost curve in Figure B. Because the fixed cost of land is not included in our set of costs, we identify the costs in this diagram as variable costs.

Figure B

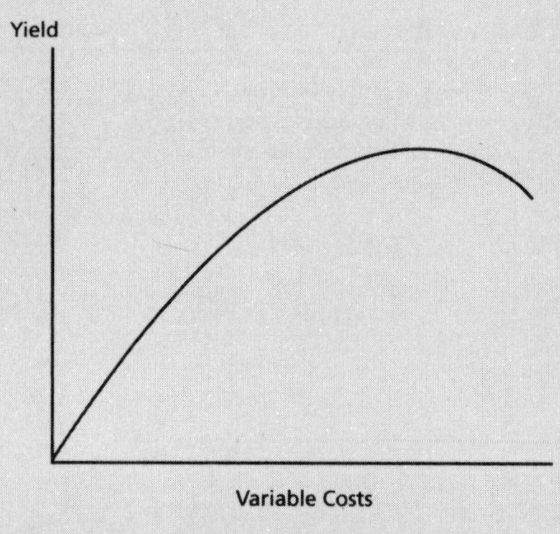

Although not specifically diagrammed, notice in Figure B that as yield increases, there are diminishing marginal returns to costs, just as there were diminishing marginal returns to seed in Figure A.

It is a convention of economics to draw cost curves with the cost on the vertical axis and the yield on the horizontal axis. So let us redraw Figure B in the conventional way, namely as shown in Figure C. (To see what happens to the cost curve when this is done, trace Figure B and then flip it over and look at it from the back side, to obtain Figure C.)

(continues)

Box 7.2 continued

Figure C

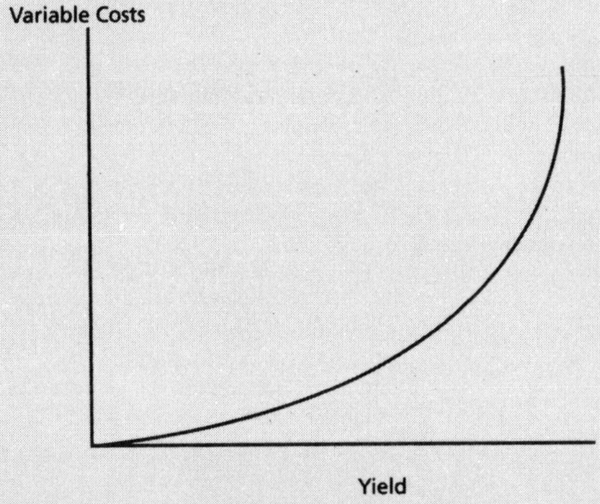

With Figure C, instead of thinking about what happens to yield as we increase costs by one unit, we can think about what happens to costs as we increase yield by one unit. In Figure C we see that we produce under conditions of increasing marginal cost. For each additional unit of yield from our hectare of land, we have to spend somewhat more on our bundle of variable costs. These increasing marginal costs are diagrammed in Figure D.

Figure D

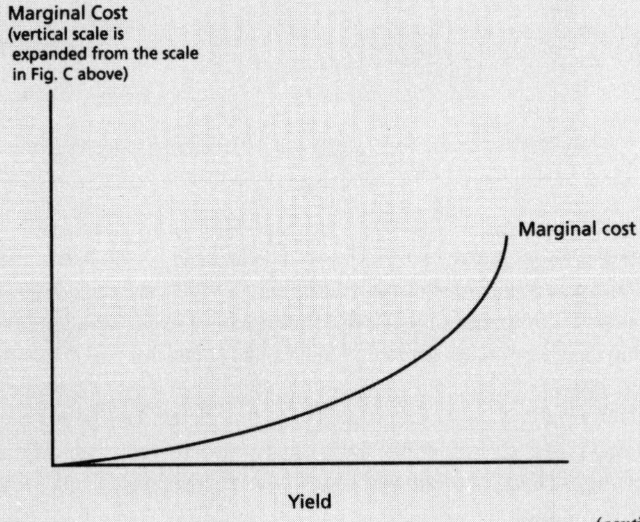

(continues)

Box 7.2 continued

Thus we see that we are producing under diminishing marginal returns, which results in producing with increasing marginal costs. The two are based on the same underlying physical relationship.

For an individual producer, the marginal cost curve (Figure D) is the price schedule at which he is willing to produce various amounts of goods for the market. He is willing to produce up to the point where the price equals his marginal cost of production. If he is producing at this point and we want to motivate him to produce more, barring a shift in technology or a reduction in the costs of some of his inputs, we will have to pay him more. This is represented in Figure E as shifting production from A to B, which is motivated by an increase in price from P_1 to P_2.

Adding up the marginal cost curves (individuals' supply curves) for all the individual producers yields the supply curve for the industry. Increasing the price of the product will motivate the industry to produce more.

Figure E

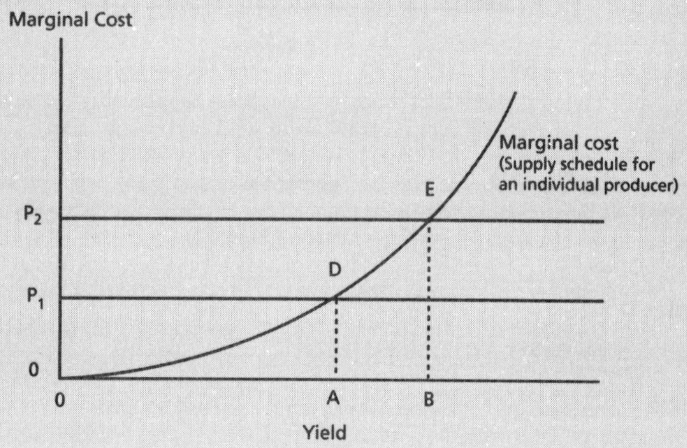

what this means, suppose that this were not the case; instead, suppose that the last dollar spent on bananas, for example, provides the consumer with very little pleasure, compared to the last dollar spent on tomatoes. In this case, the consumer would get more pleasure if he or she spent that last banana dollar on tomatoes instead.

- Each producer produces at the point that gives the highest profits (revenue minus cost). The producer does this by producing units only as long as the received price exceeds the additional cost of production, and by refusing to produce any additional units.

Figure 7.3 Aggregate Supply and Demand

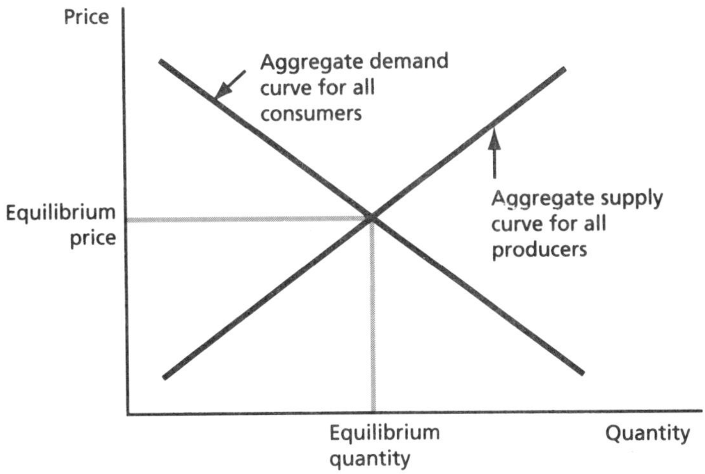

This description of how a competitive market equilibrium, in which each producer and consumer acts in his or her own self-interest, resulting in an allocation of goods that is (in the sense described above) optimal for the group as a whole, is at the heart of Adam Smith's image of an "invisible hand" guiding economic activity (see Box 7.3). This concept will be discussed more fully in Chapter 16, which explores criticisms of the "economics approach" to policy analysis.

Combining Goods into Groups

In many economic discussions, and frequently in this book, the commodities referred to are not actually single goods. We have some idea of what it means to talk about the price of a tomato, or the price of a banana, or the price of a bushel of corn or a pound of chicken wings. But when we talk about the price of "food" or the quantity of "food," that is an abstract concept. In the rest of this chapter, we refer to "aggregate supply" and "aggregate demand" as shorthand terms meaning aggregate supply and demand of food.

How Demand Changes over Time

One important aspect of our study of the world food problem is to understand how supply and demand for food have changed in the past and how they might

Box 7.3 Adam Smith's "Invisible Hand"

"[E]very individual . . . neither intends to promote the public interest, nor knows how much he is promoting it. . . . [B]y directing that industry in such a manner as its produce may be of the greatest value, he intends only his own gain, and he is in this, as in many other cases, led by an invisible hand to promote an end which was no part of his intention. Nor is it always the worse for the society that it was no part of it. By pursuing his own interest he frequently promotes that of the society more effectually than when he really intends to promote it. I have never known much good done by those who affected to trade for the public good. It is an affectation, indeed, not very common among merchants, and very few words need be employed in dissuading them from it."

Source: Adam Smith, *Wealth of Nations*, Book IV, http://socserv2.socsci.mcmaster.ca/~econ/ugcm/3ll3/smith/wealth/index.html.

be expected to change in the future. In this section we examine factors that shift the aggregate demand for food.

Recall that aggregate demand for food shows the total quantity of food demanded at every level of price. The total quantity of food demanded at any given price might increase for any of the following reasons: (1) the number of people in the economy increases; (2) people in the economy have more money to spend on all goods, so part of the additional income is spent on food; (3) people's tastes change, such that they get more pleasure out of food compared to other nonfood goods. (Economics students will recognize that we have failed to mention an additional factor: [4] prices of other nonfood items increase, such that the pleasure per dollar of additional expenditure on those nonfood goods declines.)

Population Growth Shifts the Aggregate Demand Curve

Aggregate demand for food depends on how many individual consumer demand curves we are aggregating. This is the usual departure point for discussions of the world hunger problem: Can food supplies keep up with population growth? In other words, how many people are there to feed? This will be the focus of Chapter 8, and population policy will be discussed in Chapter 18. For now, we simply note that an increase in population causes an outward shift in the aggregate demand curve for food. At any particular price level, the aggregate quantity demanded is higher because there are more people contributing to the aggregate demand. Holding all other factors constant (and in particular assuming that the aggregate supply curve remains constant), the effect of the outward shift in aggregate demand is to increase the price, and the higher price induces farmers to produce more food, so that the aggregate equilibrium quantity increases.

Changes in Income or Income Distribution Shift the Aggregate Demand Curve

Aggregate demand for food is not the same thing as quantity of food that will provide every individual a healthy diet, nor is there any implicit promise that the equilibrium price will guarantee that the average individual (or all individuals, or even most individuals) will have sufficient food. Aggregate supply and demand can describe famine conditions or conditions of great plenty.

An individual's demand curve for food shows how much food he or she will want to buy at every price level, *given his or her income.* A person with a low income will have a demand curve that reflects low quantities. For most goods (so-called normal goods), if the person gets more income to spend, his or her demand curve will shift up and to the right, as shown in Figure 7.4: at every price, the person will demand a higher quantity.

How big will the shift be? Economists use the concept of "income elasticity of demand" to measure this. This concept and some measures of the income elasticity of demand will be discussed in more detail later in this chapter, where we will show that the size of the shift is different for a poor person than for a rich person. If a poor person's income increases by 10 percent, the person will spend a lot of the increase on food and the person's demand curve for

Figure 7.4 Outward Shift in Demand

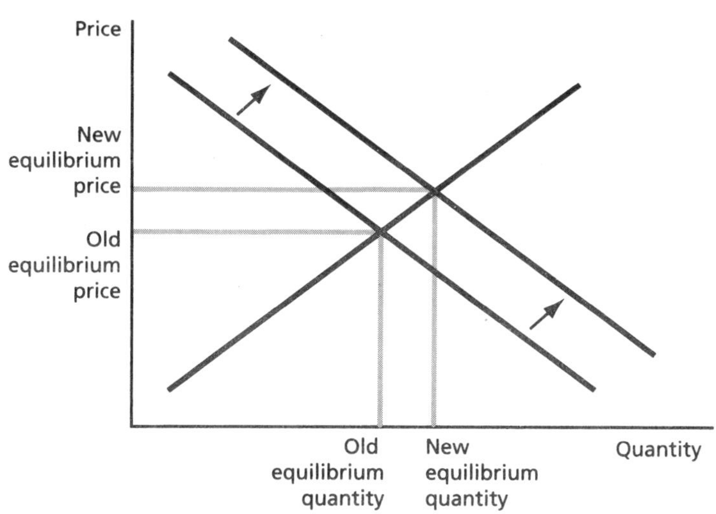

food will shift a relatively large amount. If a rich person's income increases by 10 percent, the person will spend little of the increase on food and the person's demand curve for food will shift a small amount. This fact has implications for the relationship between changes in aggregate demand for food and changes in income distribution.

If everyone's income grows, each person's demand for food will shift out and aggregate demand will also shift out. If average income stays the same, but income is redistributed from a rich person to a poor person, the aggregate demand curve for food will also shift out, since the poor person's demand for food will shift out by a large amount, while the rich person's demand for food shifts back (in the opposite direction of the arrow in Figure 7.4), but by a smaller amount.

Changes in Tastes and Preferences Can Shift the Aggregate Demand Curve

A final factor that may shift the aggregate demand curve for food is worth mentioning, even though it may seem obvious: changes in people's tastes and preferences over time. To see how this works at the level of the individual, let us return to our analogy of the boy and the raspberries and imagine that the same boy revisits the same raspberry patch the following summer, except that this year he is wearing braces on his teeth. As he begins to eat the raspberries, he discovers that the tiny seeds get stuck in his braces in an annoying way. The pleasure he derives from eating the berries is less. In economics terminology, his individual demand curve has shifted down and to the left.

A change in tastes and preferences is not the same thing as a change in quantity of food a person wants to buy. As the price of food changes, the quantity of food a person wants to buy will change as the person moves along their (constant) demand curve. As a person's income changes, the quantity of food the person wants to buy will change as their demand curve shifts. A change in tastes and preferences causes a change in the quantity of food a person wants to buy even as prices and income remain constant. For example, a person who decides to lose weight by eating less food will have a shift in demand.

In the aggregate, some trends in general behavior may cause a detectable shift in aggregate demand for food in developed countries: an increasing concern with obesity and overeating, an increasing interest in diets low in carbohydrates, an increasing interest in food produced "organically," and a shift toward a more vegetarian diet.

How Supply Changes over Time

We next look at two types of changes that may cause shifts in the aggregate supply of food over time: changes in availability (or price) of resources used to produce food, and changes in technology or in efficiency of resource use.

Changes in Availability of Resources Used to Produce Food

If the resources used to produce food become more readily available—if the price of those resources falls—then at any level of output price, farmers will be willing to produce more. In the berry patch analogy, this is equivalent to an increase in the density of berries on the bush, or a decrease in the density of the thorns. Because it is less costly to produce an additional unit, more units are produced. An increase in production at every price level means an outward shift in supply, such as that illustrated in Figure 7.5.

Some simple examples illustrate this. If a farmer obtains more land, then at every price level he will produce more output; his individual supply curve will shift out. If the farmer has more children able to work in the field, the extra labor will cause the farm's supply curve to shift out. If a dry creek begins to run with water so that the farmer can irrigate, the supply curve will shift out. If the general weather pattern changes to be more conducive to agricultural production, the supply curve will shift out. If the soil on the farm is eroded, or soil nutrients deplete, the supply curve will shift back (in the opposite direction of the arrows in Figure 7.5). All of these—land, labor, water, weather, soil—are examples of productive resources; changing their availability shifts the supply curve.

When productive resources are bought and sold on a market, changes in general availability are reflected in the level of price for the resources. If fertilizer becomes more plentiful, the fertilizer price drops, and the supply of agricultural output shifts out. Similarly (as we will see later in the book), a government

Figure 7.5 Outward Shift in Supply

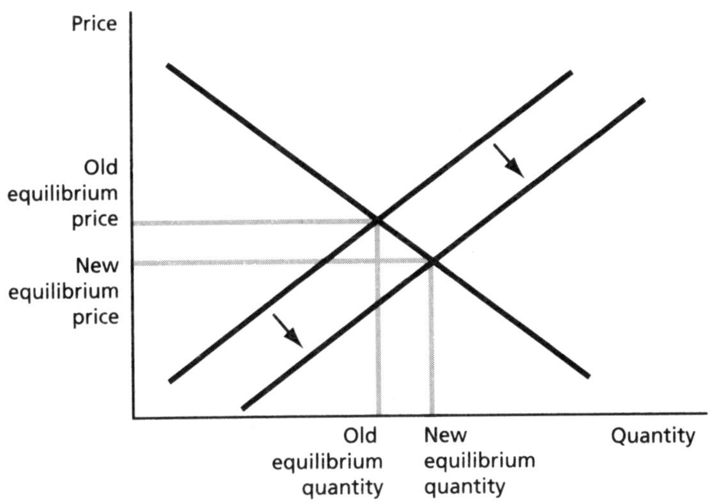

program to reduce fertilizer price (or to reduce the price of any other productive resource) has the same effect: it reduces farmers' production costs and causes an outward shift in the supply of agricultural output. During 2007–2008, oil and gasoline prices more than doubled; during that period (as noted at the end of Chapter 6) the supply curve for food shifted back (up and to the left) and contributed to the increase in food prices.

Changes in Technology or in Efficiency of Resource Use

The amount of food produced from a given set of resources also depends on the efficiency with which the resources are used. Available water can be used efficiently to irrigate, or some of it can be wasted in runoff. Labor hours of a second worker can (inadvertently or intentionally) undo the work of the first worker. Anything that systematically improves the efficiency of resource use will have the same impact as an increase in resource availability.

A technological improvement is some kind of new knowledge that allows farmers to produce more output with the same resources. Often the new knowledge is embodied in some piece of equipment (e.g., the invention of the plow) or other input (a new variety of seed). But the concept is broad enough to encompass improvements in the ability to predict weather patterns, or a better understanding of agronomic processes. A technological improvement is similar to a reduction in the price of a productive resource: both reduce the cost to the farmer of producing an additional unit of output, and both shift the supply curve out.

Using the Concept of Elasticity to Quantify Economic Changes

So far in this chapter we have focused mainly on the direction of changes: as price goes up, consumers reduce the quantity of food demanded and producers increase the quantity of food supplied; as income goes up, consumers increase the quantity of food demanded. But by how much? If the price of food increases by 3 percent, how much will quantities supplied and demanded change? If a household's income increases by 12 percent, how much will the household increase the quantity of food purchased?

Economists answer these questions using a concept called "elasticity." An elasticity measures the percentage change in quantity supplied or demanded in response to a 1 percent change in price or income. Here we will limit our attention to three types of elasticity. *Income elasticity of demand* measures the extent to which a demand curve shifts out in response to an increase in income. *Price elasticity of demand* measures the steepness of the demand curve: How much does quantity demanded fall as price increases (or how much does quantity demanded rise as price decreases) as we move along a demand curve? *Price elasticity of supply* measures the steepness of the supply curve: How does quantity supplied change in response to a change in price as we move along a supply curve?

Income Elasticity of Demand

In simplest terms, income elasticity of demand is the percentage change in the consumption of something, such as rice, when a 1 percent change occurs in income. How this elasticity will change depends on income level. If a poverty-stricken Indian villager suffers a 1 percent decrease in income, he might decrease his consumption of rice by half a percent or so. But a wealthy stockbroker who suffers a 1 percent decrease in income might not change her rice consumption at all (she might spend less on recreational travel, for instance, but this is getting ahead of our story).

Let us look at some general ways in which food consumption changes as income rises or falls. Studies of how food consumption changes in countries as their average per capita incomes grow show consistently that food consumption increases as income increases, but that the rate of increase in consumption falls off as incomes reach higher and higher levels. This is because the proportion of the household budget spent on food decreases as income increases. The first person to write about this was Ernst Engel, and the phenomenon has become known as Engel's law.

Figure 7.6 provides an illustration from Indonesia of how food is related to income. Only the families in the top half of the income groupings (the right-hand half of Figure 7.6a) consume a minimally sufficient number of calories. Figure 7.6b provides a second illustration of Engel's law: as income rises, the percentage of income spent on food declines from 75 for the lowest-income families in the sample population, to 60 for the highest-income families. Figure 7.6c shows how the food expenditure mix changes as income increases. As income grows, the East Javanese spend a smaller proportion of their food budget on starchy staples—cassava, rice, maize, and wheat flour—and a larger proportion on other items, especially animal products. This phenomenon is called Bennett's law (after agricultural economist Merrill Bennett; see Bennett 1941), which states that the *starchy staple ratio* (the ratio of starchy foods such as cereals and root crops to other foods in the diet) falls as income increases. Figure 7.6d shows how the energy derived from various food sources shifts as income rises.

One aspect of Figure 7.6d deserves special attention. Notice that for people in this survey, starting at the very lowest levels of income, additional income leads to an *increase* in cassava consumption; but for only slightly higher levels of income, increased income leads to a *decrease* in cassava consumption.

Figure 7.6 Relationships Between Income Level and Nutritional Status, East Java, Indonesia, 1977–1978

a. Energy (Kcal)

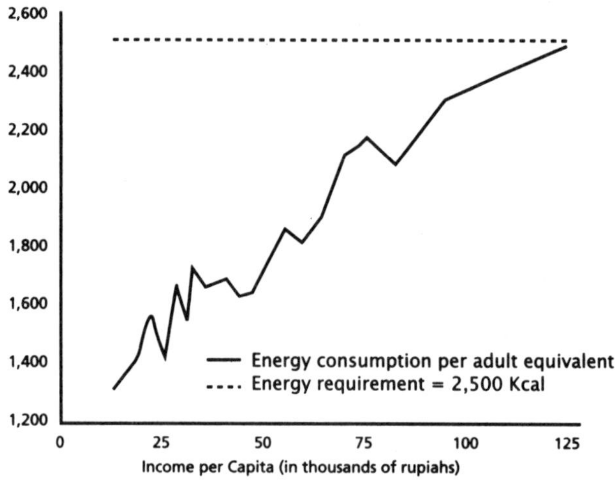

b. Food Expenditures (percentage of income)

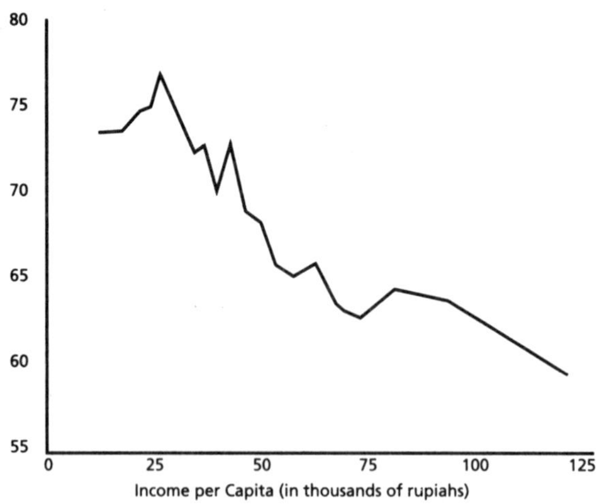

(continues)

Figure 7.6 continued

c. Percentage of Food Budget

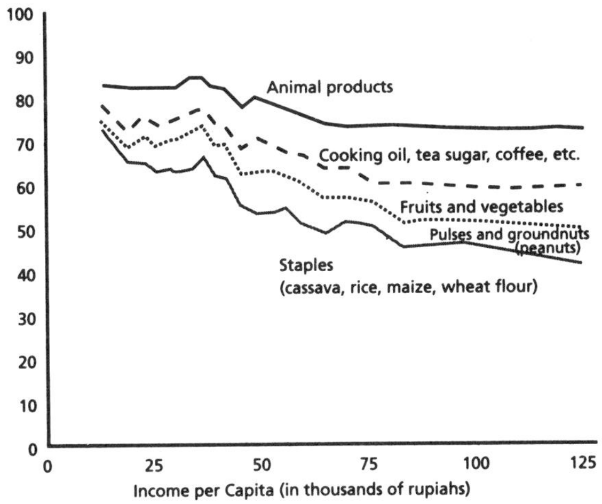

d. Percentage of Total Energy Consumption

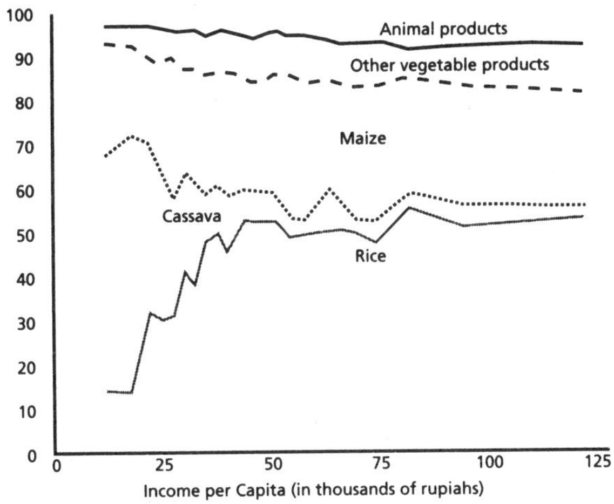

Source: Ho 1984.

Note: An income of 125,000 rupiahs in 1977–1978 was worth something over $200. The study region included Madura and the nearby regency of Sidorajo.

In a situation like this, as a person's income rises, instead of demand shifting out (as shown in Figure 7.4), the demand curve shifts in the opposite direction—quantity demanded at any given price is lower than it was before the income increase.

Commodities like this—of which people consume less as their incomes rise—are known by the ignominious name of *inferior goods.* A study of maize consumption in South Africa (Alderman and Lindert 1998:221) showed a similar pattern. At very low income levels, households used additional income to *increase* their maize consumption, but as income reached higher levels, households used additional income to *reduce* their maize consumption and replace it with consumption of other, more palatable foods. In the United States, during the recent economic slowdown, grocery stores reported large increases in sales of the meat product Spam (Martin 2008). This implies that as incomes fall (during a recession), demand for Spam shifts out; so, in the United States, Spam is an inferior good—it has a negative income elasticity of demand.

The explanation for this behavior is that inferior goods (like cassava in Indonesia or maize in South Africa) provide the most calories per dollar spent, but are not the tastiest foods available. For very poor people who have insufficient calories in their diets, increased income provides an opportunity to reduce the extent of undernutrition in the household through purchase of the most efficient source of calories. But for slightly richer households where undernutrition is less of a problem, increased income provides an opportunity to add tastier food to the household diet without significantly harming nutritional status. (An extreme example of inferior goods is discussed in Box 7.4.)

In Figure 7.7, we see that, for the world as a whole, changes in diet are consistent with Bennett's law. Carbohydrates make up about 70–75 percent of the dietary calories of the lowest-income countries, and only 45–50 percent of the dietary calories of the highest-income countries.

Quantifying income elasticities. As discussed earlier, income elasticity of demand is the percentage change in the consumption of something, such as rice or calories or "all food," when there is a 1 percent change in income, holding all other variables (such as price) constant. Reading this definition, one might think that it would be a fairly simple matter to estimate an income elasticity: observe a change (or difference) in consumption, such as "per capita milk consumption increased 2 percent between 2004 and 2008"; observe a comparable change (or difference) in income, such as "per capita income increased by 4 percent between 2004 and 2008"; divide the first number by the second number. But such a procedure would fail to give a good estimate of an income elasticity, because it would fail to hold other things constant—prices and tastes also changed between 2004 and 2008. In the real world of constantly changing incomes and prices, estimating elasticities is a lot more difficult than this. We leave descriptions of how this is done to others (e.g., Deaton and Muellbauer

Box 7.4 An Economic Experiment to Verify the Existence of Giffen Goods

Earlier in Chapter 7, we mentioned Alfred Marshall as the father of modern economics. Marshall credited his colleague, Robert Giffen, with the following insight: "[A] rise in the price of bread makes so large a drain on the resources of the poorer labouring families, . . . that they are forced to curtail their consumption of meat and the more expensive . . . foods; and, bread being still the cheapest food which they can get and will take, they consume more, and not less of it." In other words, Giffen posited that for very low-income consumers, an increase in price could actually cause an *increase* in consumption of inferior goods.

Economists have long recognized the theoretical possibility that Giffen pointed out. However, they have looked in vain (until recently) for a real-world example of a "Giffen good." Introductory economics textbooks sometimes suggest that potatoes during the Irish famine might be an example, but careful consideration (Rosen 1999) raises serious doubt. "It is unlikely that consumption of potatoes could have increased when the price rose during the famine, at least in the aggregate, precisely because the price rise was caused by a blight that destroyed much of the crop" (Jensen and Miller 2008:1553–1554).

A problem with finding real-life examples of Giffen goods is that if we look at *aggregate* changes in quantity consumed when the price of a staple good rises, then we combine in that aggregate the consumption responses of very low-income people (for whom the staple good is a normal good) with the consumption responses of higher-income people (for whom the staple good is an inferior good). One way around this problem is to construct artificial situations in which people with the same income levels face different prices. These artificial situations are know as "economic experiments."

Jensen and Miller went into the Chinese countryside (Hunan province) and conducted an economic experiment designed to test whether Giffen behavior could be observed in the real world. They handed out coupons that entitled the holder to buy rice at reduced prices. In return, the coupon recipients agreed to be interviewed about their incomes and consumption patterns. Based on this experiment, Jensen and Miller conclude: "For the group consuming at least some substantial share (20 percent) of calories from sources other than rice, i.e., the poor-but-not-too-poor, we find very strong evidence of Giffen Behavior. . . . When faced with an increase in the price of the staple good, these households do indeed 'consume more, and not less, of it'" (2008:1566).

1980; Huang 1985; Johnson, Hassan, and Green 1984). We will accept elasticity estimates done by others and concentrate on their meaning and implications for policy planners.

T. J. Ho (1984) calculated income elasticity for the East Javanese consumers whose food consumption patterns were outlined in Figure 7.6. She found the income elasticity of expenditure on food to be 0.58. That is, for this community, a 1 percent increase in income will produce a 0.58 percent increase in spending for food.

Figure 7.7 Percentage of Calories Derived from Fats, Carbohydrates, and Proteins, by Annual GNP per Capita, 2001

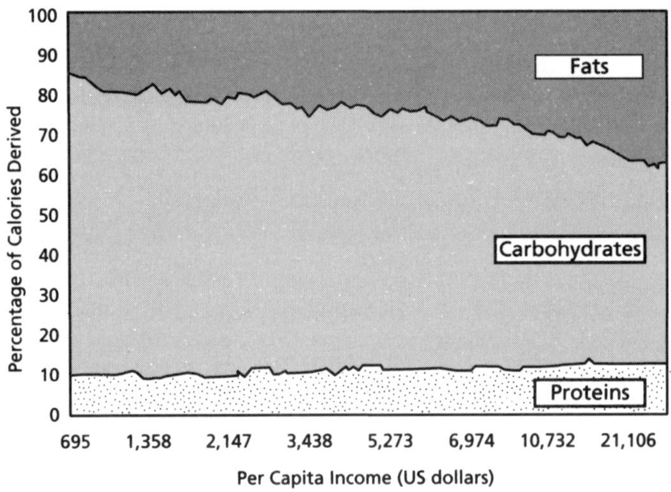

For purposes of illustration, let us hypothesize that this community spends half its income on food and, to make the computations simple, let us assume that its income is \$100. A 1 percent increase in income raises income to \$101. If food spending increases by 0.58 percent, these people will spend 29 cents more on food, for a total of \$50.29. The remaining 71 cents of increased income will be spent on nonfood items, bringing the total for that category (housing, clothing, paying off debts, etc.) to \$50.71. We can deduce, therefore, that the income elasticity of demand for nonfood items in this community is 1.42. Had this community increased its spending on food by exactly the same percentage (and in this case exactly the same dollar amount) as its spending on nonfood items, this would indicate that the income elasticity of demand for both food and nonfood items is exactly 1.0.

Bouis and Haddad (1992), in a review of the literature, report a wide range of calorie-income elasticities, from 0.01 in Nicaragua to 1.18 in India. They argue that the wide range is attributable in large part to the different methods that different studies use to collect data on calories, and to different conceptual measures of "income." They conclude that if appropriate measures are used, elasticity estimates fall in the 0.08–0.14 range.

Ho (1984) estimates the income elasticity of demand for calories to be 0.28 and for protein to be 0.52. These elasticities condense some of the information in Figure 7.6d. Review this graph and notice again how the East Javanese are substituting rice for cassava as their incomes increase. To a lesser extent, they are substituting animal products and other vegetable products for cassava at the same time. Calories from cassava are cheaper than calories from rice, other vegetables, or meat. Therefore, as the East Javanese increase their spending on food, they buy fewer calories per rupiah and more of other food properties such as protein content and flavor. This is what the income elasticity figures (low for calories and high for protein) are telling us.

Here is another way of looking at those elasticity figures for calories versus protein. A 1 percent increase in income yields only a 0.28 percent increase in consumption of calories but a more generous 0.52 percent increase in protein consumption. That is, as income changes, the East Javanese change their consumption of protein more (in percentages) than they change their consumption of calories. While it is easy to arrive at this conclusion without the aid of the elasticity figures, those figures allow us to calculate how much the people of East Java change their consumption of these two nutrients when income changes. This is the beauty—and the importance—of elasticity. It quantifies things.

Of course, people do not go to the market and buy nutrients like calories and protein. They buy food. So let us look at some income elasticity figures for particular commodities and see how these elasticities change with income. A detailed table of income elasticities in various countries for a variety of food groups and nutrients can be found at the website of the US Department of Agriculture (USDA) (see http://www.ers.usda.gov/data/internationalfooddemand).

How income elasticity changes with income. Table 7.1 shows some income elasticities for three income groups in rural Brazil. Notice that, except for cassava flour, the income elasticity figures are positive. That is, for most foods, consumption increases as income increases. Cassava flour, with a negative income

Table 7.1 Income Elasticities for Calorie Intake, Selected Foods, by Income Group, Rural Brazil, 1974–1975

	Income Group		
	Lowest 30 Percent	Middle 50 Percent	Highest 20 Percent
Cassava flour	–3.50	–1.590	–.356
Rice	1.99	.172	.173
Milk	2.27	.147	.172
Eggs	1.93	.630	.114
Mean per capita calorie intake	1,963	2,432	2,771

Source: Gray 1982:26.
Note: The 1974–1975 National Household Expenditure Survey (ENDEF) of the Brazilian Geographical and Statistical Institute was used as the database from which to calculate these income elasticities.

elasticity, is an inferior good. Notice also that the absolute values of these elasticity figures (i.e., their values regardless of negative/positive sign) tend to decrease from low- to high-income consumers. That is, food consumption among low-income consumers is considerably more responsive to changes in income than it is among high-income consumers. This phenomenon is found among all third world populations.

When the low-income consumer in the Brazilian sample receives a 1 percent increase in income, he or she tends to increase rice consumption by about 2 percent. But the high-income consumer tends to change his or her consumption of rice by less than 0.2 percent when income changes by 1 percent. We call the low-income family's consumption response own-price *elastic*. That is, a 1 percent change in its income yields a greater than 1 percent change in consumption. On the other hand, we call the high-income family's consumption response own-price *inelastic*. That is, a 1 percent change in its income produces a less than 1 percent change in consumption. If a family changed its consumption of rice by exactly 1 percent when its income changed by 1 percent, we would say it had an income elasticity of demand for rice of 1.0—neither elastic nor inelastic.

Remember that the elasticity figures show percentage change in consumption following a 1 percent change in income. The income elasticity of demand for rice in rural Brazil is about the same as the income elasticity of demand for eggs. But because rice makes up 15 percent of the total calories consumed among this income group, and eggs only 0.5 percent, a 1 percent change in income results in a far greater change in actual consumption of rice than of eggs (see Table 7.2). The nutritional significance of an elasticity, therefore, depends not only on the magnitude of the elasticity but also on the magnitude of consumption of the goods under consideration.

Table 7.2 Changes in Calorie Consumption Resulting from a 1 Percent Increase in Income, Selected Foods, Lowest 30 Percent of Consumers by Income, Rural Brazil, 1974–1975

	Kilo-Calories	Percentage of Total Kilocalories Consumed	Income Elasticity	Change in Kilocalories Consumed Resulting from a 1 Percent Increase in Income
Cassava flour	440	22.4	–3.50	–15.4
Rice	296	15.1	1.99	5.9
Milk	41	2.1	2.27	.9
Eggs	7	.5	1.93	.1
Other	1,152	59.9		
Total kilocalorie intake	1,963		.46	1,940
Total change in entire diet				9

Source: Calculated from Gray 1982:20, 26.

Price Elasticity of Demand

How food consumption changes with price. As Figures 7.2 and 7.3 show, quantity demanded normally increases in response to a price decrease. But by how much? The consumption response to a decline in the price of a food item can be complex. Let us think about what might happen to consumption among a group of poor people if the price of rice were to fall substantially.

A fall in the price of rice would likely result in the consumption of more rice. But it might well be that not all the increased purchasing power that results from a fall in the price of rice will be spent on rice. Some of it might be spent on purchasing more of other foods, such as eggs, and some of it might be spent on purchasing nonfood items, such as entertainment.

This percentage change in rice consumption as a result of a 1 percent change in the price of rice (holding all other variables constant) is its *own-price elasticity.* The percentage change in the consumption of a good resulting from a 1 percent change in the price of *some other good* is the *cross-price elasticity* of demand.

Cross-price elasticities tend to be small. A change in the price of rice, for instance, will probably have little impact on the quantity of movie tickets a person buys. Data on cross-price elasticities are harder to obtain and may be less reliable than income and own-price elasticities. Furthermore, much sound analysis of the nutritional impact of policy alternatives can be done with what we know about income and own-price elasticities. Therefore, in this book we will not deal much with cross-price elasticities.

Quantifying own-price elasticities. Remember that price elasticity of demand is the percentage change in consumption of something, like rice, when a 1 percent change occurs in its own price, holding all other variables constant.

As with income elasticity of demand, estimating these elasticities is a lot more difficult than the above equation suggests. So again we leave descriptions of how this is done to others. A comprehensive set of demand elasticities in various countries for a variety of food groups and nutrients have been calculated by the USDA (see http://www.ers.usda.gov/data/internationalfooddemand), which obtained values of between –0.2 and –0.4.

Table 7.3 shows a set of income and own-price demand elasticities for major food groups in Indonesia. These figures are for the complete spectrum of incomes, not broken down by income groups. Notice that for Indonesian society as a whole, all the income elasticities are positive and all the own-price demand elasticities are negative. This is exactly what the theory leads us to expect: income elasticities are positive, meaning that an increase in income causes an outward shift in demand, as in Figure 7.4—the goods here are "normal goods" not "inferior goods"; price elasticities are negative, meaning demand curves are downward-sloping.

Elasticities show the relative importance consumers attach to the various foods in their diets. Items considered essential or necessary tend to have elasticities below 1.0. Think of it this way: When income falls by 1 percent, consumption of necessities falls by less than 1 percent. Or, if the price of a necessity rises by 1 percent, consumption falls by less than 1 percent. From Table 7.3, it appears that Indonesians generally regard corn and cassava, spices, and rice as necessities.

Conversely, items considered luxuries tend to have elasticities above 1.0. If income falls by 1 percent, the consumption of luxuries falls by more than 1 percent, as people cut back on luxuries and concentrate what income is left on necessities. If the price of a luxury rises by 1 percent, people are more likely to cut back substantially on consumption of that luxury than if the price of a necessity rises by 1 percent. Table 7.3 shows that Indonesians generally regard livestock and livestock products as luxuries. This is commonly the case in third world countries.

How price elasticity changes with income. Pinstrup-Andersen and colleagues (Pinstrup-Andersen et al. 1976; Pinstrup-Andersen and Caicedo 1978) were the first to show that one could estimate price (and income) elasticities by income groups as well as by the community as a whole. They divided their Cali, Colombia, sample population into five income groups, and estimated elasticities for each income group as well as for their entire sample. Some of the elasticities calculated in their groundbreaking study are shown in Table 7.4.

Notice how responsiveness to change in price generally diminishes when moving from low-income to high-income consumers. From this table it appears that high-income consumers in Cali could not care less about small changes in the price of cassava, potatoes, maize, or bread. And even with regard to pork, which the average consumer considers just over the edge into the

Table 7.3 Income and Price Elasticities for Selected Foods, Indonesia

	Income Elasticity	Own-Price Elasticity
Corn and cassava	.3	–.26
Spices	.3	–.25
Rice	.7	–.63
Coconut	1.1	–.88
Tea and coffee	1.1	–.90
Vegetables and fruits	1.2	–.97
Prepared food	1.2	–1.01
Fish	1.3	–1.04
Sugar	1.4	–1.15
Drinks	2.1	–1.71
Livestock and livestock products	2.2	–1.73

Source: Boediono 1978:362.

Table 7.4 Estimated Direct Price Elasticity of Demand by Income Group, Cali, Colombia, 1969–1970

	Low Income			High Income		
	I	II	III	IV	V	Average
Cassava	–.23	–.28	–.25	–.00	–.00	–.19
Potatoes	–.41	–.42	–.31	–.00	–.00	–.26
Rice	–.43	–.40	–.40	–.26	–.18	–.35
Maize	–.63	–.55	–.44	–.00	–.00	–.44
Bread/pastry	–.65	–.56	–.32	–.24	–.00	–.31
Beans	–.82	–.78	–.64	–.45	–.25	–.60
Peas	–1.13	–1.13	–.76	–.59	–.52	–.70
Eggs	–1.34	–1.23	–1.26	–.75	–.35	–.92
Oranges	–1.39	–.96	–.79	–.64	–.29	–.69
Milk	–1.79	–1.62	–1.12	–.64	–.20	–.77
Pork	–1.89	–1.61	–1.12	–.82	–.70	–1.01
Daily calorie intake as percentage of requirement	89	99	117	132	1,718	119

Source: Pinstrup-Andersen et al. 1976:137–138.

luxury category, high-income consumers have a demand elasticity with respect to price of less than 1.0 (a 1 percent change in the price of pork will generate a less than 1 percent change in pork consumption among these consumers). In contrast, low-income consumers are fairly responsive to changes in food prices. They consider cassava, rice, potatoes, bread, and beans necessities, but they regard animal products and fresh fruit as luxuries.

Since this pioneering study, a number of other studies have been conducted that relate food price elasticities to income. A useful survey of those studies can be found in Alderman 1986.

Price Elasticity of Supply

The final elasticity we will look at is the supply elasticity, which measures the percentage change in quantity of output in response to a 1 percent change in price of output. Estimates of supply elasticities find that supply is also "inelastic": a 1 percent increase in price will induce less than a 1 percent increase in quantity supplied. López (1980), for example, estimates a supply elasticity of 0.010 for crops and a supply elasticity of 0.472 for animal products. Table 7.5 shows some short-run supply elasticities for several farm products in African countries (the "short run," in this case, refers to the fact that these elasticities measure the response of farmers when the quantity of land they are using and the quantities of other fixed inputs, such as tube wells, are not allowed to vary).

Although the evidence is not conclusive, it appears that agricultural supply elasticities are somewhat higher in the developed world than in the developing

Table 7.5 Short-Run Supply Elasticities, Selected Crops, African Countries

	Elasticity
Wheat	.31
Maize	.23
Sorghum	.10
Groundnuts	.24
Cotton	.23
Tobacco	.48
Cocoa	.15
Coffee	.14
Rubber	.14

Source: World Bank, *World Development Report,* 1986:68; data derived from Askari and Cummings 1976 and Scandizzo and Bruce 1980.

world (Askari and Cummings 1976; Herdt 1970:518–519), indicating that farmers in developing countries are somewhat less responsive to changes in prices than are farmers in the developed world. If this is the case, it is most likely explained by three characteristics of farmers in developing countries: (1) they are less involved in the market economy—they sell a smaller percentage of their production and therefore are less impressed by swings in market prices; (2) they use lower quantities of purchased inputs relative to output sold and are therefore less able to adjust their production to variations in market prices; (3) they are more risk-averse than farmers in the developed world—they do not like spending large amounts on purchased inputs when a chance exists that, because of low prices, the investment may not pay off. Nevertheless, hundreds of estimates of supply response to price show a positive relationship between price and production (supply curves do slope up).

Similar results are obtained for long-run aggregate supply elasticities (evaluations of the "long run" allow all inputs, including land and major capital items, to vary in quantity when the output price changes). Because all inputs are allowed to vary, long-run supply elasticities are generally higher than short-run supply elasticities. In the developing world, the aggregate supply elasticity of agricultural output with respect to price appears to range between 0.3 and 0.9. Among higher-income developing countries the range is 0.6 to 0.9, and among the poorest developing countries the range is 0.3 to 0.5 (Chhibber 1988).

Economic Analysis of the 2007–2008 Food Price Increase

At the end of Chapter 6, we described the rapid increase in food prices during 2007 and 2008. The tools of economic analysis can help us better understand the causes of that price increase, their relative importance, and implications for the future. A number of interrelated causes of the high prices have been identified by economic analysis (e.g., Glauber 2008a; IFPRI 2008; OECD and FAO 2007).

Weather

Bad weather caused a very poor grain crop in various parts of the world, but especially in Australia. Based on USDA estimates, the Australian wheat crop was 25 million metric tons in the 2005–2006 season, and fell to 10 million metric tons in the 2006–2007 season. Even though wheat production bounced back slightly to 13 million metric tons in the 2007–2008 season, Australia had nearly exhausted its stocks. From the Australian perspective, this change from the 2005–2006 to the 2006–2007 crop seasons represents a dramatic crop failure—a 60 percent decline. However, Australia contributes a fairly small part of total world production, which declined by 4 percent, from 621 to 596 million metric tons. A backward shift in the world wheat supply curve can be illustrated by Figure 7.5 (with "new" and "old" supply curves reversed, and the arrows describing the shift pointing in the opposite direction). A decline in equilibrium quantity of 4 percent is associated with a higher price, as we shift backward along the aggregate demand curve. Using a price elasticity of –0.3 (the midrange of USDA estimates described previously), a 4 percent decline in equilibrium price is associated with a 13 percent increase in wheat price. This weather-related supply shortfall appears to be temporary. By the 2008–2009 crop season, Australian wheat production had increased to 20 million metric tons, not quite as high as during the 2005–2006 crop season, but twice as high as during the 2006–2007 bottom.

Restriction of Exports from India and China

Another source of reduced supply on the world market during the 2007–2008 period was related to policy rather than weather. India and China instituted export restrictions that insulated their domestic prices from world prices. If Figure 7.5 is taken to show *world* supply and demand for grain, then it illustrates (with the direction of the shift reversed, so that "new" supply is above and to the left of "old" supply) this phenomenon. *The Economist* (March 27, 2008) reported that, according to one estimate, these policies could add 20 percent to wholesale food prices. As world food prices come down (as they did between summer 2008 and fall 2008), these restrictions on exports are likely to be weakened or eliminated; however, the long-term prospects for these policies are uncertain.

Energy Prices

The increase in food prices occurred simultaneously with a rapid increase in oil prices. The average gasoline price in the United States was $2.16 in January

2007 and rose to $4.12 in July 2008 (US Energy Information Administration 2008). But energy costs account for only about 8 percent of the consumer's food dollar; so a 100 percent increase in energy prices would cause (at most) an 8 percent upward shift in the supply curve (again, see Figure 7.5, with the direction of the shift reversed). With a demand elasticity of 0.20 and a supply elasticity of 0.33 (within the range of normal estimates of elasticities), the new equilibrium price would be about 5 percent above the old equilibrium price. Higher energy prices may be a fairly permanent aspect of the economy in the coming decades, but the size of the impact on food prices will be small.

(For more detail on the calculation above: [1] find the point at which supply intersects demand *before* the energy price increase; [2] draw the new supply curve reflecting the higher energy prices; [3] note that the vertical distance between the old and new supply curves is 8 percent, so—holding quantity constant—the price on the new supply curve is 8 percent above the price on the demand curve; [4] the new equilibrium quantity will be lower than the old equilibrium quantity; [5] a 1 percent decrease in quantity would raise the price on the demand curve by 5 percent [since each 1 percent increase in price is associated with a 0.20 decrease in quantity—the definition of the own-price elasticity of demand]; [6] a 1 percent decrease in quantity would reduce the price on the new supply curve by 3 percent [since each 1 percent decrease is associated with a 0.33 decrease in quantity—the definition of the own-price elasticity of supply]; [7] therefore the 8 percent disequilibrium in price has been eliminated, moving up the demand curve by 5 percent above the old price, and down the new supply curve by 3 percent.)

Increased Meat Demand in Asia

Some analysts attributed the oil price increase to increased demand in China and India. An Indian company announced a plan to produce a new inexpensive automobile for sale in India; perhaps people in India and China were just reaching a level of prosperity at which automobile purchases would become widespread; and with more than 2 billion consumers in those two countries, the potential impact on gasoline demand (and therefore oil and gasoline prices) was mind-boggling. With this analysis of energy markets fresh in mind, it seemed natural to draw an analogy of increased demand for meat and animal products as a cause of the food price increase. This explanation of a shift in demand for food—illustrated by the shift as shown in Figure 7.4—is discussed in more detail in Chapter 10.

The difficulty in understanding this as an important cause of the 2007–2008 food price increase is that the phenomenon of growing incomes and food demand in Asia is not anything sudden or new. Per capita incomes in China and India have been growing for decades; calories per capita have been growing;

percentage of calories from animal products has been increasing. Up-to-date information on food consumption during the 2007 and 2008 period is not yet available, and perhaps we will become aware of some unusually high increases during that period; but for the time being, there is nothing about Asian food consumption patterns during the 2007–2008 period that would explain a sudden price increase.

Biofuel Demand

The most widely discussed explanation of high food prices is expanded use of crops to produce biofuels. The use of food crops to produce biofuels has been pushed by policies in a number of oil-importing countries, including the United States. As a result of government policy, corn used in ethanol production in the United States more than doubled from 2005–2006 to 2008–2009 (projected), according to USDA estimates. It accounted for 5.8 percent of world corn production in 2005–2006, and 12 percent of world corn production in 2008–2009. Biofuel production using food crops also increased in Europe (biodiesel from oilseeds) and Brazil (ethanol from sugarcane). The impact of this policy-induced biofuel boom has been to shift the demand curve for food crops up and to the right (again, as illustrated in Figure 7.4).

Growing biofuel demand has had a substantial impact on food prices. For example, the USDA (see Glauber 2008b) estimates that, if there had been no growth in biofuel demand, the corn price would have risen by 48 percent instead of the actual 62 percent between April 2007 and April 2008; overall food prices would have increased 40.6 percent instead of 45 percent. What is more, biofuel demand for food crops is expected to continue to grow (see OECD and FAO 2007, for example). This will be discussed in more detail in Chapter 10.

The Organization for Economic Cooperation and Development and the Food and Agriculture Organization conclude: "Currently strong world market prices for many agricultural commodities . . . are, in large measure, due to factors of a temporary nature, such as drought related supply shortfalls. . . . But structural changes such as increased . . . demand for biofuel production . . . may keep prices above historic equilibrium levels during the next 10 years" (OECD and FAO 2007:10).

Psychology and Price Speculation

It is difficult to construct an economic analysis based on the above factors that would explain the sharp rise in food prices during the 15-month period ending in June 2008. Therefore, one must consider the possibility of a speculative price bubble, which can occur as follows. An investor notices that food prices have risen rapidly, and speculates that this trend is likely to continue. Based on this observation, our investor decides to buy futures contracts for agricultural commodities—investments that will make money if prices continue to rise. And if other investors reach the same conclusion slightly *after* our investor does, and also rush to buy futures contracts, they will bid up the price, and our investor makes money. But if other investors think that food prices are unlikely to keep rising and decide against buying futures contracts, our investor loses money. The point is that whether our investor makes money or loses money depends less on the underlying supply-demand situation than on the *psychology* of investment decisions. Is our investor leading the herd of investors, or following them? In retrospect, we can see that food prices did fall sharply during the summer and fall of 2008, suggesting that they had risen above the levels that were justified by underlying supply-demand conditions.

Economics and the Concept of Food Security

In the mid-1980s, Reutlinger and colleagues introduced the concept of food security defined as "access by all people at all times to enough food for an active, healthy life" (1986:1). By focusing on people, this concept puts more attention on the purchasing power of those families at risk for undernutrition. A household's food production deficit is the difference between the amount of food the household needs for an active healthy life and the amount of food the household produces for itself. The household is food-insecure if its income (together with other sources of money for buying food) is less than the cost of buying enough food to make up for the food production deficit. The cost of buying enough food depends on the quantity of food needed, and the price of food.

The food security framework (or the related "entitlements" framework of Amartya Sen—see Box 7.5) stresses nutritional needs, prices, incomes, and home production (more generally, any way of obtaining food other than purchasing it on the market) as the four elements that determine a household's food security. The supply-demand framework of economics explains how prices are determined. This contributes to an understanding of food security since it helps explain the factors influencing the price of food; it also helps explain income levels, to the extent that a person's income depends on the labor wage (the price of labor) or on the prices of other goods and services the household sells.

The concept of food insecurity also illuminates a possible source of confusion about use of the term "demand" in economics. An individual's demand curve for food reflects how much food the person will want to buy at various price levels, given the amount of money a person has available for all purchases. A person with low income may go hungry, or be severely undernourished, while buying the quantity of food specified on his or her demand curve

Box 7.5 Amartya Sen

The 1998 Nobel Prize in economics was awarded to Indian economist Amartya Sen, who has written extensively on the economics of hunger. Sen's organizing framework is similar to the exposition of food security as presented in this chapter. Sen uses the term "entitlements" to refer to an individual's ability to acquire food. "Since food . . . [is] not distributed freely, people's consumption depends on their 'entitlements,' that is, on the . . . goods over which they can establish ownership through production and trade, using their own means. Some people own the food they themselves grow, while others buy [food] in the market on the basis of incomes earned" (Sen 1990).

Sen points out that hunger can exist even when there are no food shortages in the aggregate. In this framework, hunger can be seen as an "entitlement failure"—the failure of a person to assert an entitlement right to a quantity of food large enough to provide adequate nutrition. Production, income, and price all work together in defining a person's entitlements; and entitlements depend on the social, political, and cultural systems in which the person lives.

at the prevailing price. This causes some people to view economics as a heartless discipline that has no sympathy for human suffering. But the concepts of supply and demand are not intended to be prescriptions for the world's problems; rather they are a model for understanding the sometimes complex ways in which markets work.

8

It's Not Food vs. Population

Land, unlike people, . . . does not breed.
—Robert Heilbroner (1953:82, paraphrasing Thomas Malthus)

The debate over food versus people started with an argument between the young reverend Thomas Robert Malthus and his father. The elder Malthus was enthusiastic about a recently published book that promised a future world devoid of "disease, anguish, melancholy, or resentment" (Godwin 1793). Young Thomas was not impressed. In fact, he was so skeptical about such a utopian future that he wrote down his objections. The father was so struck with Thomas's words that he encouraged his son to publish them (Heilbroner 1953:69–70). First issued anonymously in 1798 as *An Essay on the Principle of Population as It Affects the Future Improvement of Society,* Malthus's "essay" was never short; by its sixth edition, though still claiming to be an essay, it covered some 600 pages of detailed argument. For Malthus in his own words, see Box 8.1.

Malthus postulated that the reproductive capacity of humans must put continual pressure on the "means of subsistence." Human numbers, he wrote, could increase by "geometric" progression: 2, 4, 8, 16, 32, 64, 128, 256 (we now call this progression *exponential*). Malthus did not see how subsistence could increase any faster than an "arithmetic" progression: 1, 2, 3, 4, 5, 6, 7, 8, 9 (we now call this progression *linear*). Unlike people, land does not breed, and Malthus thought that the potential for human numbers to increase exponentially must therefore put continuous pressure on our food supply.

Malthus enumerated a long list of checks to population growth, including war, sickly seasons, epidemics, pestilence, and plague. Humans themselves, Malthus thought, would be unable to check their own population growth because, the only way to limit family size, according to him, was through "moral restraint" (the technology of contraception was next to nonexistent at the time). And in Malthus's view, given the "passion between the sexes," moral restraint was not strong enough to effectively limit human fertility. Therefore, lurking in the shadows, always ready to impose the ultimate check on population

Box 8.1 Excerpts from *An Essay on the Principle of Population* by Thomas Malthus, 1798

"It has been said that the great question is now at issue, whether man shall henceforth start forwards with accelerated velocity towards illimitable, and hitherto unconceived improvement, or be condemned to a perpetual oscillation between happiness and misery, and after every effort remain still at an immeasurable distance from the wished-for goal. . . ."

"I think I may fairly make two postulata. First, That food is necessary to the existence of man. Secondly, That the passion between the sexes is necessary and will remain nearly in its present state. . . ."

"Assuming then my postulata as granted, I say, that the power of population is indefinitely greater than the power in the earth to produce subsistence for man. Population, when unchecked, increases in a geometrical ratio. Subsistence increases only in an arithmetical ratio. A slight acquaintance with numbers will shew the immensity of the first power in comparison of the second. . . ."

"By that law of our nature which makes food necessary to the life of man, the effects of these two unequal powers must be kept equal. . . ."

"This implies a strong and constantly operating check on population from the difficulty of subsistence. This difficulty must fall somewhere and must necessarily be severely felt by a large portion of mankind. . . ."

"Taking the population of the world at any number, a thousand millions, for instance, the human species would increase in the ratio of—1, 2, 4, 8, 16, 32, 64, 128, 256, 512, etc. and subsistence as—1, 2, 3, 4, 5, 6, 7, 8, 9, 10, etc. In two centuries and a quarter, the population would be to the means of subsistence as 512 to 10: in three centuries as 4,096 to 13, and in two thousand years the difference would be almost incalculable, though the produce in that time would have increased to an immense extent. . . ."

"No limits whatever are placed to the productions of the earth; they may increase for ever and be greater than any assignable quantity. Yet still the power of population being a power of a superior order, the increase of the human species can only be kept commensurate to the increase of the means of subsistence by the constant operation of the strong law of necessity acting as a check upon the greater power. . . ."

Source: http://socserv2.socsci.mcmaster.ca/~econ/ugcm/3ll3/malthus/popu.txt.

growth, was famine. "Famine stalks in the rear, and with one mighty blow, levels the population with the food of the world" (Heilbroner 1953:83).

There was plausibility to the Malthusian argument. It was, in fact, a precursor to the now widely accepted ecological principle that any population will expand until it fills the ecological niche available to it. What Malthus did not foresee was that there would eventually be other checks to human population growth besides war, pestilence, and famine; that changing attitudes about family size—a kind of "small is beautiful" philosophy—could combine with a new technology in the form of effective and simple contraception to limit population growth. Nor did he foresee the enormous increases in agricultural production that would accompany the application of science to farming.

As important as Malthus's book was for the thesis it espoused, it was more important for stimulating thinking among people who read it. Charles Darwin, for instance, reported that he read Malthus "for amusement," yet this reading inspired the theory of natural selection and survival of the fittest that would dominate his treatise *On the Origin of Species* (Bettany 1890; Herbert 1971).

Others were not amused. As one biographer put it: "Malthus was not ignored. For thirty years it rained refutations" (James Bonner, quoted in Heilbroner 1953:76). In the storm of protest that followed the publication of his essay, Malthus was likened to Satan and denounced as an "immoral, revolutionary, hard-hearted, and cruel atheist" (Bettany 1890:ix). But the strongest refutation of the seeming inevitability of a perpetual tendency toward famine that Malthus postulated lies in what has happened since he wrote his essay.

Since 1800 the population of the world has, in fact, grown exponentially—or nearly so (see Figure 8.1). And more remarkable—and a testament to Malthus's reasoning—his 1789 treatise was published at the very beginning of this exponential growth trend. On the other hand, the growth of the world population seems destined to slow, and eventually stop, through a process demographers call the demographic transition.

Figure 8.1 World Population, 0–2000

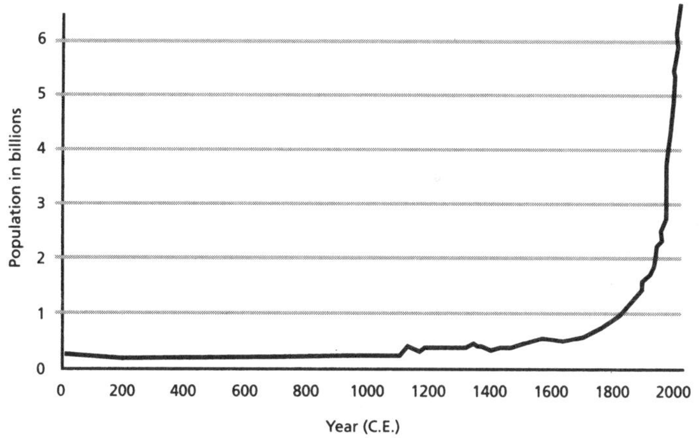

Source: US Bureau of the Census, "Collection of Historical Estimates of World Population" (10,000 B.C.E. to 1950) and "Census World Population Estimates" (1950–2008), http://www.census.gov/ipc/www/idb/worldpopinfo.html.

The Theory of Demographic Transition

Originally described by Frank Notestein (1945), the literature contains a number of ways of defining the *demographic transition.* We adopt and paraphrase from a conceptualization by Carl Haub (1987:19) of the Population Reference Bureau in Washington, D.C.

The theory of demographic transition offers a general model for the gradual evolution of a population's birth and death rates (see Box 8.2) from the preindustrial to the modern pattern, which results in an S-shaped curve of population growth through time. According to the theory, population growth goes through four stages, illustrated in Figure 8.2.

The first stage is the preindustrial period. Birth rates are high and fertility is uncontrolled, with birth rates—generally within the range of 25 to 45 per 1,000—approximately equal to death rates. Periodic famines, plagues, and wars cause brief periods of population loss. Population grows, but slowly. Population growth rates in this stage are close to zero and stay relatively constant over time.

In the second stage, mortality declines while fertility remains high. With better public health services and more reliable food and water supplies, death rates fall and life expectancy increases. If no accompanying decrease in birth rates occurs, population grows rapidly, and faster and faster over time.

Box 8.2 Terms Commonly Used by Demographers

Crude birth rate, or birth rate: The number of births per year per thousand individuals in the population.

Crude death rate, or death rate: The number of deaths per year per thousand individuals in the population.

Gross reproductive rate (GRR): The number of female children a newborn female will have during her lifetime if current levels of fertility by age of female continue through time.

Life expectancy at birth, or life expectancy: The average expected age of death of newborns who follow a given age-specific mortality schedule.

Net reproductive rate (NRR): The expected number of daughters per newborn female, after subjecting those newborn females to a given set of mortality rates. (NRR is lower than GRR, because some of the newborn females will die before completing their reproductive years.)

Total fertility rate, or fertility rate: The total number of children a female will bear during her lifetime.

Figure 8.2 The Theory of Demographic Transition

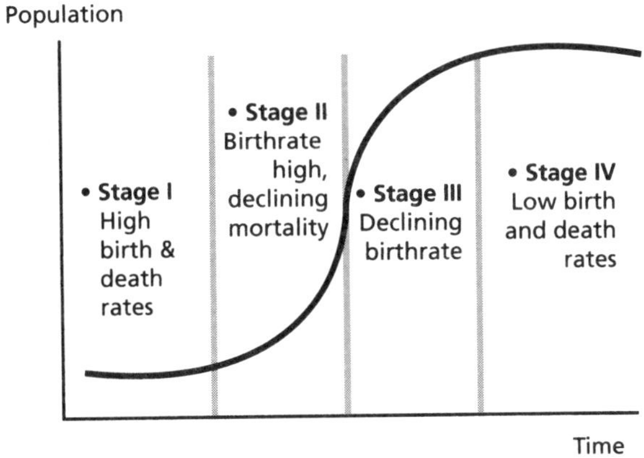

In the third stage, fertility declines. At some point, usually as a country urbanizes and industrializes, its birth rate decreases in response to desires to limit family size. Population continues to grow rapidly for a while. But eventually, birth rates approach death rates, and population growth slows. In this stage, population growth rates are still positive, but population growth slows over time. By the end of this stage, population growth may have fallen to near zero.

The fourth stage is the modern period. By this point, both birth and death rates are low. After the birth rate falls as low as the death rate, population stabilizes. Stability in a population over a number of generations implies that life expectancy is stable (each generation lives as long as the last, and as long as the next) and also that fertility is stable, meaning a "replacement rate" of 2.1 children per woman (2.1 rather than 2.0 to account for childhood mortality and other factors that keep some small percentage of the population from reproducing).

The experience of fertility rates in the advanced economies of Europe in recent decades has caused some demographers (e.g., Haupt and Kane 2004) to speculate that the theory of demographic transition will have to be expanded to include a fifth stage—one in which fertility rates fall *below* the replacement rate, and population dwindles rather than grows. According to the US Census Bureau website (www.census.gov; international database), the current fertility rate in Eastern Europe as a whole is 1.37, and the current rate in Western Europe as a whole is 1.55, with some countries (Italy, Spain, Moldova, for example) having fertility rates of 1.30 or less.

There is abundant evidence to support the validity of the theory of demographic transition. First, we can find developed countries in which all four stages of demographic transition have occurred. The experience of Sweden, for example, is shown in Figure 8.3. In the decades before 1805, Sweden was

Figure 8.3 Birth and Death Rates in Sweden, 1751–2008

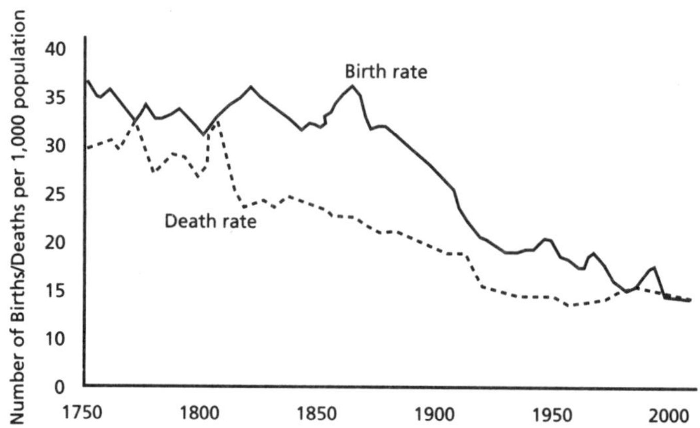

Sources: 1750–1946, Haub 1987:20; 1951–2008, US Bureau of the Census, International Database.

in the last phase of the first stage, with fertility and mortality rates approximately equal, but high. The second stage in Sweden covered the seventy years between 1805 and 1875, as the death rate began to fall, but the birth rate stayed high. The third stage in Sweden covered the hundred-year period between 1875 and 1975, as the birth rate declined faster than the death rate. Sweden is now in the fourth stage, with birth and death rates approximately equal, but low (in fact, currently, the birth rate in Sweden is slightly below the death rate).

A second source of evidence supporting the theory of demographic transition is a comparison of population growth rates in developed countries to those in developing countries. For example, in the period 2000–2008, the population of Africa (home to many of the poorest countries of the world) grew at a rate of 2.45 percent per year; the population of Asia (home to many developing countries) grew at a rate of 1.18 percent per year; and the population of Europe grew at a rate of only 0.02 percent per year. In what are today's developed countries, the demographic transition is essentially finished: for "more developed countries," according to the US Census Bureau website, birth rates are now just slightly higher than death rates; and death rates are expected to exceed birth rates in 2014.

In developing countries, the process of demographic transition is still under way. Compare the demographic transition as shown in Figure 8.3 (with Sweden as representative of the developed world) with that shown in Figure 8.4 (with Mexico as representative of the developing world). The death-rate decline in Sweden began shortly after 1800 and took approximately 150 years to fall from 30 to 15 per 1,000. The more dramatic death-rate decline in Mexico did not begin until about 1915 and took only 40 years to fall to 10 per 1,000.

Figure 8.4 Birth and Death Rates in Mexico, 1895–2005

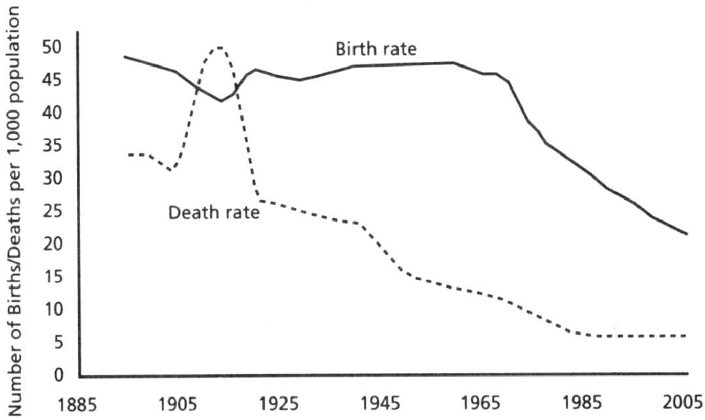

Sources: 1895–1955, Haub 1987:20; 1960–2005, US Bureau of Census, International Database.

The birth rate in Mexico remained above 40 per 1,000 until the early 1970s, when it began a rapid decline. Consequently, by the early 1970s, Mexico's population growth rate was above 3 percent, yielding a doubling time of fewer than twenty-three years.

Figure 8.5 makes this point a little more generally: population growth is higher in less developed countries. The countries with high population growth rates are bunched at the left of the figure in low per capita incomes. The rich countries generally have lower rates of population growth. This has two implications. First, economic growth, or improved economic prosperity, can be a powerful policy mechanism for reducing population growth. This will be discussed in more detail in Chapter 18. Second, since population is growing fastest among groups of people who have low incomes, the growth in demand for food may lag behind population growth. When we get to future projections in Chapter 23, we will assume that a 50 percent increase in population translates into a 50 percent increase in demand for food; in reality this may overstate the impact on demand.

A final piece of evidence supporting the theory of demographic transition is shown in Figure 8.6. For the world as a whole, the growth of population is slowing. The actual growth—numbers of people added to world's population every year—peaked in the late 1980s. The *rate* of growth—actual growth as a percentage of total population—peaked in the late 1960s. At that point in time, the world as a whole moved from the second to the third stage of demographic transition. Note that although the rate of growth is still positive—world population continues to grow—it is slowing over time. In addition, the US Census

Figure 8.5 Population Grows Faster in Countries with Lower per Capita Incomes: Evidence from 165 Countries, 2001

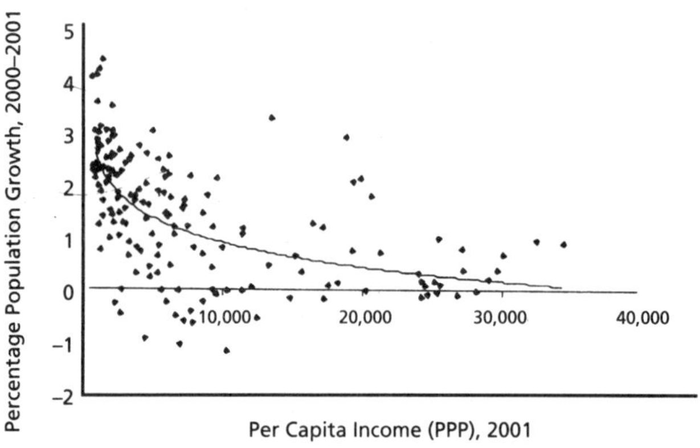

Source: World Bank 2003, *World Development Indicators.*

Bureau predicts that the world population growth rate will continue to decline for the next fifty years. (One interesting aspect of Figure 8.6: the sharp dip in world population growth rate in the late 1950s and early 1960s corresponds with the "Great Leap Forward" famine in China, discussed in Chapter 2.)

Lester Brown and Hal Kane, in criticizing the optimistic view of demographic transition, point out a less happy road to population stabilization:

> As we approach the end of the twentieth century, a gap has emerged in the [demographic transition] analysis. The theorists did not say what happens when second stage population growth rates of 3 percent per year begin to overwhelm local life-support systems, making it impossible to sustain the economic and social gains that are counted on to reduce births. Unfortunately, trends that lead to ecological deterioration and economic decline are also self-reinforcing: Once populations expand to the point where their demands begin to exceed the sustainable yields of local forests, grasslands, croplands, or aquifers, they begin directly or indirectly to consume the resource base itself. . . . This . . . reduces food production and incomes, triggering a downward spiral in a process we describe as the demographic trap. All countries will complete the demographic transition, reaching population stability with low death and birth rates, or will get caught in the demographic trap, which eventually will also lead to demographic stability—but with high birth rates and high death rates. (1994:55–56)

If the world's population growth does, in fact, stop as projected, humans will have succeeded in controlling their own numbers without war, pestilence, and famine—something Malthus did not expect.

Figure 8.6 Annual Growth Rate of World Population, Historical and Projected, 1950–2050

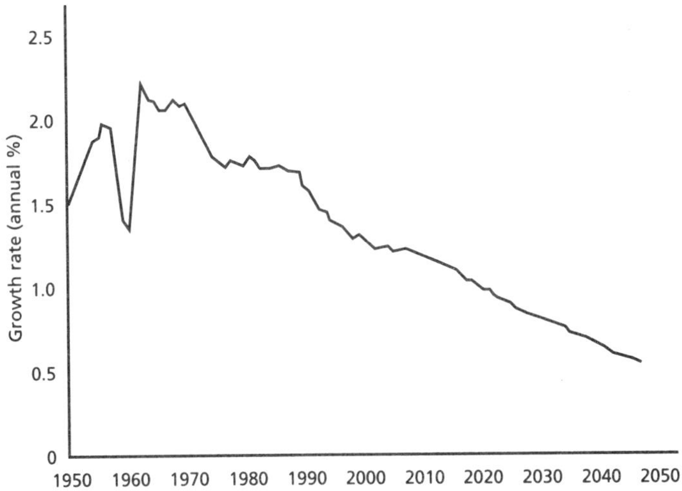

Sources: US Bureau of the Census, International Database, 2008.

Projections of Future World Population

As we consider the future prospects for world food supply and demand, we think first of population. How many people will there be to feed? The US Census Bureau, basing its statistics on the growth rates shown in Figure 8.6, projects that by 2050, world population will be about 9.54 billion, 42 percent higher than the current (2008) population of 6.71 billion (see Table 8.1). The United Nations has also made population projections into the future. Its projections are based on assumptions about how life expectancy will change, and about how fertility rates will change. See UN Population Division 2002a for a description of these assumptions.

Life expectancy at birth is assumed to increase as average nutrition continues to improve and as average incomes continue to rise. The improvements are assumed to be greatest in countries where life expectancy is the lowest, and therefore where potential for improvement is the greatest. These assumptions mean that life expectancies are projected to increase substantially in some parts of the world. Babies born in Africa in 2050 are expected to live sixteen years longer than babies of their grandparents' generation, born in 2000 (UN Population Division 2002b). This may sound like quite a dramatic increase. But life expectancy has increased by substantial amounts in other locations.

Table 8.1 Population in 2050: Four Projections

Projection	Population Size in 2050 (billions)	Percentage Increase from 2008	Average Annual Growth Rate (%)
UN low variant	7.792	16	0.36
UN medium variant	9.191	37	0.75
US Census Bureau	9.539	42	0.84
UN high variant	10.756	60	1.12

Sources: United Nations, Department of Economic and Social Affairs, Population Division, *World Population Prospects: The 2006 Revision* and *World Urbanization Prospects: The 2005 Revision,* http://esa.un.org/unpp, November 2008; US Bureau of the Census, International Database, November 2008.

For example, life expectancy in Saudi Arabia increased from 53.9 years in the early 1970s to 70.9 years in the late 1990s; in Indonesia, life expectancy increased from 49.2 to 64.5 years over the same period; and in South America as a whole, life expectancy increased from 60.5 to 69.0 years (UN Population Division 2002b). The UN's assumption is, therefore, within the range of historical experience.

It is more difficult to make reasonable assumptions about future levels of fertility. As a result of this difficulty, the UN presents three possible scenarios, or variants. The medium variant assumes that fertility will decline (from its current level of about three children born to an average woman of childbearing age) until it reaches a level of 1.85 children per woman. (This is below the "replacement level" of 2.1—the number of children each woman must give birth to, on average, to ensure that two children survive to puberty. Even if the fertility rate falls to replacement level, population would continue to grow because of increases in life expectancy.) In the high variant, fertility is assumed to be 0.5 percent above the medium variant. In the low variant, fertility is assumed to be 0.5 percent below the medium variant.

Again, to put these projections into historical context, let us examine recent experience. For the world as a whole, the fertility rate dropped from 4.5 per 1,000 in 1970–1975 to 2.8 per 1,000 in 1995–2000. In Asia, where economic growth has been exceptionally strong in the past twenty years, the fertility rate dropped from 5.1 in the early 1970s to 2.7 in the late 1990s. In Europe, fertility was at the replacement rate of 2.1 in the early 1970s, but dropped to 1.4 by the late 1990s (UN Population Division 2002b). Thus, the UN's low-variant assumption is that fertility worldwide will drop to levels currently observed in Europe.

Based on these assumptions, the UN's three population projections are shown in Table 8.1. The medium variant is quite close to the US Census Bureau's prediction. How good are the projections? Table 8.2 shows UN projections of world population in 2000 made at different points in time. Several things are notable:

Table 8.2 Past Projections of World Population in 2000: How Accurate Were They?

Year Projection Was Made	Projected Population in 2000 (billions)
1957	6.28
1963	6.13
1968	6.49
1973	6.25
1980	6.12
1984	6.12
1988	6.25
1990	6.26
1992	6.23
1994	6.16
1996	6.09
1998	6.06
Actual 2000 (US Census Bureau)	6.07
Actual 2000 (UN)	6.12

Source: National Research Council 2002.

- For obvious reasons, near-term projections are more accurate than long-term projections.
- However, the projection made in the early 1960s was remarkably close to the actual population that existed in 2000; the 1963 projection was off by only 3 or 4 percent.
- All of the projections (except the last one) overestimated the actual population. For the most part, projections of future population have diminished over time. But this is not a hard and fast rule. For example, the second edition of this book reported the 1996 projections of population in 2050: the US Census Bureau projected 9.350 billion and the UN, in its medium variant, projected 9.850 billion. The third edition of the book reported a 2002 Census projection of 9.084 billion and a UN medium variant of 8.919 billion. And presently (2008), the US Census Bureau projects 9.539 billion and the UN projects 9.191 billion.

A report published in the journal *Nature* (Lutz, Sanderson, and Sherbov 2001) criticized the UN's projection methods and made its own set of projections to the year 2100. The median projection of that report tracked closely with the UN's medium variant through 2050, calling for world population to peak at about 9 billion during the decade of the 2060s and then decline.

The specter of AIDS hangs over any discussion of future population. The projections shown in Table 8.1 include estimates of increased mortality as a result of the disease. The United Nations estimates that at the end of 2007, about 33 million people were infected with HIV, the virus that causes AIDS. The

AIDS epidemic has stabilized or even begun to decline in some areas. However, the problem remains severe in parts of sub-Saharan Africa (see Table 8.3). Although the world hopes that medical science can make progress on AIDS treatment, it is still the prognosis that many of these 33 million people will die from AIDS during the next five to twenty years. In 2007, an estimated 2 million people died from AIDS worldwide. To put this number into perspective, consider that over 55 million people died in 2007 from all causes. If all HIV-infected people died within five years, that would increase the number of deaths during that five-year period by about 10 percent. One review of the literature concludes:

> At the world level, AIDS is unlikely to suppress population size or growth rates. However, the impact of AIDS may be felt by some individual countries. For example, the US Bureau of the Census predicts that in some countries, populations in 2020 will be considerably smaller as a result of the AIDS pandemic—45 percent smaller in Uganda, 35 percent in Rwanda, and 30 percent in Malawi. (Brown 1997)

Current Trends in per Capita Food Production

Not only did Malthus not expect humans to be able to control their own population size, but he also did not expect our food supply to keep up with a dramatic, exponential growth in our population. The latter half of the twentieth century ex perienced the most rapid growth of population in the entire history of the world, yet during this period of breakneck population growth, food production grew even faster, such that per capita food production gradually increased.

The factors contributing to growth in food production will be discussed in Chapters 11–14. Here we simply note the fact that food production has continued to grow faster than population. As Table 8.4 shows, worldwide food production per capita has increased steadily over the past several decades, growing

Table 8.3 Percentage of Adult Populations (Age 15–45) Infected with HIV, 2007

Swaziland	26.1
Botswana	23.9
Lesotho	23.2
South Africa	18.1
Namibia	15.3
Zambia	15.2
Mozambique	12.5
Malawi	11.9
Sub-Saharan Africa	5.0

Source: UNAIDS 2008.

Table 8.4 Index of Net Food Production per Capita, Selected Areas and Years (1961 = 100)

	Sub-Saharan Africa	Developing Countries	Developed Countries	World
1961	100.00	100.00	100.00	100.00
1971	103.91	106.10	113.70	106.48
1981	88.90	114.97	123.69	111.78
1991	90.62	132.84	123.28	116.99
2001	91.38	163.95	121.34	129.44
2005	88.25	173.84	124.58	134.91

Source: FAOSTAT 2008a.
Note: See Box 14.3 for a description of index numbers.

about 35 percent between 1961 and 2005. Food production per capita in the developing world has grown much more rapidly, nearly 75 percent between 1961 and 2005. In the developed world, food supply grew faster than population from the early 1960s until the 1980s. Since then, food production per capita has remained roughly constant. The situation in sub-Saharan Africa is an exception to the worldwide trend: food output per capita has declined steadily from the 1960s to the 1980s and has remained roughly constant since then. In 2005, food production per capita in sub-Saharan Africa was about 12 percent lower than it was in 1961.

A World Bank study (Mundlak, Larson, and Crego 1996:3) concludes that worldwide food supply is growing faster than food demand: "Has supply lagged demand? If that were so, agricultural prices would have risen, but this has not happened." The study finds that median prices to farmers dropped by 0.61 percent during the 1967–1992 period, and that 71 percent of world production from 1967 to 1992 came from countries in which (inflation-adjusted) farm prices fell (though recall the more recent experience regarding food prices described in Chapter 6). The study finds that the median growth rate in agricultural production was 2.25 percent per year (meaning that half of the world's food production takes place in countries with agricultural growth rates of less than 2.25 percent). The study confirms that in most countries, per capita agricultural production grew—food became more plentiful.

FAO data on nutrient availability per capita also indicate steady improvement. As Figure 8.7 shows, since 1961, worldwide calories per capita have increased over 20 percent, from 2,253 calories per person per day to 2,808 calories per person per day in 2003. Similarly, protein availability has increased about 20 percent, to 75 grams per person per day in 2003. These levels of nutrients are sufficient for an adequate diet for the average person. One interpretation of this fact is that *if the world's food supply were evenly divided among all people, there would be enough food for everybody.*

Figure 8.7 Worldwide Calories per Capita per Day, 1961–2003

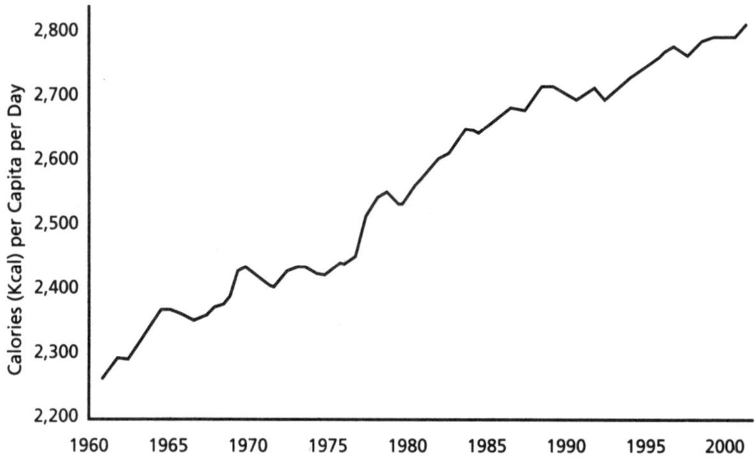

Source: FAOSTAT 2008a.

So an important question in any discussion of the incidence and permanence of world hunger is how the world's food supplies are allocated among the people of the world. In the next chapter, we look at purchasing power—and its components, income and food prices—as the immediate problem explaining why certain people in the world are unable to afford adequate nutrition.

9

Income Distribution and Undernutrition

"Get off this estate!"
"What for?"
"Because it's mine."
"Where did you get it?"
"From my father."
"Where did he get it?"
"From his father."
"And where did he get it?"
"He fought for it."
"Well, I'll fight you for it!"

—Carl Sandburg (1936:75)

Income is central to the problem of undernutrition. A family is hungry because the family is poor. In this chapter we look at two closely related issues. First we look at the distribution of income in the world. What countries are poor? How poor are they? How is income distributed within countries? How are these things changing over time? Second, we look at how changes in income affect the aggregate demand for food.

Who Are the Poor?

The World Bank (2008b) lists thirteen countries with per capita incomes of less than \$365 and thirty-one countries with per capita incomes of less than \$730. It is hard for most of us to imagine how a person could feed him- or herself on a little over \$2 per day, let alone \$1 per day. And people have needs besides food.

Who are the poorest of the poor? They live in the third world. They are landless or nearly so. If they do have a bit of land, typically they earn more than half their livelihood working for others. Whether they live packed tightly into city slums or sprinkled across the countryside, they are poorly educated, often illiterate, and commonly superstitious. When employed, they accept the most menial of jobs. Some are subsistence fishermen. Some live in relative isolation in remote farming areas. Often they are squatters, neither owning nor renting the land on which they put up their huts. Their food larder is usually almost empty.

Their households are often fragmented, with one or more members away trying to find work so they can send money home. They may be in debt—to wealthier relatives, to friends, to employers, to local moneylenders. The household head is often young, not yet having found good employment, but already burdened with the responsibility of raising children. (For a good essay on the poor of a particular region, see Carner 1984.)

Comparing Average Incomes in Different Countries

In an international economy, prices are influenced by consumption in all countries. The affordability of food for poor people in one country is influenced by the consumption patterns of affluent people in other countries. Looking at the distribution of income in the world, we see that average per capita incomes are very different from one country to another (see Table 9.1). To pick the most dramatic example, in 2007, per capita income in Burundi was $110 per year, while per capita income in Norway was $76,450. The average Norwegian

Table 9.1 GNP per Capita, Selected Countries, 2007

	GNP per Capita ($)	
	Exchange Rate Comparison	Purchasing Power Parity Comparison
Burundi	110	230
Congo, Dem. Rep.	140	290
Liberia	150	290
Ethiopia	220	780
Eritrea	230	520
Malawi	250	750
India	950	2,740
China	2,360	5,370
Brazil	5,910	9,370
Singapore	32,470	48,520
Japan	37,670	34,600
Germany	38,860	33,820
United States	46,060	45,850
Switzerland	59,880	43,080
Norway	76,450	53,690

Source: World Bank 2008a, Appendix table 1.

earned nearly 700 times the amount earned by the average person in Burundi (World Bank 2008a).

Income, as the term is used here, refers to gross national product (GNP) or to the closely related concepts of gross domestic product (GDP) or gross national income (GNI), which measure the total value of goods and services produced in the economy. The richest ten countries in the world have less than 10 percent of the world's population but produce over 60 percent of the world's goods. The poorest fifty-six countries support more than 50 percent of the world's population but produce less than 5 percent of the world's goods.

An Alternative Way of Comparing Incomes in Different Countries: Purchasing Power Parity

Some economists have expressed doubt about whether the usual method of comparing GNP per capita in different countries gives an accurate view of the quality of life in those countries. The usual method—as reflected in the numbers in Table 9.1—translates the value of goods and services in a country from the local currency to US dollars by using the market exchange rate. In effect, this measure translates local currency into dollars by considering how many units of the local currency it would take to buy a dollar on the foreign exchange market.

An alternative method translates local currency into a dollar equivalent through the use of purchasing power parity (PPP)—that is, by comparing the purchasing power of the local currency to the purchasing power of the dollar. In effect, this measure translates how much it would cost in the local currency, spent in the local market, to buy the same quantity of goods that could be purchased in the United States with one dollar.

These two conversion methods can change the relative positions of countries in the ranking of income per capita. For example, using the exchange rate conversion, the per capita incomes of Eritrea and Ethiopia are almost the same; but using the PPP conversion, Ethiopia's per capita income is higher than Eritrea's by 50 percent. Or, using the exchange rate conversion, Switzerland's per capita income is higher than that of the United States; but using the PPP method, US per capita income is higher.

With only a few exceptions (see Singapore in Table 9.1), using the PPP conversion makes poor countries appear to be a little less poor and makes rich countries appear to be a little less rich. In effect, the cost of living is relatively high in high-income countries and is relatively low in low-income countries. To illustrate: Using either conversion method, Norway is the richest country and Burundi is the poorest. However, with the exchange rate comparison, average income in Norway is 695 times the average income in Burundi, while with the PPP conversion method, average income in Norway is a mere (!) 233 times the average income in Burundi. This relationship between per capita income and cost of living is documented clearly in the World Bank's *China Quarterly Update* (World Bank 2008a). It shows that countries at the very low end of the income scale have costs of living that are about 70 percent lower than the cost of living in the United States.

It can be difficult for people who grew up and live in the United States or other developed countries to understand how people in poor countries can live on such tiny incomes. Box 9.1 describes the life of a poor person who lives in Malawi, where average income per person per year is only $750.

Some may dismiss the description of a single Malawi household in Box 9.1 as "anecdotal." However, "data is the plural of anecdote," as political scientist Raymond Wolfinger once said (Shapiro 2004). Data from fourteen surveys of households in the poorest parts of the world were reviewed by Banerjee and Duflo (2007) to see if any general lessons could be drawn. We cannot hope to describe here all of the results of their review, but a number of things are worth noting:

Box 9.1 Life in Rural Malawi

The figures in Table 9.1 tell us that Malawi is one of the poorest countries in the world; average income per capita (PPP method) is about $2 a day. But what does this really mean? How can a person live on $2 a day? It is difficult—perhaps impossible—to convey the depth of poverty experienced by people in rural Malawi. Reporter Barry Bearak (2003), writing for the *New York Times Sunday Magazine,* visited the Malawian countryside and describes the life of Adilesi Faisoni, a grandmother in the village of Mkulumimba.

A photograph accompanying the article shows her "worldly goods": a cup, three bowls, a cooking pot, some hearth stones and stirring sticks, a cleaning rag. That's it. No electricity, no furniture, nothing. She owns a single set of clothes. Her grandchildren wear used T-shirts from the United States with pictures of Power Rangers and Teenage Mutant Ninja Turtles. The clothes are ragged and worn. Her house is a 9-by-12-foot mud hut. She lives here with her daughter and ten grandchildren. How do they all fit? "We squeeze like worms."

Her diet is almost exclusively *nsima,* a thick porridge made from maize meal. During the hungry months, she may only eat a single bowl each day. "There is no way to get used to hunger," she tells the reporter. "All the time something is moving in your stomach. You feel the emptiness. You feel your intestines moving. They are too empty and they are searching for something to fill up on."

Last year, the maize meal ran out, and her family had to eat pumpkin leaves and wild vegetables. Her husband starved to death. "There was nothing to do but beg, and you were begging from others who needed to beg."

- Poor households are large and young. Median household size is between seven and eight persons. There are six children under age 18 for each adult over age 51.
- The poorest households spend between 56 and 78 percent of their money on food. If household income increases by 1 percent, expenditure on food increases by two-thirds of 1 percent. About half the increased expenditure on food is devoted to increasing available calories (by buying more of the cheapest form of calories) and about half is devoted to increasing food variety and palatability.
- Almost all of the poorest households spend some money on things like tobacco, alcohol, and festivals and celebrations, but very few households spend anything at all on entertainment such as movies and theater.
- There is considerable variation in the different countries surveyed in the ownership of assets by poor households. In Mexico, only 4 percent of poor households own land; in one area of India, 99 percent of households own land; in all countries, the amount of land owned is small. In Peru, 70 percent of households own a radio; in one area of India, the corresponding number is 11 percent.

According to a survey of extremely poor people in Udaipur, India, Banerjee and Duflo found the following:

- Nearly half of the poorest households report that at some time during the year, adults have to reduce meal size, and in one-third of the households, adults sometimes have to go for entire days without eating. About 12 percent of households report that children are forced to cut meals. "Happiness" (see Box 9.2) was also measured in this survey, and was found to have a strong negative correlation with cutting meals.
- About 65 percent of adults are undernourished (have a body mass index of less than 18.5), and about 55 percent have anemia.
- Most households have a bed or cot; about half have a watch or clock; 10 percent have a stool or chair; 5 percent have a table. Almost no households have an electric fan, a sewing machine, a bullock cart, a motorized bicycle, or a tractor.

Does Income per Capita Measure What Is Important?

Some readers may at this point be a little skeptical about exactly what these average per capita numbers mean. And there are some legitimate concerns. In the following sections we will take up the issue of income distribution. Related to that, other indicators of quality of life, or depth of poverty, in different countries have been developed.

The World Bank tabulates data on the number of people in each country who live in extreme poverty as measured by incomes of less than $1 or $2 per day. To make these comparable over time, the $1 poverty line is adjusted for inflation. For the numbers shown in Table 9.2, the $1 standard refers to 1985 dollars; adjusted for inflation, it is equivalent to about $1.38 in the year 2005. The United Nations Development Progamme (UNDP) compiles an index of indicators of quality of life, known as the Human Development Index (HDI). This index includes per capita income, but also includes measures of health and education.

Table 9.2 allows us to compare income per capita with these other indicators. Several points emerge:

- Very high-income countries also have high HDI scores and virtually no people living on less than $1 per day.
- Countries with low HDI scores and high poverty rates also have very low incomes per capita.
- Countries can have similar HDI scores even though they have very different incomes (compare South Africa to Indonesia).
- Countries can have very different HDI scores even though they have the same incomes (compare Georgia to Lesotho).
- Countries can have similar HDI scores and incomes and still have very different poverty rates (compare Nicaragua to Indonesia).

Despite the fact that rankings of income per capita differ from rankings of HDI or poverty rate, income per capita is an important indicator of quality of life in a country. Figure 9.1 shows that countries with higher per capita incomes for the most part have better quality of life—their citizens live longer, poverty and undernutrition are less prevalent, and (not shown in Figure 9.1;

Table 9.2 Alternative Quality-of-Life Measures, Selected Countries, 2005

	GNI per Capita (PPP conversion, $)	Human Development Index	Percentage of Population Living on Less Than $1 per Day
Norway	41,420	.968	0
United States	41,890	.951	0
Indonesia	3,843	.728	7.5
Nicaragua	3,674	.710	45.5
South Africa	11,110	.674	10.7
Georgia	3,365	.754	N/A
Lesotho	3,335	.549	36.4
Niger	781	.374	60.6
Sierra Leone	806	.336	57.0

Source: UNDP 2007.

Figure 9.1 Income Matters

Poor people are better off in high-income countries

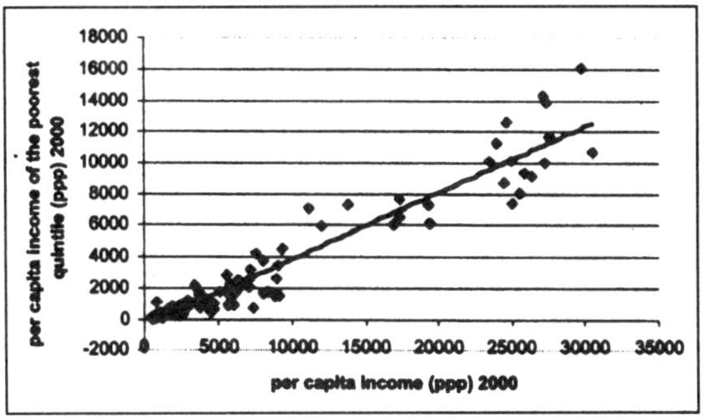

People live longer in high-income countries

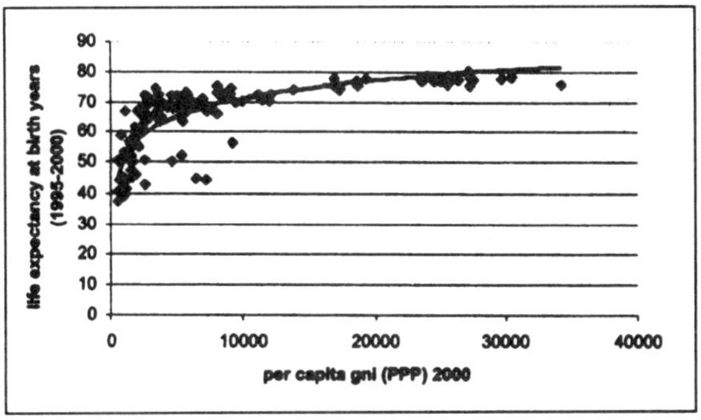

Fewer people are undernourished in high-income countries

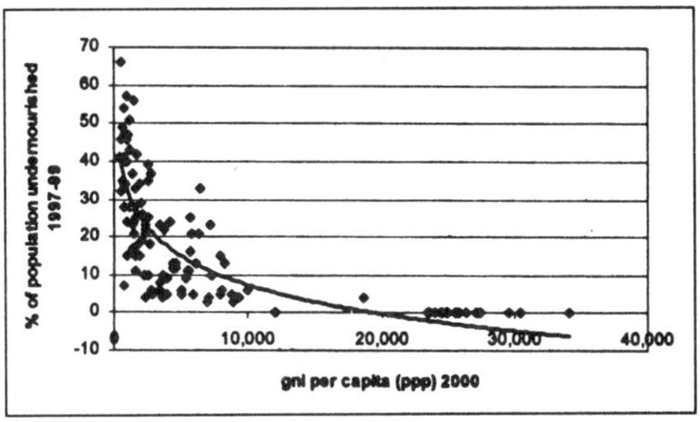

see Hayward 2006: fig. 8) environmental quality is better. Box 9.2 explores some recent research about whether people in higher income countries are "happier" than people in poor countries.

Measuring Income Distribution Within a Country

Another legitimate criticism of average per capita income as an indicator of quality of life is that it is an average. Some people in the country have incomes above the average and some have incomes below the average. This point is made by the Nicaragua-Indonesia comparison in Table 9.2. The two countries have nearly identical *average* per capita incomes, but Nicaragua has a much greater poverty problem, because income is distributed much more equally in Indonesia than in Nicaragua. Next we look at measures of how income is distributed within a country.

Box 9.2 Income and Happiness

Money can't buy happiness (or as American humorist Leo Rosten modifies it: "money can't buy happiness, but neither can poverty"). In the 1970s, economist Richard Easterlin investigated this proposition empirically, and his conclusion—that there is no link between income and happiness—became known as the "Easterlin paradox." In a recent paper, Betsey Stevenson and Justin Wolfers (2008) of the University of Pennsylvania reinvestigated the issue using information from opinion polls in which people were asked questions about how happy they were. Their conclusion: "[W]e establish a clear positive link between average levels of subjective well-being and GDP per capita across countries, and find no evidence of a satiation point beyond which wealthier countries have no further increases in subjective well-being. We show that the estimated relationship is consistent across many datasets and is similar to the relationship between subject well-being and income observed within countries. Finally, examining the relationship between changes in subjective well-being and income over time within countries we find economic growth associated with rising happiness. Together these findings indicate a clear role for absolute income and a more limited role for relative income comparisons in determining happiness."

The issue of whether poverty is an absolute or a relative situation continues to engage economists. If one accepts the notion that there is an absolute standard, then one is led to conclude that there is virtually no poverty problem in the United States (see Boxes 9.1 and 9.4 especially, but also Box 9.3). On the other hand, if one accepts the notion that poverty is a problem of relative wealth, then the problem is defined in such a way that it is impossible to solve—some households are always in the bottom 10 or 20 percent of income, no matter what happens to the overall distribution of income.

Pareto's Law

In the late 1800s, Vilfredo Pareto, an Italian mathematician, economist, and sociologist, examined income distribution in a number of countries and found a remarkably consistent pattern. Furthermore, he was able to fit this pattern to a mathematical formulation that soon became known as Pareto's law. Pareto's law had its problems, and, as it turned out, one of the main benefits of Pareto's work on income distribution was to stimulate others to think about alternative methods of measuring it. (For a concise discussion of Pareto's work on distribution, see Steindl 1987.)

The Lorenz Curve

In 1905, US statistician Max Lorenz proposed a method of comparing distributions of income and wealth through a cumulative income or wealth curve, known as the Lorenz curve, an example of which is shown as the dashed line in Figure 9.2. The vertical axis, OC, represents percentage of total income for the group under analysis. The horizontal axis, OE, represents the percentage of individuals (or families) in the group.

To conceptualize how a Lorenz curve is constructed, imagine a group of 100 individuals, each with a different income. Find the total amount of income

Figure 9.2 A Lorenz Curve

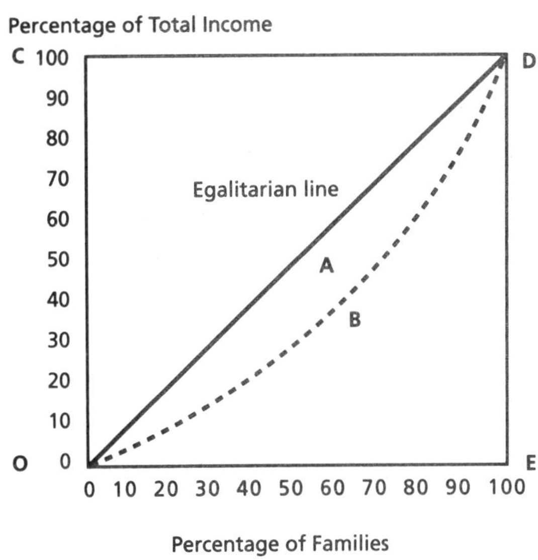

Source: Adapted from Kakwani 1987:244.

for the group as a whole. Then line everyone up in order of income from lowest to highest. Starting from the person with the lowest income, add up the total incomes of the poorest 10 people, and calculate that total as a percentage of total income for all 100 people. (The poorest 10 percent of the population will have less than 10 percent of the total income, because, after all, they *are* the poorest.) Then repeat the process for the poorest 20 people, the poorest 30 people, and so forth. The "poorest 100"—that is, everyone in the entire group—will, as a group, have 100 percent of the group's income. We now have 10 data points: the poorest 10 percent have *x* percent of the total income; the poorest 20 percent have *y* percent of the total income; and so forth. Connect the 10 data points with a smooth line and you have a Lorenz curve (Cowell 1977: 23).

The straight diagonal line, OD, in Figure 9.2 is called the *egalitarian line.* If everyone in the group under analysis had exactly the same income, the Lorenz curve would correspond to the egalitarian line. (The "poorest" 10 percent in this case would have 10 percent of the income; the "poorest" 20 percent would have 20 percent, etc.—because everyone has the same income.) At the other extreme, if all the income accrued to one individual, the Lorenz curve would be the right angle represented by OED in Figure 9.2. (The poorest 10 percent would have 0 percent of the income; the poorest 20 percent would have 0 percent of the income; even the poorest 90 percent would have 0 percent of the income; but the "poorest" 100 percent would have 100 percent of the income, again, by definition.) Examining these two extremes shows that the more the Lorenz curve bends away from the egalitarian line, the greater the inequality of income.

Figure 9.3a shows a Lorenz curve for the distribution of land owned in the Indian village of Bagbana during 1968 and 1981. More than 30 percent of the families in this village owned no land, which is why both Lorenz curves in Figure 9.3 track the zero line of land owned for more than a third of the way across the horizontal axis. Notice that inequality in land distribution in this village increased during the thirteen years from 1968 to 1981.

Figure 9.3b shows the Lorenz curve for farm income in this same village over the same period; more people had farm income than owned land (many were landless farm workers). During the thirteen years under consideration, the inequality of farm income decreased.

The Gini Coefficient

The search for a better method of measuring income and wealth distribution did not end with Lorenz. In 1912, Italian economist Corrado Gini proposed yet another measure of inequality, the Gini ratio, often called the Gini coefficient (Dagum 1987). Gini used the Lorenz curve as the basis of his ratio. He simply compared the area of the triangle OED (see Figure 9.2) with the area of the

Figure 9.3 Lorenz Curves for Farm Land and Farm Income, Bagbana Village, India, 1968 and 1981

a. Cumulative Land

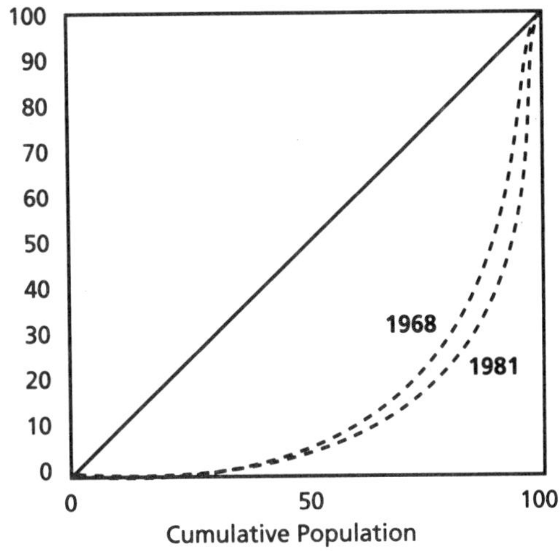

b. Cumulative Farm Income

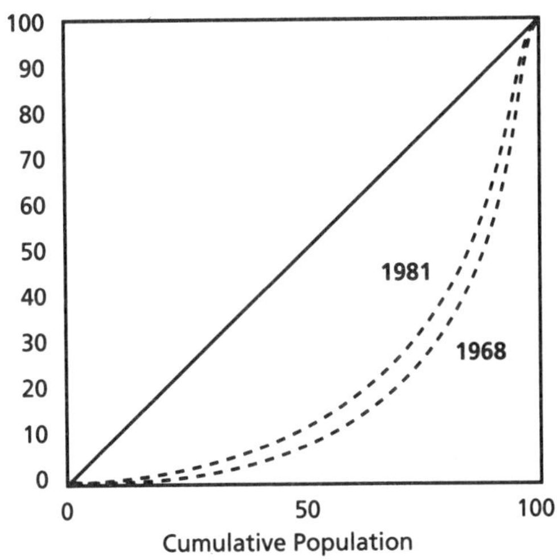

Source: Field surveys by Phillips Foster and his assistants. Lorenz curves by Paul Fishstein.

lens-shaped piece taken out of that triangle by the Lorenz curve. Labeling the lens-shaped piece as A and the remainder of the triangle as B, the Gini ratio is:

$$\frac{A}{A+B}$$

If one individual in the group has all the income, the Gini ratio becomes 1. If the size of A approaches 0, the Gini ratio approaches 0. The range of the Gini ratio is thus from 0 to 1. Often the ratio above is multiplied by 100 and the Gini coefficient is reported on a scale of 0 to 100.

Although popular, the Gini coefficient is open to criticism (Paglin 1974). It shows nothing about the location of the concentration of income inequality among high- versus low-income groups. The shape of the Lorenz curve could conceivably change (showing the poor as better-off and the middle class as worse-off, for instance) without any change in the Gini ratio.

Furthermore, neither the Lorenz curve nor the Gini ratio take account of expected variation in income as people age. Imagine a country consisting of working adults with ages distributed evenly from 20 to 50 years and with everyone in their 20s paid the same, everyone in their 30s paid the same, but more than those in their 20s, and everyone in their 40s paid the same, but more than those in their 30s. In this case, the Gini ratio would be greater than 0, even though every individual in the country has exactly the same income pattern over their lifetime.

Now imagine the same country twenty years in the future. Income by age is still exactly as it was twenty years ago, but now there are more people in their 20s and 40s, and fewer people in their 30s. The Gini ratio for the country would be higher than before, indicating greater inequality, even though lifetime income patterns have not changed at all—every individual's lifetime earning pattern is exactly the same as that in the preceding generation. Lemieux (2006) estimates that demographic changes may account for 69–95 percent of observed changes in income distribution.

Relative Income Share

Lorenz curves and Gini ratios make for useful comparisons and appear frequently in the literature of income and wealth distribution, but they may be hard for nonspecialists to understand and time-consuming to explain to politicians. The underlying data may provide a more intuitively appealing description of income distribution. For example, from Table 9.3, rather than saying, "The Gini coefficient of expenditure distribution is 30.0 for Ethiopia and 50.6 for Nigeria," one could say, "The richest 10 percent of the population in Ethiopia accounts for 25.5 percent of total expenditures, while in Nigeria the richest 10 percent accounts for 40.8 percent of total expenditures." Sometimes the ratio of the percentages earned by the richest and poorest 10 percent is reported as a

Table 9.3 Income, Population, and Income Distribution, Selected Countries

	2007 GNI ($ billions)	% of World Total	2007 Population (millions)	% of World Total	Survey Year	Gini Index	Percentage Share of Income or Consumption: Poorest 10%	Poorest 20%	Richest 20%	Richest 10%
US	13,877.00	26.39	302	4.57	2000[c,d]	40.8	1.9	5.4	45.8	29.9
Japan	4,813.00	9.15	128	1.94	1993[c,d]	24.9	4.8	10.6	35.7	21.7
Germany	3,197.00	6.08	82	1.24	2000[c,d]	28.3	3.2	8.5	36.9	22.1
China	3,121.00	5.93	1,320	19.96	2001[a,b]	44.7	1.8	4.7	50.0	33.1
UK	2,609.00	4.96	61	0.92	1999[c,d]	36.0	2.1	6.1	44.0	28.5
France	2,447.00	4.65	62	0.94	1995[c,d]	32.7	2.8	7.2	40.2	25.1
Italy	1,991.30	3.78	59	0.89	2000[c,d]	36.0	2.3	6.5	42.0	26.8
Spain	1,300.00	2.51	45	0.68	1990[c,d]	32.5	2.8	7.5	40.3	25.2
Canada	1,133.00	2.47	33	0.50	1998[c,d]	33.1	2.5	7.0	40.4	25.0
Brazil	1,071.00	2.15	192	2.90	2001[c,d]	59.3	0.7	2.4	63.2	46.9
Russia	1,069.30	2.04	142	2.15	2002[a,b]	31.0	3.3	8.2	39.3	23.8
India	955.80	2.03	1,123	16.98	1999–2000[a,b]	32.5	3.9	8.9	43.3	28.5
Korea, Rep.	878.00	1.82	49	0.74	1998[c,d]	31.6	2.9	7.9	37.5	22.5
Mexico	246.54	1.67	105	1.59	2000[a,b]	54.6	1.0	3.1	59.1	43.1
Australia	755.80	1.44	21	0.32	1994[c,d]	35.2	2.0	5.9	41.3	25.4
Netherlands	750.70	1.43	16	0.24	1999[c,d]	30.9	2.5	7.6	38.7	22.9
Turkey	592.90	1.13	74	1.12	2000[a,b]	40.0	2.3	6.1	46.7	30.7
Indonesia	373.12	0.71	226	3.41	2002[a,b]	34.3	3.6	8.4	43.3	28.5
Iran	246.54	0.47	72	1.09	1998[a,b]	43.0	2.0	5.1	49.9	33.7
Egypt	229.40	0.44	75	1.13	1999–2000[a,b]	34.4	3.7	8.6	43.6	29.5
Philippines	142.62	0.27	88	1.33	2000[a,b]	46.1	2.2	5.4	52.3	36.3
Pakistan	141.01	0.27	162	2.46	1998–1999[a,b]	33.0	3.7	8.8	42.3	28.3
Nigeria	137.09	0.26	148	2.24	1996–1997[a,b]	50.6	1.6	4.4	55.7	40.8
Bangladesh	75.05	0.14	159	2.40	2000[a,b]	31.8	3.9	9.0	41.3	26.7
Vietnam	67.24	0.13	85	1.29	2002[a,b]	37.0	3.2	7.5	45.4	29.9
Ethiopia	17.56	0.03	79	1.19	1999–2000[a,b]	30.0	3.9	9.1	39.4	25.5

Source: World Bank.

Notes: a. Refers to expenditure shares by percentiles of population. b. Ranked by per capita expenditure. c. Refers to income shares by percentiles of population. d. Ranked by per capita income.

measure of income dispersion: "In Ethiopia the richest 10 percent spend (on average) 7 times as much as the poorest 10 percent, while in Nigeria the richest 10 percent spend 25 times as much as the poorest 10 percent."

Income Distributions in Different Countries and Changes over Time

Information on income distribution is collected and published by the World Bank. Some of this is presented in Table 9.3 for the world's seventeen largest economies and eighteen most populous countries. Together, these countries account for 82 percent of the world's economic output and 74 percent of the world's population.

We can see from the table that there is quite a wide variation in the degrees of income equality. Of the countries reported, the most equal distribution, as reflected by the lowest Gini coefficient, is in Japan (Gini = 24.9); the most unequal distribution is in Brazil (Gini = 59.3). The poorest 20 percent of the population controls 10.6 percent of the income in Japan, but only 2.4 percent in Brazil. The richest 20 percent of the population controls 35.7 percent of the income in Japan, but 63.2 percent of the income in Brazil.

Over the past three decades, there appears to be a trend toward greater inequality of income in many countries. The increase in inequality has been especially pronounced in formerly socialist (Soviet-bloc) countries since the end of the Cold War and the breakup of the Soviet Union. For example, the Gini coefficient in Poland increased from 25.5 in 1991 to 33 in 1993; in Bulgaria the Gini coefficient increased from 24 to 34 over the same period. Table 9.4 illustrates changes in the Gini coefficient over time for a sample of countries. The increase for China probably represents reforms in the general economic system. However, the increases in inequality were not limited to countries that were replacing a socialist system with a more market-oriented economy. In both the United States and the United Kingdom, inequality increased gradually but substantially over a period going back to the late 1960s. From 1968 to 1991, the Gini coefficients increased from 33 to 38 in the United States, and from 24 to 32 in the United Kingdom. This trend does not appear in every country. For example, Japan and India showed little change. Italy showed a substantial increase in equality, with the Gini coefficient dropping from about 40 in the mid-1970s to about 32 in the early 1990s. (See the World Bank's website on income distribution, http://www.worldbank.org/poverty/inequal/index.htm, for an excellent overview of the literature and data on income distribution.)

One additional aspect of Table 9.4 is worth noting. Gini coefficients in all countries are quite stable over time; even when there is a distinct trend in the coefficients, the changes over a five-year period are not large. This should be reassuring to the reader who noticed that our income distribution data in Table 9.3 is for different years in different countries.

Table 9.4 Changes in Gini Coefficients over Time, Selected Countries, 1968–1991

	United States	United Kingdom	Taiwan	Japan	Italy	India	China
1968	33.50	24.10	28.90	34.90		31.86	
1969	33.64	24.90		35.70		31.47	
1970	34.06	25.10	29.42	35.50		30.38	
1971	34.30	25.70		36.90			
1972	34.46	26.00	29.02	33.40		31.85	
1973	34.42	25.10	33.60	32.50		29.17	
1974	34.16	24.20	28.09	33.60	41.00		
1975	34.42	23.30	31.20	34.40	39.00		
1976	34.42	23.20	28.40	33.90	35.00		
1977	34.98	22.90	28.00	33.70	36.30	32.14	
1978	35.02	23.10	28.43	32.90	35.98		
1979	35.06	24.40	27.70	33.90	37.19		
1980	35.20	24.90	27.96	33.40	34.29		
1981	35.62	25.40	28.15	34.30	33.12		
1982	36.48	25.20	28.51	34.80	32.02		
1983	36.70	25.70	28.45		32.87	31.49	27.20
1984	36.90	25.80	28.81		33.15		25.70
1985	37.26	27.10	29.20	35.90			31.40
1986	37.56	27.80	29.29		33.58	32.22	33.30
1987	37.56	29.30	29.65		35.58	31.82	34.30
1988	37.76	30.80	30.02			31.15	34.90
1989	38.16	31.20	30.41	37.60	32.74	30.46	36.00
1990	37.80	32.30	30.11	35.00		29.69	34.60
1991	37.94	32.40	30.49		32.19	32.53	36.20

Source: Deininger and Squire 1997.

Because of data availability, income distribution in the United States has been studied in considerable detail. There are several notable findings of that research. For example, income distribution is becoming more unequal over time. Piketty and Saez (2003) show that percentage of income earned by the top 1 percent of earners in the United States increased from about 8 percent in the early 1970s to about 14 percent in the late 1990s.

Changes in the distribution of *wealth,* however, are more ambiguous. Kennickell (2006) finds that the Gini coefficient of wealth increased from 78.4 in 1995 to 80.5 in 2004, indicating an increase in concentration of wealth. At the same time, however, he finds that the percentage of wealth held by the wealthiest 1 percent declined from 34.6 percent in 1995 to 33.4 percent in 2004.

But distribution of expenditures is considerably less unequal than is distribution of income. People go into debt to finance consumption during low-income years of their lives (in their 20s and 30s, for example) and pay off that debt by constraining consumption during high-income years of their lives (in their 40s and 50s). Cox and Alm (2008) show this difference for the United States. If we split households into groups based on *income,* we find that the richest fifth earn average incomes that are fifteen times higher than the average incomes in the poorest fifth. However, if we split households into groups based on *expenditure,* we find that households in the richest fifth spent (on average) about four times the amount spent by households in the poorest fifth.

The overall income distribution, and changes in that distribution, mask the fact that there is considerable movement of individuals from one level of the income distribution to another level over time. The US Department of Treasury (2007) found that among people in the bottom 20 percent of income distribution in 1996, half had moved out of that group by 2005. Among people who were in the top hundredth of 1 percent of income distribution (the richest of the rich) in 1996, 75 percent had dropped out of that group by 2005. The median incomes of these 1995 "superrich" actually declined over the next decade.

Factors Influencing Income Distribution

The causes of income distribution are complex and not fully understood. In this section, we review some of the explanations that have been advanced.

The "Lucky Rich" or the "Worthy Rich"?

People are more likely to favor policies to redistribute income if they believe that the rich are just "lucky" and the poor "unlucky." Wages and salaries in the general economy are set by a process that is opaque and sometimes seemingly arbitrary.

One may be inclined to view Tiger Woods's high income as justified, since television ratings rise when he is playing, and at the same time be suspicious of high compensation for a corporate executive, if it seems that the executive could be easily replaced with any of hundreds or thousands of other, lower-paid candidates. Tiger Woods is worth his millions, in this view, while the executive is simply the lucky one who got the plum job.

In the world of economic models of perfect competition, every worker is paid a wage equal to his or her marginal productivity, so in some rudimentary way, every person is "worth" exactly what they earn. But even the most doctrinaire free-market economist would recognize that compensation for rare-skill jobs is not set in a competitive market. What is more, even a competitively determined income can be perceived as having an element of luck, insofar as it depends on the worker's physical capital (tools and equipment, for example) and human capital (health and education, for example) and on whether innate or acquired skills are the skills currently in high demand.

Technology Adoption and the Kuznets Curve

In his 1954 presidential address to the American Economic Association, Harvard economist Simon Kuznets (1955) hypothesized that during the early phases of development, third world countries might experience increasing income inequalities before "leveling forces become strong enough to first stabilize and then reduce income inequalities." His idea that the path of income equality through time in the third world would trace a U-shaped curve became known as the Kuznets curve (see Figure 9.4). Subsequent studies have lent support to Kuznets's hypothesis (Adelman and Morris 1973; Ahluwalia 1976b; Chenery, Robinson, and Syrquin 1986).

Using cross-sectional data of a sample of sixty countries, including forty third world countries, fourteen developed countries, and six socialist countries, Ahluwalia estimated relative income shares as per capita income changed. The result of one of his multiple regressions, which estimates percentage income share for the lowest 40 percent of the population from income variables, is shown in Figure 9.4. In this diagram, the income share of the lowest 40 percent of the population declines from about 17 percent at around $150 per capita, to about 12 percent at around $400 per capita, and then increases back to about 17 percent as income increases to $3,000 per capita.

It seems reasonable to postulate that the poor may benefit less from development than the rich. As development takes place, those in the more advanced sector of the economy are likely to be the first to take advantage of it. Therefore, they reap the first gains. After all, when the new productive techniques come along, they often require new knowledge and substantial amounts of capital. The

Figure 9.4 Estimated Relationship Between Income Share and per Capita GNP, 60 Countries, Various Years Prior to 1975

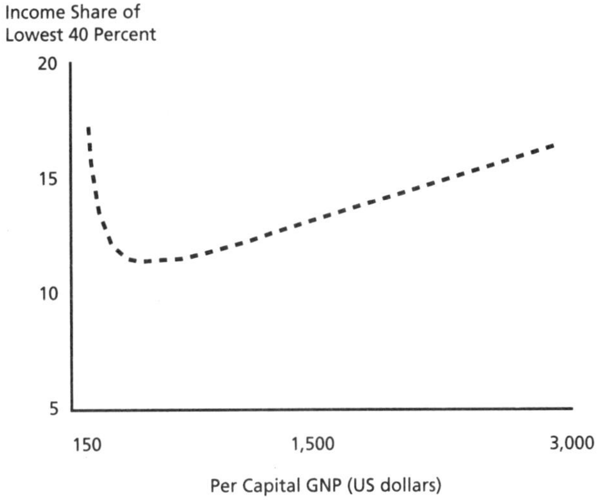

Source: Adapted from Ahluwalia 1976a:133.

railroads and canals and electric companies, which were originally privately owned, are a case in point. It was difficult for the poor even to imagine "making a killing" in these areas.

Modernization could conceivably make the poor worse-off. As Ahluwalia puts it: "An aggressively expanding technologically advanced, modern sector, competing against the traditional sector for markets and resources (and benefiting in this competition from an entrenched position in the institutional and political context) may well generate both a relative and absolute decline in incomes of the poor" (1976b:330–331). He concludes from his research, however, that though the initial stages of development are likely to make the poor worse-off relative to the rich, these same initial stages are not necessarily inclined to make the poor worse-off in absolute terms.

An Empirical Analysis of Factors Influencing Income Distribution

Ahluwalia's study, which estimated a type of Kuznets curve of income distribution, also produced evidence of how some variables besides per capita income affect income distribution. Table 9.5 shows the results of two of his multiple regressions on sixty countries, mentioned above. In these regressions, the variables in the left-hand column are hypothesized to have an impact on relative income shares.

The numbers in the middle column (equation 1) show how the variables in the left-hand column influence percentage income share among the top 20 percent of the population. The numbers in the right-hand column (equation 2) show how the variables in the left-hand column influence percentage income share among the bottom 40 percent of the population.

The variables on GNP per capita and GDP growth rate are included to account for the Kuznets curve. The variable at the bottom of the table, "dummy for socialist countries," is included to account for the fact that, generally, socialist countries have a more equal income distribution than nonsocialist countries. The remaining variables are included to see how they influence relative income share when the influence of the Kuznets curve and socialism are accounted for.

The signs on the numbers are of particular interest to us in examining the regression, because they tell us the direction of influence that the related variable has on percentage income share when the other variables are at their average value. (Because the units of measurement of each of the variables influence the size of the regression coefficients—the numbers not in parentheses—and because we do not have these units of measurement, the size of the numbers is not of particular interest to us here. The size of the numbers in parentheses is important because they are the results of a test of significance—the bigger the number, the more likely its variable is to be of significant influence.)

Table 9.5 Cross-Country Regressions That Explain Income Shares

	Dependent Variable: Percentage Income Share			
	Richest 20 Percent		Poorest 40 Percent	
Explanatory Variable	Direction of Influence of Variable	Equation 1	Direction of Influence of Variable	Equation 2
Constant	–	9.07 (0.27)[a]	+	77.93 (4.11)
Log per capita GNP	+	50.35 (2.13)	–	47.28 (3.50)
(Log per capita GNP)^2	–	8.16 (1.98)	+	7.65 (3.35)
Growth rate of GDP	–	0.11 (0.32)	+	0.11 (0.55)
Literacy rate	–	0.09 (2.21)	+	0.06 (2.56)
Secondary school enrollment	–	0.14 (2.48)	+	0.02 (0.74)
Growth rate of population	+	3.59 (4.29)	–	1.19 (2.56)
Share of agriculture in GDP	–	0.25 (2.23)	+	0.04 (0.65)
Share of urban population	–	0.10 (1.68)	+	0.06 (1.79)
Dummy for socialist countries	–	9.41 (3.27)	+	8.57 (5.35)
R^2		.76		.69
F		22.31		6.21
SEE		4.6		2.6

Source: Ahluwalia 1976a:131.

Note: For each explanatory variable, the signs switch between the richest 20 percent and the poorest 40 percent of the population.

a. Values in parentheses are T-ratios. For this sample, a T-value of 1.68 indicates significance at the 10 percent level for a two-tailed test.

First look at signs for the "dummy for socialist countries." In equation 1, the sign for this variable is negative, meaning that socialism tends to lower the percentage income share of the top 20 percent of the population. In equation 2, the sign is positive. That is, socialism tends to raise the relative income share of the lowest 40 percent of the population.

Now examine the signs for the other variables. Increasing the literacy rate, the rate of secondary school enrollment, the share of agriculture in GDP, and the share of urban population appear to increase the relative income share of the poor and to reduce the relative income share of the rich.

The share of agriculture in GDP is not significant for the poor ($t = 0.65$, or the coefficient is different from zero at only a 25-percent level of significance), so we will not consider this variable important. As the percentage of the urban population increases, relative income share of the poor increases. This seems reasonable, because urban people generally enjoy higher incomes than rural people, and because, as rural people move to the city in search of better jobs, the total income distribution may become more equal. However, because of the problems associated with growth in third world cities (crowding, pollution, crime, etc.), it is hard to argue for urbanization as a means to reduce income inequality.

Literacy rate and secondary school enrollment are good indexes of overall education among the poor in third world countries. The wealthy see to it that their children get an education; the poor are usually illiterate and lack secondary education. Increasing the rate of literacy and enrollment in secondary education makes the poor potentially more productive and gives them better access to employment and better-paying jobs, thus reducing income inequality.

The influence of population growth rate on relative income share is harder to understand. It is the only variable among consideration here for which increasing its value makes the poor worse-off relative to the rich. The relationship between income and population is discussed more fully in Chapters 8 and 18.

Why Income Distribution Matters to the Problem of Undernutrition

In the world as a whole, low-income people tend to underconsume food, while high-income people tend to overconsume it. Reutlinger and Selowsky (1976) pointed out that increased inequality of income hurts the nutritional status of the poor in two ways. The first, and obvious, way is that increased inequality means that the poor have less income, so their food consumption declines and their chances of being undernourished increase. The second way that increased income inequality hurts nutritional status of the poor is that increased incomes for the rich translate into increased demand for meat and dairy products. As discussed in more detail in the next chapter, the greater the demand of the wealthy for animal products, the more the price of grain increases and the harder it is for the poor to buy the grain they need for minimal nutrition.

The Redistribution-Incentive Paradox

The observation that income distribution can influence undernutrition among the poor by increasing the price of staples like grain raises a related possibility: that income redistribution can influence the price of food through the supply side. Just as income inequality can hurt the nutritional status of the poor as described by Reutlinger and Selowsky, a *more equal* income distribution—specifically, a government policy to redistribute income from the rich to the poor—could also increase the incidence of undernutrition, because income redistribution might decrease the monetary incentives for farmers to produce, thereby reducing supply and increasing the price of food.

During long periods of the twentieth century, two major and vastly different countries—China and the Soviet Union—experimented with general economic policies that greatly reduced differences in income. As they succeeded, they reduced differences in food consumption, but at the same time they experienced difficulties with food production.

Before their respective socialist revolutions, both countries enjoyed healthy agricultural economies. In 1917, the year of the Russian Revolution, the world's leading agricultural geographers wrote that "the Russian Empire leads the world in both acreage and production of wheat. . . . Nearly one-fifth of the average harvest is exported" (USDA 1917:13). China's socialist revolution was completed after World War II. Before that war, the success of China's agriculture is reflected in the fact that it accounted for 93 percent of the world's soybean exports.

Both Russia and China went from being major food exporters before their socialist revolutions to major food importers afterward. And during the 1980s, both countries tried to reintroduce market-oriented incentives, in part to bolster their flagging agricultural productivity.

So, while there may be nutritional gains from reducing income inequality, income redistribution has the possibility, when carried to extremes, of introducing incentive problems that lead to low levels of food consumption for all. In reducing undernutrition, then, the appropriate debate with reference to income redistribution should center on how much and by what methods.

Does Income Equality Promote Growth?

A study in the late 1970s answered this question with a tentative yes. Ahluwalia, Carter, and Chenery (1979) examined twelve countries for which they had data on growth and income shares for a ten-year period. They found that the countries that were most successful in reducing income inequality were also the countries with the highest rates of growth in per capita incomes.

However, a more comprehensive study by the World Bank (Deininger and Squire 1997) failed to make a strong correlation between growth and income inequality. Of the eighty-eight countries whose per capita GDP grew for a decade, income inequality improved slightly in about half the cases and worsened slightly in the other half. However, the study found that the distribution of *wealth* (as measured by landownership) does strongly affect growth. Countries having great inequality of wealth grow more slowly than countries with less inequality. For example, fifteen developing countries have a Gini coefficient (for land distribution) higher than 70. Of these fifteen countries, only two showed growth rates higher than 2.5 percent per year during the period 1960–1992.

Global Redistribution: Do Incomes Grow Faster in Poor Countries?

No matter how large the income disparities that exist within a country, they are dwarfed by the differences in income between rich and poor countries. Recall the comparison of average incomes in Norway and Burundi earlier in this chapter. How can such enormous differences exist between countries? One obvious answer is that much more investment in productive capital occurs in rich countries; people in those countries are more productive because they have more and better capital, better equipment in their workplaces, better roads and communications, and better education. Many economists believe that as time passes, income per capita in poor countries will catch up with income in rich countries; in other words, incomes per capita will *converge.* Investors will discover that investing in poor countries having low levels of capital has a higher payoff than investing in rich countries having high levels of capital; this investment will increase the capital stock in poor countries and therefore increase production per person in those countries.

Looking at two centuries of experience in currently developed (rich) countries, we see clear evidence of convergence. The Maddison (2001) data provide the best evidence. In 1820, the United Kingdom and the Netherlands were, by a considerable margin, the richest countries in the world (measured by GDP per capita). Average per capita income in those two countries was $1,718, almost three times higher than the average income of $676 in five countries near the bottom of the per capita income list: Canada, Australia, Japan, Finland, and New Zealand. By 1998, those five countries that were "relatively poor" in 1820 had more than caught up with the leaders: average income in the Netherlands and the United Kingdom was $19,030, and average income in Canada, Australia, Japan, Finland, and New Zealand was higher—$20,034.

For developing countries in modern times, there is mixed evidence about whether this convergence theory is correct. In 1980, per capita income in high-income countries was $9,433—about 8.4 times higher than the $1,127 average income in low- and middle-income countries. By 1999, this ratio had increased to 9.4—incomes were *diverging.* However, in the years since 1999, this ratio has fallen, until by 2007 it was 7.4.

Table 9.6 gives more insight into the convergence (or lack thereof) of incomes. Incomes in China and India have increased faster than those of rich countries since 1973. Between 1973 and 1998 (Maddison data), average income in high-income countries grew by about 60 percent, while India's per capita income more than doubled and China's nearly quadrupled. Between 1998 and 2007 (World Bank data), rich-country income rose by 20 percent, while India's rose by about 65 percent and China's more than doubled. The rest of Asia outperformed India during the 1973–1998 period, but lagged somewhat

Table 9.6 Evidence of Convergence: Per Capita Incomes in Constant 1990 Dollars

	Maddison Data				World Bank Data		Annual Percentage Growth	
	1870	1950	1973	1998	1998	2007	1973–1998	1998–2007
Rich countries	1,894	5,663	13,141	21,470	20,791	24,725	1.96	1.93
China	530	439	839	3,117	1,658	3,678	5.25	8.85
India	533	619	853	1,746	1,142	1,877	2.87	5.52
Other Asia	620	757	1,486	3,328	1,548	2,170	3.23	3.75
Latin America	698	2,554	4,531	5,795	5,354	6,385	0.98	1.96
Africa	444	852	1,365	1,368	1,054	1,281	0.01	2.17
World	867	2,114	4,104	5,709	5,313	6,747	1.32	2.65

Sources: Maddison 2001; World Bank, Data and Statistics, 2008.

Note: Maddison numbers for 1998 differ from World Bank numbers for a variety of reasons: country groups are slightly different; methods of conversion to dollar equivalents are different; standards for adjusting to constant 1990 dollars may be different; World Bank numbers are later numbers and underlying information about GNI may have been revised.

behind India in the 1998–2007 period. Overall, the pace of per capita income growth in Asia provides strong evidence for the convergence hypothesis.

However, other parts of the developing world have not fared as well when it comes to income growth. Latin America showed a period of convergence with high-income countries in the period 1870–1950. However, in the years since 1950, this convergence has reversed. Incomes in Africa show no signs of convergence. The ratio of average incomes in rich countries to average incomes in Africa was about four to one in 1870; the gap widened to about ten to one by 1973, and is now about twenty to one.

Paul Collier, in his book *The Bottom Billion,* describes the situation this way:

> These differences between the bottom billion [mostly in sub-Saharan Africa] and the rest of the developing world [China, India, and other success stories in Asia] will rapidly cumulate into two different worlds. Indeed, the divergence has . . . already pushed most of the countries of the bottom billion to the lowest spot in the global pile. It was not always that way. Before globalization gave huge opportunities to China and India, they were poorer than many of the [bottom billion] countries. . . . But China and India broke free . . . whereas other countries . . . didn't. For the last two decades this has produced a growth pattern that appears confusing. Some initially poor countries are growing very well, and so it can easily look as if there is not really a problem. Over the next two decades the true nature of the problem is going to become apparent. . . . By 2050, the development gulf will no longer be between a rich billion in the most developed countries and the five billion in the developing countries; rather it will be between the trapped billion and the rest of humankind. (2007:10–11)

World Distribution of Income and How It Has Changed over Time

So far we have discussed two sources of income disparity: income disparity within countries, which appears to be growing, and income disparity between countries, which may be declining due to convergence. What does this mean to the overall distribution of income in the world?

A description of world distribution of income is found in Box 9.3. Notice that most incomes in the United States are at the top end of world distribution.

Box 9.3 World Distribution of Income: Who Is Rich?

Who is rich? Would you say that all people in the top half of the income distribution are rich? Or do you have to be in the top quarter, or the top 10 percent, or the top 5 percent? Or would you say that only the richest 1 percent of the population are "the rich"? This is a subjective judgment, and well-intentioned, well-informed people can come to different conclusions. Once you have decided how you would define "rich," use the table below to find out how much a rich person earns (according to your definition). We predict you will be surprised.

Using data such as those shown in Tables 9.1 and 9.3, we constructed an estimated world distribution of income. In the table below, we show the world distribution of income in 2007 using the PPP comparison.

As described in the text, this method of comparing incomes attempts to adjust for different costs of living, so that a person making $5,000 in one country has approximately the same standard of living as a person making $5,000 in another country. The countries used for our exercise account for 93 percent of the world's population and 96 percent of the world's income in 2007. Therefore, the countries omitted are relatively poor countries, and our estimate of the distribution may be slightly biased in the upward direction—perhaps people are not as rich as we make them out to be.

So . . . are you one of the rich?

World Distribution of Income 2007

This percentage of the world's population . . .	Earns less than this annual income
50	$3,650
65	$6,200
75	$10,150
80	$13,800
85	$19,150
90	$27,500
95	$42,000
98	$63,500
99	$83,000
99.5	$105,000

For example, if a college student spends $42,000 a year on tuition, room, and board, that student spends more than the annual incomes of 95 percent of the world's population. A single person at the poverty line in the United States earns more than about 75 percent of the world's population. See Box 9.4 for a more detailed picture of poverty in the United States.

Sala-i-Martin's estimate (2002) of world distribution of income and how it has changed over time is very similar to the distribution reported in Box 9.3. He concludes that income is becoming more equally distributed. He estimates that the worldwide Gini coefficient fell from about 66 in the 1970s to 65 in the 1980s to 63 in the 1990s. This has translated into a reduction in poverty worldwide (measured by the percentage of people living on incomes of less than $2 per day, adjusted for inflation). Sala-i-Martin finds that the worldwide poverty rate declined from 40 percent in 1970 to less than 20 percent by the late 1990s. China's poverty rate fell from about 75 percent in 1970 to 20 percent in 1998. The reduction in poverty was found in every geographical area except Africa, where the poverty rate increased from 53 percent in 1970 to 64 percent in 1998.

Box 9.4 Poverty in the United States

Based on the data presented in Box 9.3, a person at the poverty line in the United States would have a higher income than about 75 percent of the people in the world. How does the life of a poor person in the United States compare to the life of Adilesi Faisoni in Malawi, described in Box 9.1?

Single individuals are categorized as "poor" in the United States if their income (2007) is below $10,210 a year; for a family of four, the poverty line is $20,650. In the United States:

97.7 percent of poor households live in houses or apartments that have complete indoor plumbing.
80.7 percent of poor households have air conditioning.
99.4 percent of poor households have a refrigerator.
97.3 percent of poor households have a color television; 54.6 percent have more than one color television; 62.7 percent have cable or satellite television.

Sources: US Bureau of the Census 2008; US Energy Information Administration 2001.

10

Other Factors Influencing Demand for Food

Chapter 8 presented some projections about the size of the world's population—it could grow by more than 50 percent over the next fifty years. Does this mean that if food supply grows by 50 percent, we will have enough extra food to feed those extra people? To answer that question we must first examine the factors that influence food consumption per person. In Chapter 9, we discussed how demand for food increases as income increases, and we examined some trends in per capita income levels and the distribution of income. In this chapter, we ask the question: What factors influence the amount of food and the types of food consumed per capita?

Population Characteristics and Calorie Requirements

How does average demand for food change as a result of characteristics of the population? We saw in Chapter 3 that nutrient requirements depend on age, sex, pregnancy and breast-feeding, and physical activity.

Age Structure

The age composition of a population reflects the underlying demographic conditions of the past and at the same time is an important determinant of demographic conditions of the future.

Population pyramids. The most convenient way to visualize the age structure of a population is through a pyramidal graph of population distribution according to age and sex. Conventionally, population pyramids represent age cohorts by five- or ten-year intervals, and place males on the left of a vertical line and females on the right, with the youngest cohort at the bottom. The graphic representation of the age cohorts can reflect either actual numbers or percentage distribution. Figure 10.1 shows a numerical population pyramid for the industrialized nations versus the third world in 1985, with projections to 2025. The

Figure 10.1 Population Pyramids for Less and More Developed Countries, 1985 and Projections to 2025

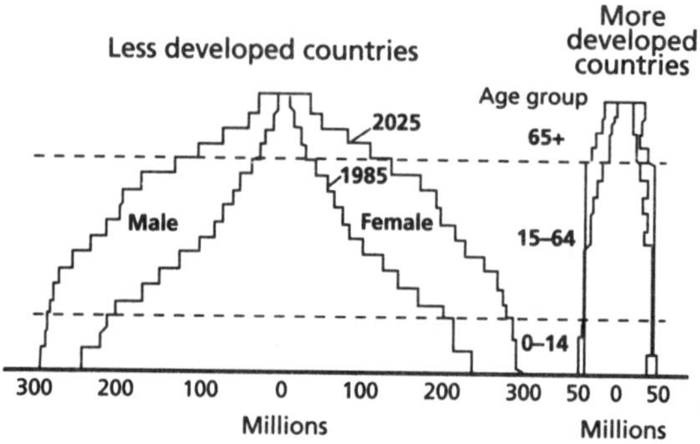

Source: Adapted from Merrick et al. 1986:19.

horizontal lines across both sets of pyramids mark the dividing lines that are commonly, but arbitrarily, placed to separate the dependent age categories (in this case, younger than age 15, and 65 or older) from the working-age population. Such numerical population pyramids do a nice job of illustrating the differences in actual population size of the third world versus the developed world. They can also show demographic features such as the higher survival rate of older women (notice the difference between the numbers of men and women in the oldest cohort for the 2025 projection for the developed world).

Age structure is of interest to demographers because, as noted above, it provides clues about past and future demographic patterns. It is important to policymakers because of its impact on two things: momentum in population growth, and dependency ratios.

Momentum in population growth. Over a long period, a population would simply reproduce itself if individual couples produced exactly the right number of children to replace themselves, accounting for the fact that some children die before they arrive at childbearing age. In most populations this number comes to 2.1 children per couple (or per woman). If a population has remained constant for a couple of generations and then its fertility rate rises higher than 2.1, it will grow. If, on the other hand, its fertility rate falls below 2.1, it will shrink. Thus a fertility rate of 2.1 is considered the replacement level (Merrick et al. 1986:6). (In this discussion, we assume that no net immigration or emigration increases or decreases population size, and that no improvements in healthcare raise average life expectancy.)

One might expect that when the fertility rate of a rapidly growing population falls to 2.1, births and deaths would be in balance, and population growth would stop. This is not the case, at least not immediately. The reason is demographic momentum. A population that has had high fertility in the years before reaching replacement-level fertility will have a much younger age structure than a population with low fertility before crossing the replacement threshold. Consider a developed country versus a developing country (see Figure 10.1). For the developed country, the number of females expected to enter their childbearing years over the next five years (the female cohort aged 10–15, roughly) is almost exactly the same as the number of females expected to exit their childbearing years (the female cohort aged 45–50, roughly). Therefore the number of childbearing women will remain about the same, and a replacement fertility rate would imply a stable population size. For the developing country, however, the 10- to 15-year-old cohort is much larger than the 45- to 50-year-old cohort. Therefore, the number of childbearing women is expected to grow over the next five years. Even if fertility fell to a replacement rate (2.1 children per woman of childbearing age), population would continue to grow because the number of women of childbearing age continues to grow. This is the phenomenon of demographic momentum.

On a worldwide basis, population momentum means that "even if there were a sudden reduction of fertility to the level strictly needed to replace the population, the world population would still increase by more than 2 billion" by the year 2050 (FAO 1996e: paper no. 4).

Dependency ratios. Population pyramids also illustrate the dependency ratios in a population. A dependency ratio is usually defined as the ratio of dependents to working-age adults. Working-age adults are generally identified as those who are 15 to 64 years old. The adult dependency ratio is the percentage of the population who are 65 and older divided by the percentage of the population who are working-age; the child dependency ratio is the percentage of the population who are younger than 15 divided by the percentage of the population who are working-age.

Dependency ratios can influence overall nutritional status in a population because the young and the elderly are more likely to be food-insecure, since they rely on others to provide them with food. In a rapidly growing population, the child dependency ratio is far greater than in a slowly growing population.

Future age structure of the population. What kind of age structure will we see in the future? The theory of demographic transition described in Chapter 8 predicts that as living conditions improve, life expectancy will increase, followed by a decline in fertility rates. This implies that in the future, there will be a smaller percentage of children and a larger percentage of adults. Future population pyramids will be narrower.

If there are proportionately more adults than children, the need for food will grow faster than the population. The simple example in Table 10.1 illustrates this

Table 10.1 How Changing Age Structure of a Population Can Affect Food Requirements

	Present	Future
Number of children (requiring 1,800 calories per capita per day)	4 million	5 million
Number of adults (requiring 2,700 calories per capita per day)	6 million	10 million
Total population	10 million	15 million
Total calorie requirements	23,400 million	36,000 million
Calories per capita per day	2,340	2,400

point. Although population increases by 50 percent, calorie requirements increase by 54 percent because the future population is 66 percent adult rather than the present 60 percent.

Other Demographic Characteristics and Food Requirements

Average per capita food requirements in a population also depend on other characteristics of the people, such as pregnancy, physical activity, and height.

Pregnancy

Pregnant and breast-feeding women require higher caloric intake. Two opposing trends exist here. As the base of the population pyramid contracts, we see an increase in the ratio of women of childbearing age to total population. On the other hand, the drop in fertility rates means that each woman of childbearing age is becoming pregnant fewer times during her lifetime. The combined effect is expected to be close to zero for the period 1995–2025 (FAO 1996e: paper no. 4).

Physical Activity

Anyone who has ever exercised in an attempt to control his or her weight knows that physical activity burns calories. On a worldwide scale, what is likely to affect average per capita calorie requirements is not "average visits to the gym," but the average physical activity of adults on the job. Farming (especially in developing countries) requires more physical activity than many city jobs. Therefore, as urban populations grow faster than rural populations in the future, we should expect to see a decline in the average activity level. The Food and Agriculture Organization projects that this change will cause a reduction in per capita food requirements of from 1 to 4 percent for the period 1995–2025 (FAO 1996e: paper no. 4).

Height

Good nutrition during infancy and childhood can cause a person to become a taller adult. But taller adults need more calories to maintain their bodily functions. The FAO's projections are based on an underlying assumption that food supplies will continue to grow faster than food demand; thus the incidence of undernutrition will decline, and there will be less stunting. This is expected to add 2 percent to energy requirements in developing countries between 1995 and 2025, and to have no substantial effect in the developing world (FAO 1996e: paper no. 4).

Income Growth and per Capita Food Demand

The total effect of demographic changes on food demand is expected to be negligible; however, growth of per capita income is likely to have a much bigger effect on food demand. As we saw in Table 9.6 for the 1998–2007 period, per capita income grew at an annual rate of 2.65 percent for the world as a whole. If average income per capita continues to grow at a rate of 2.65 percent per year, then in forty years it will increase by 189 percent. As discussed in Chapter 7, estimates show that the income elasticity for food is between 0.1 and 0.3. This means that a 1 percent increase in a person's income will increase the quantity of food demanded by the person by from 0.1 to 0.3 percent. If we use an income elasticity of 0.2, a 189 percent increase in income would be associated with a 38 percent increase in food demand. Using the changes projected in Table 10.2, an income increase of 49 percent (1 percent annual growth), combined with a income elasticity of 0.1, would mean an increase in per capita food consumption of 4.9 percent over the forty-year period; an income increase of 232 percent (3 percent annual growth), combined with an income elasticity of 0.3, would mean an increase in per capita food consumption of about 70 percent.

The most likely scenario is at the low end of this range, for two reasons. First, although average growth of per capita incomes in the 2–3 percent range have been observed over a decade, it may be hard to sustain this high growth level for forty years. Second, income elasticities decline as income increases; therefore the 0.1 elasticity is probably more realistic than the 0.3 elasticity.

Table 10.2 The Power of Compound Growth

Annual Growth of Income per Capita	Total Growth of per Capita Income After 40 Years
1%	49%
2%	123%
3%	232%

So a good guess is that food demand per capita will grow by 5–20 percent over the next forty years. Is this a reasonable guess based on historical experience? Growth of this magnitude would mean that (assuming constant prices) the worldwide average intake of calories per capita might grow from the current level (2003) of 2,808 per day to between 2,946 calories per day (approximately equivalent to the current average diet in China) and 3,508 calories per day (approximately equivalent to the current average diet in Hungary). For an additional comparison, consider that between the early 1960s and the early 2000s, worldwide consumption of calories per person grew about 24 percent.

Growth in Population and Growth in Food per Capita: A Multiplicative Effect

Notice that growth in per capita food consumption magnifies the impact of growing population. Imagine a country in which 1,000 people consume 2,500 calories per day—total food consumption in the country is 2.5 million calories per day. Now suppose the population grows to 2,000, and per capita calorie consumption grows to 3,000; now total food consumption in the country is 6 million calories per day. The population has grown by 100 percent, but food consumption has grown by 140 percent. Not only are there more people to feed, but each person is eating more. The practical effect of this is illustrated in Table 10.3. If population grows by 40 percent over the next forty years, and if per capita food demand grows by between 5 and 20 percent, then total food demand will grow by between 47 and 68 percent.

Dietary Diversification and Demand for Food

As average incomes increase, people do not simply eat more food; they eat different kinds of food. In particular, they eat more meat and animal products, and they consume fewer calories from cereals. To illustrate this, consider the diets of various countries and country groups shown in Table 10.4.

Table 10.3 Growth in per Capita Food Consumption Magnifies the Effect of Population Growth

Growth of per Capita Food Demand over 40 Years	Growth of Total Food Demand After 40 Years of 40% Population Growth
5%	47%
10%	54%
15%	61%
20%	68%

In 2003, average calorie consumption per capita per day worldwide was 2,808, with about 477 of those calories (17 percent) coming from animal products and about 2,331 from plant sources. But it takes more than 1 calorie of grain produced and fed as animal feed to produce 1 calorie of meat. See Box 10.1 for a discussion of the number of plant-derived calories needed to produce a human-consumed calorie from various animal products. In the calculations below, we will assume that it takes 6 plant-derived calories to produce 1 animal-product calorie.

Table 10.4 Calories per Capita per Day from Animal and Vegetal Sources, Various Countries, 2003

	Total Calories	Vegetal Calories	Animal Calories
Brazil	3,145	2,470	675
China	2,940	2,296	644
India	2,472	2,270	202
Nigeria	2,713	2,627	86
United States	3,753	2,708	1,045
Developed countries	3,331	2,453	878
Developing countries	2,668	2,299	369

Source: FAOSTAT various years.

Box 10.1 Number of Plant-Derived Calories Needed to Produce 1 Calorie from Animal Products

In a background paper for the 1996 World Food Summit, the FAO (1996e) published the following estimates:

11 plant-derived calories are needed to produce 1 calorie of beef or mutton.
4 plant-derived calories are needed to produce 1 calorie of pork or poultry.
8 plant-derived calories are needed to produce 1 calorie of milk.
4 plant-derived calories are needed to produce 1 calorie of eggs.

Time magazine (Usher 1996) cited a conversion rate of 16 to 1 for cattle.

Fitzhugh (1998) argues that these numbers are too high. Animals can eat grass, crop compost, waste, and byproducts that are not part of the human diet. He estimates that the correct conversion rate is 2.3–3.4 plant calories for each calorie of animal products.

If we convert the figures on the world's food balance sheet for 2003 from kilograms to calories, we find that the ratio of calories used in animal feed to calories of animal products eaten by humans is about 2.2 to 1. Although we will use the figure of 6 plant calories for each animal-product calorie in our projections and discussions in the text, it should be noted that this may overstate the impact of dietary diversification. This issue comes up again in Chapter 20 where conversion rates of 2.4–3.5 are used.

In 2003, people in developing countries consumed about 14 percent of their 2,668 calories per day as animal products. Suppose this percentage rises to 17 percent by 2050. Even if caloric intake remained constant, this shift to animal products would mean that each person would need 9.4 percent more plant-derived calories. The calculation is as follows: The current developing-country diet is 369 calories from animal sources, and 2,299 from plant sources. If 6 plant-derived calories are needed to produce 1 animal-product calorie, this means that the current developing-country diet requires 4,513 plant-derived calories. If calories from animal products rise to 17 percent of 2,668 (453 calories), then the future diet would require 4,933 plant-derived calories, or a 9.4 percent increase. To put it somewhat differently: if diets in developing countries were slightly more diversified (17 percent animal products instead of the current 14 percent)—even with no increase in calories per capita—the effect on food demand would be the same as a 9.4 percent increase in population.

Consider a different scenario—suppose that by 2050, people in developing countries have incorporated meat into their diets to such an extent as to resemble the diets in developed countries today. In this case, the future developing-country diet would be 26 percent animal-product calories (694 out of 2,668). Converting this to plant-derived calories, and adding the 1,974 calories consumed directly from plant sources, yields 6,138 plant-derived calories—a 36 percent increase from the present diet.

Delgado (2003) projects that per capita consumption will grow between 1997 and 2020 by 2.1 percent per year for meat, and 1.7 percent per year for milk. These growth rates are considerably higher than in the preceding two scenarios. If animal calories do grow at these rates, then by 2020 or 2025, many people in the world will have reached some kind of satiation point where they no longer react to increased income by adding more animal calories to their diets.

The same multiplicative effect between growth in population and growth in food consumption per capita applies more generally to all the effects described in this chapter. To review, using an example from Table 10.3:

- If population grows by 40 percent over the next forty years, and if per capita food demand increases by 15 percent, then total food demand will grow by 61 percent. The computation is: $[(1 + 0.40) \times (1 + 0.15)] - 1$.
- If diversification of diets has an additional impact on effective (or plant-equivalent) demand of 10 percent, then total food demand will grow by 77 percent. The computation is: $[(1 + 0.40) \times (1 + 0.15) \times (1 + 0.10)] - 1$.

The numbers in this example were chosen to be within the "reasonable" range. They are by no means intended as exact projections.

Notice the impact of accounting for changes in income-induced food demand. Population grows by 40 percent, but total food demand grows by 77 percent—nearly double the growth in population. Even though the income effects are relatively small—15 percent growth in calories consumed, 10 percent growth due to diet diversification—their cumulative effect is large.

Agricultural Production for Nonfood Uses

It is easy to make a false equivalence between "food" and "agriculture." In fact, a growing proportion of agricultural production is used for nonfood purposes. Fiber crops (such as cotton and jute), tobacco, and coffee and tea are grown on a land area equal to about 3.6 percent of total arable land (arable land plus permanent cropland). This percentage has declined slightly, from 4.1 percent, since the early 1980s (FAOSTAT 2008c, 2008d). Production of illegal drugs accounts for about half of 1 percent of total arable land (UN Office of Drug and Crime 2007). There is no evidence that this percentage is growing.

One non-human-food use that is growing is pet food, spurred by increased pet ownership in China. Worldwide spending on pet food is expected to more than double between 2004 and 2009 (Chaney 2008). However, despite this rapid growth, the total impact of pet-food demand on total food demand is fairly small: if pet ownership in China were to reach the levels that currently exist in the United States, that would have an impact equal to about 2 percent greater total growth in human population.

The nonfood use likely to have a substantial impact on food demand during the next decades is biofuels. As noted in Chapter 7, the increase in food prices during 2007–2008 was caused in part by increased use of food crops for biofuel production. This trend is expected to continue and to accelerate during the coming decades. The USDA (2008a) projects that US ethanol production will increase from 8 billion gallons a year in 2008 to 30 billion gallons a year in 2025. The OECD and FAO (2007) project that biofuel production in the United States, European Union, Brazil, China, and Canada will more than double during the 2006–2016 period. Based on the numbers in the OECD and FAO report, we calculate that in 2006, use of crops for biofuels was equivalent to 150 calories per person per day; by 2016, this is expected to increase to 315 calories per person per day.

As we try to project further into the future, several uncertainties arise. If all energy and transportation technology were to be frozen at 2008 levels, then use of food crops for ethanol will undoubtedly continue to grow. However, two types of possible technological developments could slow or reverse the growing use of food crops for ethanol. First, the coming decades may see development of alternatives to transportation powered by gasoline-ethanol: cars powered by electricity or hydrogen, for example. Second, there is significant current research into producing biofuels without using food crops. For example, it is possible to produce ethanol from cellulose derived from nonfood plants like "switchgrass" (US Department of Energy 2007), and this may become a more profitable source of ethanol than corn. It is also possible to produce ethanol and biodiesel from algae (Haag 2007). Either of these options would allow production of biofuels with considerably less impact on the food market. In the scenarios below, we will examine biofuel growth in the range of 5–15 percent. A 10 percent growth in demand would be equivalent to 280 calories per person per day compared to the growth of 165 calories per capita per day projected by the OECD and FAO for the 2006–2016 period.

Scenarios for the Future Demand for Food

In making projections about the future, the word "if" is used frequently. It is useful to construct alternative scenarios to see how much variation results. Table 10.5 shows a number of scenarios about possible growth in demand for food. Scenario building allows us to investigate differences of opinion about how things will change. For example, scenario 2 in table 10.5 differs from scenario 1 only in the population growth number—scenario 1 assumes that population will grow about 40 percent between now and 2050 (similar to the UN's medium variant, or the US Census Bureau projection); scenario 2 assumes that population will grow only 15 percent (similar to the UN's low variant). Because of the magnifying effects of growing demand per capita, this 25 percent difference in population growth translates into a 35 percent difference in total demand growth between scenario 1 and scenario 2.

However, scenarios need to be internally consistent. For example, it is inconsistent to assume that incomes per capita will be stagnant and while a large amount of dietary diversification occurs; dietary diversification is a result of growth in incomes. Because of the link between income and fertility rates (from the theory of demographic transition in Chapter 8), high population growth is likely to be associated with low income growth (and vice versa). Low income growth reduces total calories per capita (the effect of income growth) and reduces the expansion of animal-derived calories (the effect of dietary diversity).

Table 10.5 Scenarios for Growth in Food Demand to 2050 (percentage increases)

Scenario	Effect of Population Growth	Effect of Increasing Income	Effect of Dietary Diversification	Effect of Growth in Biofuels	Total Growth in Food Demand
1	40	20	10	10	95
2	15	20	10	5	57
3	15	40	15	10	95
4	60	10	5	10	95
5	40	20	10	15	100

Therefore, the requirement of internal consistency carries with it a certain element of self-correction of projections. This is illustrated by comparing scenario 1 to scenarios 3 and 4. Scenario 3 has lower population growth, but higher growth in per capita income (3 percent per year, rather than 2 percent per year assumed in scenario 1). Scenario 4 has higher population growth (along the lines of the UN's high variant), but lower per capita income growth (1 percent annual growth). In these scenarios, the higher (or lower) growth in income is exactly offset by the lower (or higher) growth in population, so that in all of scenarios 1, 3, and 4, total food demand grows by about 95 percent. Roughly speaking, demand would double between now and 2050. We will return to these scenarios and others in the concluding final chapter.

Finally, scenario 5 shows the potential impact of a policy-induced growth in ethanol demand (as translated into per capita demand for food) of 15 percent. As discussed above, this would result from a continuation and intensification of the trends expected over the next decade. The impact is to increase total demand for food by an additional 5 percent.

11

Agricultural Land and Water

The previous few chapters helped us analyze questions such as: How much food is enough? How many people will need to be fed? How much food will the average person eat? How do the average level and distribution of income in the world influence the answers to these questions? We now turn to the supply side: Will there be enough food? How can we increase the supply of food to make food more affordable? Prices of food are not etched in stone and handed down from on high. They are determined in the market day-by-day and week-by-week through the interplay of supply-and-demand forces. When the supply of food increases, prices drop, food becomes more affordable, and hunger decreases. In this chapter we begin an examination of the factors that determine food supply.

The Basic Equation of Food Supply

The typical way to analyze the food supply is to focus on crops, and to split up output according to the following simple equation:

$$\text{Total output} = \frac{\text{Output}}{\text{Acre}} \times \text{Number of acres}$$

Typically "output per acre" is referred to by the shorthand term "yield." Of course, one might think that this equation ignores the possibility of obtaining food from animal products. But as described in Chapter 10, animal food products require animal feed, and animal feed comes from crops. And only 17 percent of calories worldwide comes from meat products. Almost 50 percent comes from cereal crops (rice, wheat, maize, etc.). Therefore we will discuss food production primarily from the perspective of crop production.

The output equation obviously splits into two factors: land area and yield. In our discussion in Chapters 11 to 14, we will explore four principal influences on current and future agricultural output:

- Quantity of available agricultural resources—land and water.
- Quality of agricultural resources.
- Intensity of input use on the land.
- Technological change.

This chapter deals with the first of these influences. Chapter 12 deals with the relationship between agricultural production and environmental quality, Chapter 13 deals with input use, and Chapter 14 discusses technological change.

Available Land

One way to increase food production is to increase the amount of land devoted to agricultural production. Data in Table 11.1 show that agricultural land has increased slowly but steadily. Worldwide, total land in agriculture increased 2.7 percent during the 1960s, 2.1 percent during the 1970s, 4.0 percent during the 1980s, and 2.6 percent during the 1990s.

The growth in total agricultural land masks some changes of land use within agriculture. As Figure 11.1 shows, total arable land ("arable land" is land on which crops are grown) grew more rapidly than pastureland during the 1960s, but less rapidly than pastureland during the 1980s and 1990s. This (and the declining amount of land devoted to cereal production) is consistent with the addition of meat and dairy products to the average diet, described in Chapter 10.

But can the rate of increase in agricultural land use continue into the future? An FAO report estimates that there are 2.57 billion hectares of "rainfed land with crop potential" in developing countries, excluding China (Alexandratos 1995:162–163). Of these 2.57 billion hectares, about 0.75–0.85 billion are currently in use. At first blush, this would appear to be very good news. But much of this potential land (67 percent) is hilly, or has poor soil or drainage. As Table 11.2 shows, the potential for increased production on good-quality land is considerably more limited. Nevertheless, there does appear to be potential to

Table 11.1 Worldwide Agricultural Land Use, 1961–2005 (thousand hectares)

	Agricultural Land	Arable Land	Permanent Crops	Permanent Pasture
1961	4,455,455	1,280,680	89,369	3,085,406
1971	4,574,962	1,326,527	97,200	3,151,354
1981	4,672,066	1,352,448	102,583	3,217,105
1991	4,858,507	1,401,035	120,891	3,336,581
2001	4,983,411	1,412,399	135,980	3,428,008
2005	4,967,580	1,421,169	140,511	3,405,898

Source: FAOSTAT 2008d.

Figure 11.1 Indexes of Land Use, 1961–2005

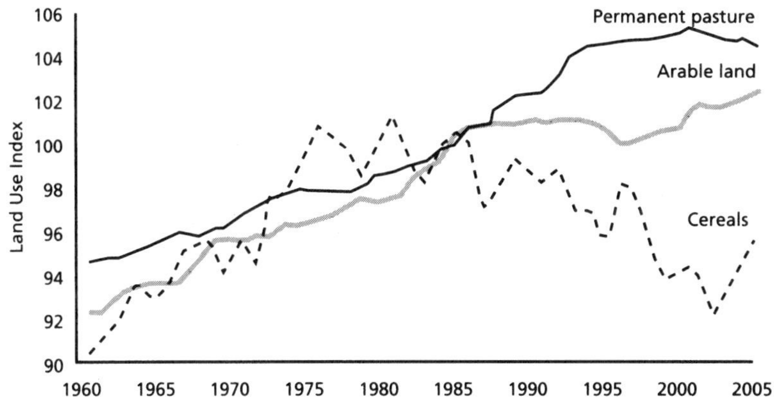

Table 11.2 Land Use and Availability in Developing Countries (Excluding China), 1996 (million hectares)

	Good Soil and Terrain	Poor Soil and Terrain	Total
Currently used for crops	547	213	760
Land with crop potential	848	1,722	2,570
Use as percentage of potential	64%	12%	30%

Source: Derived from FAO 1996a: paper no. 1.

bring new land under cultivation; the area of this potential agricultural land is perhaps 30 percent of the current area. The potential differs significantly from region to region. In South Asia, the Near East, and North Africa, there is little potential for expansion. In Brazil and the Democratic Republic of Congo (and surrounding African countries), there are large tracts of land that could be brought under agricultural production. (See Grigg 1993: chap. 6, for additional discussion.) Box 11.1 describes a large area of currently unused land in Brazil. A report by the African Development Fund estimates that in the Democratic Republic of Congo, "agricultural land is very vast (80 million ha) but only [one-tenth] of the land is being developed at the moment. The hydrographic network . . . offers tremendous water resources that can be mobilized" (2004: para 2.7.3).

The previous paragraph seems to imply that it is possible to increase the amount of land under cultivation significantly, perhaps by 30 percent or more. However, this calculation requires an additional assumption: that no land currently under cultivation is lost to agriculture. There are three reasons to be concerned that land currently used for agricultural production might not be usable

Box 11.1 Expanding Agricultural Land in the Brazilian Cerrado

One place that there is substantial opportunity for adding land to agricultural use is in the Brazilian savannah. About one-quarter of the area of Brazil (200 million hectares) is in an ecological system known as the *cerrado*. The cerrado has ample rainfall, and its climate is warm enough to grow two crops per year in most areas. Yet the cerrado remains largely underutilized for agricultural production. One study estimates that 137 million hectares of the cerrado are "well suited to large-scale mechanized farming" (Wallis 1997). As of 1990, only 12 million hectares were being used to grow crops and another 35 million hectares were being used as pasture (Schnepf, Dohlman, and Bolling 2001).

Two things have held back development of the cerrado: soil quality and transportation capacity. The soil of the cerrado is deep, but has chemical properties that are not conducive to crop growth: soil acidity is high, aluminum content is high, and soil availability of nitrogen and phosphorus is low. But these shortcomings can be overcome. Spreading lime will reduce the acidity of soil. Applying fertilizer will increase nitrogen and phosphorus. And new plant varieties have been developed that can tolerate soils with high aluminum content. Transportation infrastructure that provides a way to ship output from the cerrado to urban or international markets is limited. There is a "chicken and egg" problem here: farmers will not expand production if there is no way to move the output to market; roads and railroads will not be expanded until the need for commercial transport exists.

for agriculture in the future. First, as population grows, and urban areas expand, some farmland is paved over. The impact of suburban sprawl is obvious in parts of the United States. However, worldwide, urban areas and other human settlements take up only 3 percent of the land mass. Therefore, even significant urbanization will have a small quantitative effect on agriculture worldwide. Second, as we will discuss in more detail in the next chapter, there is concern that global warming may result in expansion of ocean areas and flooding of coastal areas. Third—and this is the factor of most concern to experts—land currently in production may be "degraded" through soil loss or contamination to such a degree that it can no longer be used to grow crops. This aspect will also be discussed in more detail in the next chapter.

On the other hand, there is undoubtedly some potential for increasing production without increasing yields *or* land area. Most significant here is likely to be an expansion of double- or triple-cropping—by which a single plot of land is planted with two or three crops sequentially during a year. Double-cropping becomes more feasible if agronomists develop crop varieties that require a shorter growing season. Other new technologies may increase the importance of non-land-based food production—most immediately, fisheries and aquaculture, but potentially including hydroponics, food from the sea, and food from space.

Taking all of these factors into account, some experts are notably more pessimistic than the FAO regarding the future potential for adding or maintaining land devoted to food production. Kendall and Pimentel cite lack of arable land as "one of the most urgent problems facing humanity . . . [and also] perhaps the most neglected" (1994:205). In another study, Pimentel calculates that nearly one-third of the world's cropland (1.5 billion hectares) has been abandoned during the past forty years because erosion has made it unproductive (Pimentel et al. 1995). Gary Gardner of Worldwatch Institute also concludes that there is little room for large-scale expansion of cropland:

> Replacing lost land is likely to be more difficult than many officials think. . . . Optimistic officials often overestimate the potential for expansion by including marginal land, where cultivation may not be sustainable. Indeed, the world's major grain producers have all overexpanded into marginal land in recent years, damaging large areas of land in the process. Many are now pulling back to the land that can be [sustainably] cultivated, with a resulting loss of grain production. (Worldwatch Institute 1996a)

The Importance of Water

Of course, finding new agricultural land is useless unless we also have sufficient water for agricultural production. As Table 11.3 shows, agriculture is a huge user of water, especially in the developing world. The amount of land that is irrigated has grown over time, though the rates of growth have declined since the 1980s (Postel 2003:60). Irrigation is especially important in growing rice. This is evidenced by the fact that irrigated land produces 80 percent of food in Bangladesh, 70 percent in China, and over 50 percent in India and Indonesia (FAO 1996b). Worldwide, irrigated land provides about 40 percent of total food. Yields on irrigated land range from 30 to 200 percent higher than on nonirrigated land. Irrigation raises corn yields from 1.7 to 3.9 metric tons per hectare in Latin America and from 1.2 to 3.1 in Africa; raises wheat yields

Table 11.3 Water Use by Country Group, 2000

	Developing Countries, Percentage	High-Income OECD Countries, Percentage	World	
			Percentage	Cubic Kilometers
Agriculture	81	41	69	2,630
Domestic	8	15	10	400
Industry	11	44	21	800
Total	100	100	100	3,830

Source: UNDP 2006:138.

from 1.8 to 4.1 in Latin America and from 1.4 to 2.4 in North Africa and the Near East; and raises vegetable yields from 5.1 to 14.2 in East Asia.

Will it be possible to continue increasing irrigation? The FAO is quite optimistic on this question: "Half or even two-thirds of future gains in crop production are expected to come from irrigated land" (1996b). A study by the World Bank and the UNDP (1990) estimated that an additional 110 million hectares of land could be brought under irrigation, producing enough more grain to feed 1.5 to 2 billion people. Worldwide, total water use in agriculture is about 2,200 cubic kilometers, amounting to 0.0088 cubic kilometers of water per 1,000 hectares of irrigated land. If 110 million hectares of newly irrigated land were irrigated at this rate, the total additional water requirement would be 968 cubic kilometers, about 30 percent of current water usage.

However, other experts are more pessimistic about future water availability. Alan Wild concludes, "Shortage of water . . . is probably the biggest biological and physical limitation to agricultural development in developing countries" (2003:217). Sandra Postel (1997) concludes that agriculture cannot increase water use much beyond current levels without causing substantial environmental problems. She argues that the figure of 44,500 cubic kilometers of water in total runoff greatly overstates available water. First, she argues, about 20 percent of that is geographically so remote that it is not available for human use. Of the remaining 32,900 cubic kilometers, about 75 percent occurs during floods, and therefore is not available for irrigation during dry periods. The actual quantity of usable water is then around 12,500 cubic kilometers. "The problem is that water use tripled between 1950 and 1990 as world population soared by some 2.7 billion. . . . Worldwide demand for water cannot triple again without causing severe shortages for crop irrigation, industrial use, basic household needs, and critical life-supporting ecosystems" (Postel 1997). She notes that water shortages are already appearing in the form of depleted groundwater. Postel also cites evidence that water tables are falling 20 centimeters a year in India's Punjab. (See Box 11.2 for a discussion of irrigation in China.) Other infamous examples of the effects of overirrigation are found in the decline of the Aral Sea in the former Soviet Union, and of Lake Chad in central Africa.

Others belittle this talk of water shortages as "doomsaying." Julian Simon of the University of Maryland was one of the most outspoken optimists about future resource availability. He based his optimism on a confidence in human ingenuity:

> Usable water is like other resources, however, in being a product of human labor and ingenuity. People "create" usable water, and there are large opportunities to discover and utilize new sources. Some additional sources are well-known and already in partial use: transport by ship from one country to another, deeper wells, cleaning dirty water, towing icebergs to places where water is needed, and desalination. . . . [In addition,] huge new supplies of

Box 11.2 Irrigation in China

Evidence of the impact of irrigation is striking in China. The Yellow River flows through a dry part of China, and farmers draw water out of the river to irrigate their crops. In the years prior to 1972, the Yellow River always had sufficient water flow to reach the sea. Then in 1972, as more and more water was drawn for irrigation, the river ran dry, failing to reach the ocean for fifteen days. Since 1986, the Yellow River has run dry every year. In the drought year of 1997, the river failed to reach the sea for 227 days. In places like this, irrigation has grown to its limit; irrigation might continue at current rates, but is unlikely to grow any further.

In other parts of China, water is used for irrigation at rates that cannot be sustained indefinitely. In many parts of the North China Plain where irrigation water is pumped from below ground, the water table was dropping by 5 feet per year over a five-ear period in the mid-1990s. However, in February 2003, scientists reported discovering a large new aquifer under the Taklamakan desert in northwest China.

Sources: Brown and Halweil 1998; US Water News Online 2003.

> groundwater have been found in the Red Sea Province of eastern Sudan, Florida, and elsewhere. (Simon 1996: chap. 6)

Efficiency of water use is the most effective way to "create" new water. Water expert Peter Gliek cites the US example: "It is a little-known fact that the United States today uses far less water per person, and less water in total, than we did twenty-five years ago. . . . It's a shocker. People don't believe it, but it's true. This is an indication that things are not the way people think they are. It is not really because we are trying to cut our water use. . . . But we have changed the nature of our economy, and we have become more efficient at doing what we want to do" (quoted in Specter 2006:34).

The FAO makes some small-scale, practical recommendations of ways that water can be used more efficiently, such as water harvesting (collecting runoff and saving it for periods of need) and drip irrigation (delivering irrigation water directly to the roots of plants) (see the FAO's contribution on water use in agriculture in UNESCO 2006). In addition, agricultural scientists have developed crop varieties that require less water to thrive, as well as chemicals that promote water retention in soil.

As we will see in the next chapter, water use in agriculture is a major source of environmental concern related to agricultural production. Increased irrigation carries the threat of increased soil erosion, increased chemical runoff and resulting water pollution, and increased threat of global warming from paddy-rice production.

Overall, when it comes to soil and water resources, the World Resources Institute gives this "bottom line" assessment:

> At a global level there is little reason to believe that crop production cannot continue to grow significantly over the next several decades. That said, the underlying condition of many of the world's agroecosystems, particularly those in developing countries, is not good. Soil degradation data, while coarse, suggest that erosion and nutrient depletion are undermining the long-term capacity of agricultural systems on well over half of the world's agricultural land. And competition for water will further magnify the issue of resource constraints to food production. Although nutrient inputs, new crop varieties, and new technologies may well offset these declining conditions for the foreseeable future, the challenge of meeting human needs seems destined to grow ever more difficult. (2000:64)

12

Agricultural Production and the Environment

The previous chapter discussed the potential for increasing land and water use for future food production. In this chapter, we explore the issue of the degree and significance of environmental damage. The issue has two faces: environmental quality is an important determinant of agricultural output; and agricultural production has a significant impact on the environment. We deal first with the interaction of agricultural production and the local environment (the environment near to the place where the agricultural production takes place). Then we discuss the interaction between agricultural production and the global environment—especially global warming.

Agricultural Production and the Local Environment

Agriculture uses natural resources—soil and water—to produce food. This can lead to deterioration in the quality of natural resources and in their ability to support food production.

Land Degradation

The environmental issue with the greatest potential for influencing future food production is land degradation. Land can become unsuitable for agricultural production in the following ways (UN Population Information Network 1995):

- Soil can disappear from land through erosion.
- Soil can become chemically unsuitable for agricultural production.
- Land can be come physically unsuitable for agricultural production.

Erosion. Wind or water can pick up soil particles from one area and move them. This can harm agricultural production in four ways. The eroded soil may contain nutrients needed for plant development. The remaining soil may be so dense that it is difficult for plant roots to develop. Erosion may reduce the capacity of the soil to retain water needed for plant growth. Finally, erosion may result in uneven terrain that makes cultivation more difficult.

Soil erosion is to a degree caused by agricultural production. Land used for agricultural production may be bare of vegetation for months at a time. The absence of roots to hold the soil in place makes the soil more easily erodible. Plowing the soil in preparation for seeding exposes it to wind and rain and increases the rate of erosion. Irrigation can contribute directly to water erosion.

Chemical characteristics of the soil. Land may become chemically unsuitable for agricultural production for several reasons. The nutrients of the soil may be depleted because of past agricultural production, especially if the same crop is grown year after year. "Salinization" of soil occurs when the salt content of the soil increases to levels unsuitable for agricultural production. Salinization can be caused by irrigating land with water that contains low levels of salts, which are left on the soil when the water evaporates. In some areas, this problem occurs because irrigation depletes the naturally occurring fresh groundwater, and causes seawater to intrude into the groundwater system. A third chemical problem with soil is "acidification." This can occur when too much fertilizer of certain types is applied, or when there are drainage problems on certain soils. Finally, other pollutants such as oil or excessive pesticides can reduce the ability of soil to support agricultural production.

Physical characteristics of the land. Agricultural land can also become unsuitable for production because of changes in the physical characteristics of the land. Soil can become less porous through compaction—when heavy machines or animals pack the soil down—or through the action of raindrops that seal the soil. Nonporous soil makes it difficult for seeds to emerge. Waterlogging occurs when water sits in the root zone of plants, and thus impedes their development. Waterlogging occurs when drainage is poor, or when a field is overirrigated.

The Extent and Impact of Land Degradation Worldwide

The Global Land Assessment of Degradation (GLASOD) study, conducted by the United Nations (ISRIC and UNEP 1991), estimated that 22 percent of agricultural land worldwide (and 38 percent of cropland) has been subject to one or more of the kinds of degradation described above. Of the 2 billion hectares of degraded land, according to this study, 83 percent was degraded by erosion, 12 percent by chemical degradation, and 5 percent by physical degradation. Seventy million hectares are so badly degraded that the damage cannot be repaired. Other studies report that the amount of degraded land increases each year by an additional 5 to 10 million hectares (Scherr and Yadav 1997). Pimentel and colleagues (1994) estimate that "each year, more than 10 million hectares

(24.7 million acres) of once-productive land are degraded and abandoned." This bears directly on the question of how much land can be devoted to agricultural production in the future. However, as we saw in the previous chapter, agricultural land has increased steadily despite these reports of degradation. Stanley Wood gives the following assessment of the current state of scientific opinion: "As a global problem, soil loss is not likely to be a major constraint to food security" (quoted in Kaiser 2004:1616).

A related possibility is that land degradation will reduce yields per hectare. This can occur for two reasons. First, when land becomes so severely degraded that it is no longer capable of supporting agricultural production, new land may be added to agricultural production to take the place of the degraded land. The new land is likely to be of relatively poor quality—otherwise, it would already have been in use. Second, when the land is degraded, but remains in agricultural use, yields on that land drop.

There is a lack of agreement among agricultural scientists about the severity of the drop in yields attributable to land degradation. Pimentel and Giampietro (1994) point to evidence that corn yields are about 20 percent lower on severely eroded lands in many parts of the United States. Mitchell, Ingco, and Duncan (1997:54) cite other studies estimating that soil erosion was responsible for yield declines of 3–4 percent over a hundred-year period. Scherr and Yadav (1997) report yield losses of 5–15 percent attributable to land degradation. (See Crosson 1996a for a review of the debate.) Den Biggelaar and colleagues (2004) estimate that yields grew at a 0.3 percent slower annual rate than they would have in the absence of land degradation, and even this might overestimate the true impact of degradation, since farmers can take steps (choosing different crops, or using more fertilizer, for example) to counteract the degradation (see Kaiser 2004:1616).

Water Quality

In the previous chapter, we discussed the importance of water in agricultural production, and cited one report that expanded irrigation will be a substantial source of increased food production in the future. Expanded irrigation requires a supply of usable water. But agricultural production can lead to degradation of water quality.

Irrigation itself is the main culprit. Irrigation in China and India has caused water tables to drop significantly, by withdrawing water faster than it is replenished. In coastal areas, depletion of groundwater reserves can result in saltwater intrusion into the groundwater system. In some soils, irrigation leeches certain salts from the soil, and carries those salts back into the groundwater, contaminating it and making it unsuitable for future irrigation. Irrigation or rainwater runoff can also carry residues from fertilizers and chemical pesticides. This also creates water-quality problems. In addition, as noted, irrigation can contribute to land degradation, exacerbating erosion, waterlogging, salinization, and acidification.

The use of surface water for irrigation also affects the ecology of rivers, lakes, and even oceans. Diversion of water for irrigation has caused the volume of water in the Aral Sea in Uzbekistan to drop by 75 percent since 1960. This huge loss of water has changed the chemical composition of the remaining water and resulted in large decline in the fish population. Total fish catch dropped from 50,000 metric tons in 1959 to 5,000 metric tons in 1994 (see the contribution by the United Nations Environment Programme [UNEP] on coastal and freshwater ecosystems in UNESCO 2006). Even without intensive irrigation, water runoff from agricultural land, carrying residues of agricultural chemicals and animal waste, can damage fisheries ecology. In the Chesapeake Bay, oyster populations are now at 2 percent of the levels common from the 1950s to the 1970s (Chesapeake Bay Foundation 2008).

Problems Associated with Agricultural Input Use

We have already discussed how use of agricultural chemicals can lead to land degradation and water pollution. In addition, chemical use can create health problems for farm workers. The manufacture of agricultural chemicals can also create environmental hazards. The 1984 explosion at a chemical plant in Bhopal, India, provided a tragic example of this. The poison gas released by the explosion is used primarily in production of insecticides. Thousands were killed and tens of thousands were seriously injured (Baylor 1996). In laboratory experiments, some pesticides have been shown to affect hormone levels—which could cause cancer, abnormalities in newborns, or reproductive problems. However, evidence is weak that this effect can be found outside the laboratory (Kamrin n.d.).

Given the increased use of mechanization and petrochemicals in agriculture, some are concerned that energy use in agriculture will become an environmental problem. However, R. S. Chen (1990) reports that agricultural production accounts for only 3.5 percent of commercial energy use in developed countries and 4.5 percent in developing countries. In developed countries, food processing and distribution use more energy than food production.

Other Environmental Problems

Water quality is a concern not just when it impinges on food production. Because people drink water, of course, reduced water quality can directly harm public health. Of special concern here is the possibility that water can become contaminated with pesticides—chemicals that are deliberately developed to be toxic. Rachel Carson's *Silent Spring* (1962) pointed out the impact that agricultural chemicals can have on the environment. This affects not only humans, but also birds, fish, and other wildlife.

A second environmental concern associated with agricultural production is the issue of maintaining genetic diversity. Especially with the increasingly widespread use of improved varieties of cereals, there is concern that the genetic material contained in traditional varieties will be lost. For example, in 1949 there were 10,000 wheat varieties in use in China; by the 1970s only 1,000 remained in use. The loss of genetic diversity can make the food supply more susceptible to disease, and may foreclose the option of future technological improvements based on genetic characteristics of the "lost" varieties. In an effort to protect future generations from lost genetic diversity, the Global Crop Diversity Trust has begun the Svalbard Global Seed Vault, which hopes to store and preserve usable seeds for every crop on the planet (Walsh 2008).

Environment and Future Prospects for Agricultural Production

Are these interfaces between the environment and food production likely to create critical constraints on future food production? Some ecologists are very alarmed (for a review, see Cohen 1996a, 1996b). David Pimentel of Cornell University (Pimentel and Giampietro 1994) states that the world's resources can support a high standard of living for fewer than 2 billion people (compared to today's actual population of over 6 billion). This pessimistic view of the future is based on the belief that the world has already expanded agricultural production into areas that cannot sustain it and has increased yields per hectare through production methods that cannot be sustained. Pimentel cites studies showing that agricultural methods that do not use chemical fertilizers result in cereal yields of between 0.5 metric tons per hectare (in semiarid regions with no fertilizer) and 2 metric tons per hectare (in humid regions using animal manure for fertilizer) (Pimentel and Giampietro 1994). Compare these figures to the average current yields of about 3 metric tons per hectare worldwide and over 5 metric tons per hectare in the United States.

Technology and the Trade-off Between Production and the Environment

If David Pimentel represents one extreme in the debate about how many people the world can feed, Julian Simon represents the other. He describes the experience of a company called PhytoFarm, which grows vegetables indoors, and estimates that these techniques "could feed a hundred times the world's present population—say 500 billion people—with factory buildings a hundred stories high, on one percent of present farmland" (Simon 1996: chap. 6). By 2008, high energy prices, and the potential to locate indoor agriculture in urban areas and thereby reduce food transportation costs, led to a resurgence of interest in this concept (see Vertical Farm 2008).

Simon's attitude reflects an enormous confidence in the ability of technology to solve problems. Technological progress is all about getting more from less. A good deal of agricultural research in the past decade has been devoted to the problem of maintaining or improving agricultural yields while doing less damage to the environment. For example, new plant varieties are being developed that are naturally resistant to pests and thus require less pesticide use. Tilling and landscaping methods to reduce soil erosion have been widely adopted in parts of the world. Drip irrigation, which delivers water directly to the plant roots, reduces water used in irrigation without causing any reduction in the effectiveness of irrigation. The new technology of aquaculture has made fish farming a rapidly growing source of food, as Table 12.1 shows. Nonmarine fish production has increased tenfold over the period shown. The huge increase during the 1980s and 1990s reflects in part the introduction of aquaculture. (Table 12.1 also shows a source of environmental concern: natural ocean—or marine—fisheries have increased production to such a degree that they are in danger of being overfished; if the breeding stock is depleted, the total ocean fish population will begin to fall.)

Another aspect of agricultural research is to develop technology that relaxes the constraints that environment imposes on agricultural output. For example, scientists are working to develop plants that can survive in brackish water: "Researchers have transferred a gene for salt tolerance from an Old World ice plant into three plants lacking salt tolerance . . . all of which then displayed significantly increased capability to grow with their roots exposed to salt. . . . [This] will contribute to the effort to engineer plants with improved ability to withstand adverse growing conditions such as under seawater irrigation" (National Science and Technology Council n.d.). Other examples of technological ways to relax the environmental constraints on agricultural production are chemicals that increase the ability of soils to retain moisture, and seed varieties that are more drought-resistant (so that crops can be grown in more arid regions). Box 12.1 describes an effort to reclaim degraded land in China.

Table 12.1 Fish Production Worldwide, 1950–2005 (metric tons)

	Freshwater Fish	Marine Fish	Total Fishery Production
1950	1,929,001	17,293,233	19,222,234
1960	3,078,619	31,625,151	34,703,770
1970	4,628,870	59,026,319	63,655,189
1980	5,101,627	62,846,261	67,947,888
1990	6,439,475	79,463,411	85,902,886
2000	8,850,086	87,863,452	96,713,538
2005	9,716,346	85,661,769	95,378,116

Source: FAOSTAT 2008a.

Box 12.1 Reclaiming Degraded Lands in Southern China

The five southernmost provinces of China have been farmed for over a thousand years, primarily with slash-and-burn techniques that cleared forests but left the soil exposed to severe erosion. The process of land degradation here is described by Parham (2001): "When vegetation is removed in these regions, the exposed soil . . . reaches temperatures so high that seeds and sprouts are killed. . . . Since new vegetation cannot be established easily, soil organic matter is reduced, and the soil becomes desiccated. . . . Even small decreases in soil organic matter have a pronounced negative effect on the soil's fertility. . . . When the original topsoil is removed by erosion, the surface becomes a mixture of aluminum-rich clays and quartz sand that contain very few minerals useful to plant life. . . . The loss of vegetative cover and soil organic matter leaves the soil subject to damage from intense tropical rainfall. With little organic matter in the soil, clay particles are moved by raindrops and plug soil pores, thus inhibiting water infiltration and increasing runoff and erosion. . . . The finer-grained eroded sediments damage aquatic productivity and bury what were once freshwater and near-shore marine aquatic breeding grounds. The remaining coarser, sandy material of the weathered granite yields soils of low fertility. Stripped of vegetation that would otherwise have absorbed or slowed the flow of water, the water pours rapidly into streams and rivers, cutting deep ravines in the soft, deeply weathered granite."

By the end of the twentieth century, an estimated 45 million hectares (over 20 percent of agricultural land in southern China) had been degraded. Use of commercial fertilizers was unsuccessful in replacing the nutrients lost through eroded soil, because the remaining coarse soil was a poor medium for holding the nutrients provided by the fertilizers.

Recent research suggests that it may be possible to restore most of the degraded land to agricultural production within two years. One research project planted fast-growing ground cover amid alternating rows of rubber trees and tea bushes. The ground cover shields the soil from the hot sun and reduces evaporation of soil moisture. The roots of the plants help reduce soil erosion, and the plants provide organic compost for the soil. The rubber trees provide shade for the tea bushes, and the tea bushes help moderate the temperatures near the roots of the rubber trees. Experimental plots indicate that this type of agriculture is profitable and can restore soil quality while reducing water runoff and flooding.

Agricultural Production and the Global Environment

Agricultural production also interacts with the environment on a global scale, especially in terms of global warming. Global warming refers to the phenomenon by which water vapor, carbon dioxide, methane, and other trace gases in the atmosphere trap heat on the surface of the planet. As the quantities of these gases (the so-called greenhouse gases) in the atmosphere increase, the amount of heat trapped will increase, and the average temperature of the planet will increase. On these matters there is a high degree of consensus among scientists (see IPCC 2007a), although there is some disagreement about the extent to which global warming has already occurred, and about the extent to which human activities are responsible for the buildup of greenhouse gases (see Leggett 2007 for a recent review).

The Impact of Agriculture on Global Warming

Agricultural production is a significant source of greenhouse gas emissions worldwide. In fact, although the industrial revolution (with its widespread burning of coal and oil products) is usually thought of as the beginning of anthropogenic (human-made) climate change, some (e.g., see Ruddiman 2005) point to the advent of agriculture as the first chapter in global warming.

There are three greenhouse gases produced by human activity that may cause global warming: carbon dioxide, methane, and nitrous oxide. Carbon dioxide comprises nearly 75–80 percent of greenhouse gases, nitrous oxide 15 percent, and methane 6 percent. Agriculture—primarily deforestation in tropical areas to clear land for agricultural production—and forestry are thought to contribute about 25–30 percent of the anthropogenic carbon dioxide released into the atmosphere (FAO 1997a). Agriculture is the primary source of anthropogenic methane and nitrous oxide. The digestive processes of ruminants (cows, sheep, goats, and other animals that chew cuds) contribute 37 percent of anthropogenic methane (Steinfeld et al. 2006). Paddy-rice production is responsible for another 13 percent. Methane forms in paddy-rice production when manure used as fertilizer decomposes in an oxygen-free environment (underwater in the flooded rice paddies). Methane production from paddy rice declines as rice farmers use production methods that substitute commercial fertilizer for animal manure, that conserve water use, and that use improved rice varieties that store more carbon in the rice plants (see Neue 1993; Lashof and Tirpak 1990; Graham 2002; Casey 2007). Livestock production is also responsible for about 65 percent of anthropogenic nitrous oxide emissions (Steinfeld et al. 2006). Release of nitrous oxide into the atmosphere can be promoted by the use of nitrogen fertilizers.

Agricultural production is a substantial source of greenhouse gases; if agricultural production grows as fast as food demand, that will add noticeably to the greenhouse gas problem. Furthermore, any realistic program to reduce greenhouse gas emissions must address the agricultural component, but there is an unavoidable tension between raising agricultural output and reducing greenhouse gases. Perhaps this tension can be reduced by technological innovations, but it cannot be eliminated.

The Impact of Global Warming on Agriculture

Obviously, when the average temperature in an area changes, the agricultural capacity of the area changes—it may become better for some crops and worse for others. Unfortunately, global warming—if it does become a real problem—is not expected to result in a gradual increase in temperature in every area of the globe. Some areas may become much warmer, some only a little warmer, some possibly colder. And global climate change does not simply mean changes in temperature. It is almost certainly associated with changes in rainfall patterns, making some areas dryer and some wetter. And it also changes the incidence of severe weather, making some areas more prone to hurricanes, tornadoes, and droughts.

There are huge uncertainties about future climate change. Will the changes be dramatic or nearly imperceptible? Even if the *average* global temperature increases significantly, what will that mean for climates in different geographical areas? Because of these uncertainties, the opinions among scientists about the possible future impacts of global warming on agricultural production vary widely. This debate has centered on two questions: whether global warming will cause flooding of coastal areas and loss of agricultural land, and the degree to which climate change will impact average crop yields worldwide. (See Box 12.2 for a related debate—200 years old—on the impact of sunspots on wheat prices.)

Box 12.2 Sunspots, Crop Yields, and Wheat Prices

In 1801, British astronomer William Herschel had a theory: sunspots influenced wheat prices. Sunspots are vortices of gas on the surface of the sun. Herschel hypothesized that sunspots would result in "copious emission of heat and therefore mild seasons" on the earth. Mild seasons would improve crop yields, which would in turn lead to lower crop prices. Herschel reported that the facts supported his theory: during five prolonged periods of low solar activity (few sunspots), wheat prices were higher. The Royal Society ridiculed Herschel's theory as a "grand absurdity." By the 1840s, astronomers had discovered that sunspots followed a cycle that peaked every eight to seventeen years, with an average cycle length of about eleven years. In the late 1800s, economist William Jevons suggested that this might be an explanation of business cycles (alternating periods of economic growth and recession).

In 2003, Pustilnik and Din examined the link between wheat prices and sunspot activity, and discovered that for all of the ten solar cycles in the 1600s and 1700s, high sunspot activity was associated with low wheat prices. The explanation, according to Pustilnik and Din, is somewhat different than Herschel's "mild seasons" theory: in periods of high solar activity, it is more difficult for charged particles from deep space to reach the earth's atmosphere; since these charged particles contribute to cloud formation, skies over England are less cloudy; this reduces threats to wheat production by frost and extended rainfall; better wheat harvests result in lower wheat prices. (See Baliunas 1999 for a report relating this phenomenon to the current global warming debate.)

Climate Change and Agricultural Area

If the average temperature of the planet increases, the volume of water in oceans will increase. This is not primarily (as popular belief has it) because of melting polar ice caps, but because the volume of water expands as its temperature increases. According to a recent report by the Intergovernmental Panel on Climate Change (IPCC 2007a), global temperature has increased by 0.5–1.0 degree centigrade since the 1870s, and average sea level has increased by about 200 millimeters (8 inches) during the same period. The report projects that increases in greenhouse gas emissions will likely increase global temperatures by 1.8–4.0 degrees centigrade over the next century (assuming all other climate-related factors remain constant and assuming that no additional policies are adopted to reduce growth of greenhouse gas emissions). Under these temperature-increase scenarios, average sea level is expected to rise by 180–590 millimeters (7–23 inches) by 2100. This estimate confirms the earlier projection of Rosenzweig and Hillel (1995) that average sea level is likely to rise by 4–20 inches by the middle of the twenty-first century.

Although this could make some areas uninhabitable and increase the threat of flooding for others (see Parry, Magalhaes, and Nih 1992; Ibe and Awosika 1991), it is not expected to have a significant impact on the worldwide availability of agricultural land. In the United States, for example, the Federal Emergency Management Agency (FEMA) estimates that a 1-foot increase in average sea level would increase the size of the "100-year flood plain" by about 20 percent, from 19,500 square miles currently to 23,000 square miles. But this increase is only one-tenth of 1 percent of the entire land mass of the United States. And as the dikes of the Netherlands remind us, human behavior can adapt to, as well as cause, rising sea levels. In addition, Rosenzweig and colleagues (1993) point out that in some areas, global warming may result in new land becoming suitable for agricultural production, because of an extended growing season or changes in rainfall pattern. (Also, the Meteorological Service of Canada [2006] estimates that 10 million hectares of agricultural land in that country are currently not being utilized because of climate constraints—though the service notes that soil quality is poor on much of this land.) Overall, the above opinions fall in line with Wittwer's: "Although important for localized regions, [cropland loss from rising sea levels] would be relatively insignificant on a worldwide basis" (1995:165–166). Of course, if sea levels were to rise much higher than the IPCC projection mentioned above, impacts on agriculture would be much greater (Rowley et al. 2007). However, estimates of cropland reductions of 10 to 50 percent (Lemons et al. 1995:122) are outliers in this debate.

The risk from rising sea levels is not only that land will be flooded, but also that drainage problems will increase, and seawater intrusion into freshwater sources will occur. The IPCC (2007a) considers salinization of irrigation water to be the most likely impact of rising sea levels on agriculture.

Climate Change and Agricultural Yields

Global warming can affect crop yields in a variety of ways (Rosenzweig and Hillel 1995), from increasing atmospheric carbon dioxide to increasing the incidence of pests and diseases.

Increase in atmospheric carbon dioxide. Atmospheric carbon dioxide (CO_2) comprises about 80 percent of the greenhouse gases, and increases the efficiency of photosynthesis and thereby boosts plant growth. Wheat, rice, and soybeans are especially responsive to increased atmospheric CO_2. High CO_2 levels also significantly increase water-use efficiency. This fact may actually increase demand for irrigation water, as farmers discover that irrigation has a larger impact on yields. In addition, high CO_2 levels increase plants' resistance to salinity and drought, and increase nutrient uptake. Finally, noxious weeds are (for the most part) less responsive to CO_2 than are crops (Wittwer 1995).

Higher temperatures. Because higher temperatures will on average increase the length of the growing season, agricultural production may become feasible in areas (closer to the North and South Poles) that are currently too cold. Soils in some of these areas (Canada and Russia) are less fertile than other soils; thus, bringing these lands under cultivation could cause a drop in average yield. In addition, some crops (notably rice) show yield declines when the temperature is too high. Finally, increased temperatures make plants mature faster. But plants that mature faster have lower food yields. On the other hand, the faster maturation, combined with the longer growing season, may extend the areas in which double- or triple-cropping is feasible.

Change in rainfall patterns. As noted above, climate change entails not just changes in temperature, but also changes in rainfall. As described in Chapter 11, soil moisture is essential to crop growth. Rainfall also can influence soil erosion. According to predictions reported by the Environmental Protection Agency (EPA 2008), precipitation will increase in areas close to the poles, and decrease by as much as 20 percent in tropical regions.

Extreme meteorological events. Global warming may increase the incidence of hurricanes, tornadoes, heavy rainstorms, and droughts, thus disrupting crop production and reducing yields. (See Box 12.3 for discussion of a meteorological event not related to global warming.)

Increase in tropospheric ozone. The EPA (2008) reports: "Since ozone levels in the lower atmosphere are shaped by both emissions and temperature, climate change will most likely increase ozone concentrations. Such changes may offset any beneficial yield effects that result from elevated CO_2 levels." Depletion of the ozone layer permits more ultraviolet radiation to reach the

Box 12.3 El Niño and Food Production

Every few years, the surface of the Pacific Ocean becomes warmer. Peruvian fishermen named this phenomenon "El Niño" (the boy-child), because it coincided with Christmas (or the coming of the Christ-child). In 1997, the warming was especially large, and this caused worldwide changes in weather patterns. In Washington, D.C., for example, the winter of 1997–1998 was exceptionally mild, with virtually no snowfall. In Los Angeles, rainfall in early 1998 was nearly twice the usual level. The winter also saw March blizzards in the Midwest, and ice storms in New England that left people without electricity for days. The United States was not the only country to feel the effects of El Niño. South America and East Africa experienced heavier rainfall than usual; parts of South Asia were unusually dry.

Because of this "weird" weather, thirty-seven countries were facing food emergencies in early 1998. But the problems were limited to certain areas. Worldwide, cereal production for 1997–1998 was slightly above the record levels of 1996–1997. FAO scientist Rene Gommes and associates (Gommes, Bakun, and Farmer 1998) conclude: "It is important not to minimize risks but also to remember that there have been El Niños without any catastrophes and catastrophes without any El Niños."

earth. This in turn can affect agricultural production. Pimentel (1993) reports a resultant potential drop in soybean yields of 30 percent (soybeans are particularly sensitive to ultraviolet radiation levels). According to studies reported by Wittwer (1995), increased ultraviolet radiation can result in substantial yield declines. However, in studies that simultaneously increased both CO_2 levels (the greenhouse gas effect discussed above) and ultraviolet radiation levels, yields were basically unchanged.

Pests and diseases. Insect pests and crop diseases thrive in higher temperatures. Thus global warming may increase the incidence of pests and disease, and thereby reduce yields.

* * *

An important determinant of the impact of climate change on crop yields is the extent to which farmers adapt their production decisions to the new climatic conditions. For example, suppose we are analyzing production in an area where rice is the predominant crop. Suppose, further, that we conclude that global warming will result in rice yields dropping by 30 percent in this area. A simplistic analysis would conclude that food production would drop by 30 percent in the area. A more sophisticated analysis would recognize the likelihood that many farmers in the area will stop growing rice, and will switch to some alternative crop. The actual drop in production may be much less than 30 percent, depending on the alternative crops available and the extent to which farmers change their cropping decisions.

Rosenzweig and colleagues (1993) estimate that if temperatures increase by 2 degrees centigrade, wheat and soybean yields will increase by 10–15 percent, and maize and rice yields will increase by about 8 percent. However, if temperatures increase by 4 degrees centigrade, yields will decline. (According to UNEP [1990], climate models predict an increase of 1.5 to 4.5 degrees centigrade over the next hundred years.) Wittwer (1995) believes that these estimates may underestimate the yield growth from global warming because they ignore some of the possible benefits from increased levels of atmospheric CO_2.

The IPCC concludes: "Many studies . . . have confirmed . . . projections [that] indicate potentially large negative impacts in developing regions, but only small changes in developed regions, which causes the globally aggregated impacts on world food production to be small" (2007b). Consistent with this conclusion, Deschenes and Greenstone (2007) have recently shown that US agriculture is likely to benefit from global warming due to increases in yields and profits.

Other prognoses for crop yields are less optimistic. For example, UNEP reports: "Mid-latitude yields may be reduced by 10–30 percent due to increased summer dryness," though "higher yields in some areas may compensate for decreases in others" (1990). Pimentel reports: "Under the projected warming trend in the United States, farmers can expect a 25 to 100 percent increase in losses due to insects, depending on the crop. . . . US crop losses due to weeds are projected to rise from 5 to 50 percent. . . . In North America, projected changes in temperature, soil moisture, carbon dioxide, and pests associated with global warming are expected to decrease food-crop production by as much as 27 percent" (1993:54–57). However, Pimentel does see some reason for hope. He projects that yields in North Africa may improve by 10–30 percent as a result of global warming.

Finally, it should be noted that if global warming causes agricultural production to shift from one geographical area (say, North America) to another (say, North Africa), there may be problems establishing the institutions and infrastructure needed to move the food from the production areas to the consuming areas. "While the overall, global impact of climate change on agricultural production may be small, regional vulnerabilities to food deficits may increase, due to problems of distributing and marketing food to specific regions and groups of people" (Rosenzweig and Hillel 1995).

13

Increasing Yields Through Input Intensity

In Chapter 11, we introduced the basic equation of food supply. In Chapters 11 and 12, we discussed one aspect of that equation: the availability of land and water, and the interactions between resource quality and agricultural production. Now we turn to the second part of the equation: yields.

Crop Yields Since 1960

The increase in crop yields in the latter half of the twentieth century is one of the great accomplishments of human history. To put the yield growth into context, consider the historical record of wheat yields in Britain shown in Table 13.1. It took 350 years for wheat yields to triple from their 1450 levels; it took 300 years to triple from their 1550 levels; it took 250 years to triple from their 1700 levels; then they nearly tripled again in the past 50 years. An acre today produces fifteen times as much wheat as it produced 500 years ago.

Worldwide, cereal yields have more than doubled since the early 1960s. As shown in Table 13.2, this growth is seen in all three of the most important cereal crops, and in most geographical areas of the world (a "percentage increase" of more than 100 signals that yields have more than doubled). Yield growth is lowest for maize and rice in Africa. Yields in Asia have grown faster than worldwide yields in all three crops.

Worldwide cereal yields are illustrated in Figure 13.1, and the consistent growth is demonstrated by the upward trend. Figure 13.2 looks at these data in a little different way; it shows growth rates in yields rather than the yields themselves. To eliminate year-to-year fluctuations caused by weather and other temporary conditions, we show average annual growth rate in yields over a seven-year period, rather than for a single year. The dotted trend line in Figure 13.2 is a logarithmic curve fitted to the data. That trend line suggests that yields are growing (notice that average growth rates are positive at every point), but are growing at a slower rate. The greater than 3 percent growth rate of the 1960s has been replaced with average growth in the 0.7–1.6 percent range.

Table 13.1 Wheat Yields in Britain, 1600–2000

Approximate Year	Approximate Yield (kilograms per hectare)
1450	500
1550	600
1600	750
1650	900
1700	1,100
1750	1,300
1800	1,500
1850	1,800
1900	2,100
1950	3,100
2000	8,000

Sources: Overton 1996:77; Cooke 1967:191, 463; FAOSTAT various years.

What Makes Yields Increase?

Three things cause yields per acre to increase:

- Productive inputs (labor, fertilizer, machinery, for example) are used more intensively on each acre of land.
- New technology increases the output obtainable without increasing inputs.
- Farmers increase their efficiency so that less potential output is lost to poor farming practices.

These three things are illustrated in Figure 13.3, which shows a "production function." The production function shows the maximum quantity of output that is obtainable from any given quantity of input (to review why production functions are shaped this way, see Box 7.2). Here, we define output as food production per acre, and input as input use per acre, and illustrate with the example of a farmer growing corn on a plot of land using nothing except his own labor.

An increase in input will move production up along an upward-sloping production function, such as from point B to point C in Figure 13.3. As the farmer puts more and more hours into caring for the field, weeding more frequently, or removing insect pests from the plants, more corn plants will survive, the ears will be bigger, and the yield of that plot of land will increase. This source of increasing yields is discussed in the first part of this chapter.

A new technology—broadly defined as new information about how to obtain output from the input—will cause an upward shift in the production function. For any given level of input, a larger output can now be achieved, such as the move from point B to point D in Figure 13.3. Perhaps the farmer is using an improved seed variety that produces bigger ears of corn, or perhaps the farmer

Table 13.2 Maize, Wheat, and Rice Yields, Various Geographical Areas, 1961 and 2007 (hectograms per hectare)

	Maize			Wheat			Rice		
	1961	2007	Percentage Increase	1961	2007	Percentage Increase	1961	2007	Percentage Increase
World	19,434	49,709	155.8	10,888	27,918	156.4	18,692	41,524	122.1
Africa	10,449	17,611	68.5	6,929	20,537	196.4	15,520	25,018	61.2
North America	39,229	94,176	140.1	13,257	25,344	91.2	38,227	80,538	110.7
South America	13,733	41,868	204.9	11,358	24,829	118.6	18,245	44,534	144.1
Asia	11,387	43,677	283.6	7,486	28,535	281.2	18,584	42,175	126.9

Source: FAOSTAT 2008c.

Figure 13.1 Worldwide Cereal Yield, 1961–2008

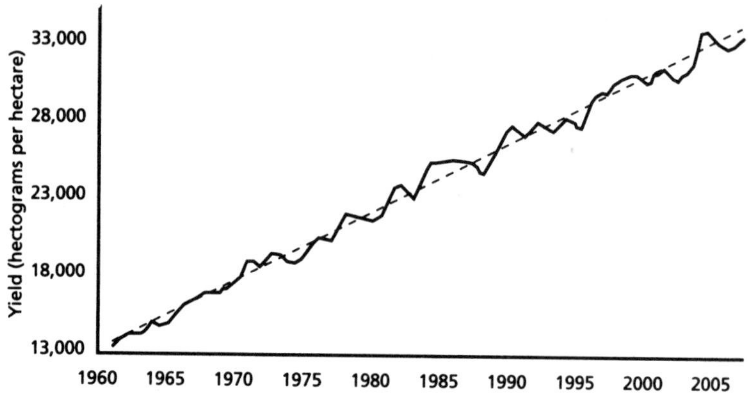

Figure 13.2 Growth Rate of Worldwide Cereal Yields, 1968–2008

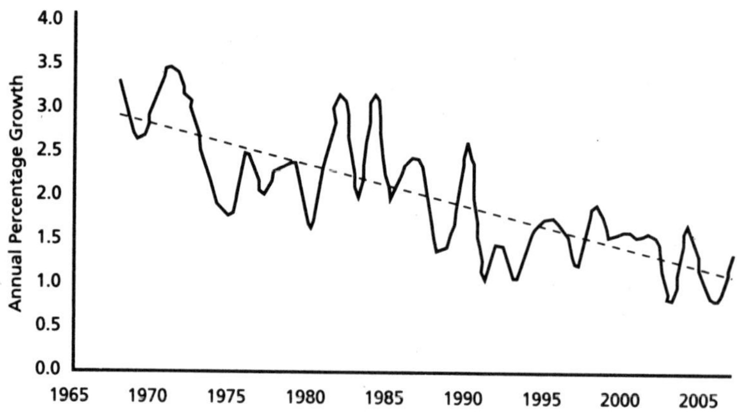

Source: FAOSTAT 2008c.

Note: Average annual growth rate in cereal yields for seven-year period ending in the year shown.

has learned that weeding in the first month of growth is especially effective, so that without increasing the number of hours of weeding, he can increase the effectiveness of weeding, and therefore the crop size. This source of increasing yields will be discussed in Chapter 14.

An improvement in efficiency is illustrated by the move from point A to point B in Figure 13.3. At point A, the farmer was failing to obtain the maximum

Figure 13.3 Different Ways of Increasing Output per Acre

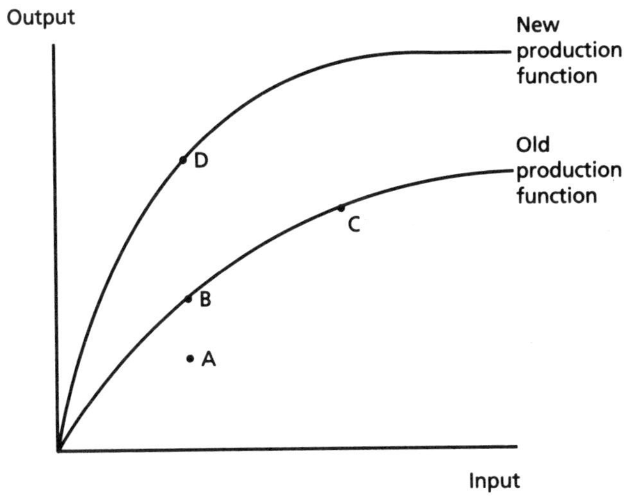

Note: Eliminate inefficiency and move from A to B. Increase input and move from B to C. Advance technology and move from B to D.

possible output using the old technology. Perhaps the farmer was stepping on his corn plants while working in the field, or perhaps his hours of work were less effective because he was weakened by hunger or sickness. This source of increasing yields will not be discussed in detail here; we discussed in Chapter 5 how undernutrition can contribute to reduced economic productivity; in Chapter 15 we will examine the interaction between undernutrition and health in more detail.

Yields and Input Use: Purchased Inputs

In Chapter 11, we saw that yields on irrigated land can be twice as high as yields on similar unirrigated land. This is a dramatic example of how yields can be increased by adding more inputs to the land. An "input," as economists use the term, is something that contributes to production. Water, labor, chemicals, and machinery are all examples of inputs. As a general rule, when we increase the intensity of input use on a plot of land, we increase the output of that plot. (See Federico 2005 for an in-depth description of trends of input use over the 1800–2000 period.)

Fertilizer

Fertilizer use—especially the application of nitrogen fertilizer—has mushroomed around the world over the past forty years, growing to over three times its 1961 level (see Table 13.3). The growth has been especially rapid in Asia (2001 use is nineteen times 1961 use) and in Latin America. Growth of fertilizer use in Africa has been much slower than in Asia or Latin America. Fertilizer use in developed countries (represented by the United States and Canada, as North America, in Table 13.3) was already well developed by 1961; therefore, growth over the successive decades has been much more modest. This is a pattern that applies to purchased inputs in general (see Federico 2005:55–56).

One way to see the impact of fertilizer is to compare countries based on their fertilizer use and their cereal yields. Figure 13.4 shows a scatter plot of fertilizer use in 160 countries and yields for 2001. The line fitted to these data points is upward-sloping; especially notable is the large grouping of countries near the origin—countries that have low fertilizer use per hectare and low yields.

Many experiments have verified the effectiveness of fertilizer in increasing crop yields. Two of these are described in Box 13.1. Several general conclusions can be drawn from these experiments:

- Fertilizer use can have a dramatic impact on yields, ranging from 20 percent improvement to over 1,000 percent improvement.
- For low levels of use, increasing fertilizer application increases yield, but at a decreasing rate, giving a yield response curve that has the same general shape as the production function illustrated in Figure 13.3.
- At some point, further increases in fertilizer application actually reduce yields.
- The impact of fertilizer on yields depends on a number of other factors, such as whether the crop is irrigated, the timing of fertilizer applications, and other farming practices used (weeding, types of crop rotation).

Table 13.3 Growth in Fertilizer Use by Continent, 1961–2001 (metric tons of total fertilizer consumption)

	Africa	Asia	Latin America	North America	World
1961	716,141	3,809,371	979,127	8,048,099	30,454,244
1971	1,850,152	12,671,942	3,061,228	16,459,951	68,743,842
1981	3,534,962	31,656,890	6,362,770	21,385,088	99,994,724
1991	3,533,961	58,786,676	7,740,283	20,957,084	104,947,291
2001	3,946,152	73,546,412	12,839,766	22,088,582	102,775,013
2001 as multiple of 1961	5.51	19.31	13.11	2.74	3.37

Source: FAOSTAT 2008d.

Figure 13.4 Higher Fertilizer Use Increases Crop Yields: Evidence from 160 Countries, 2001

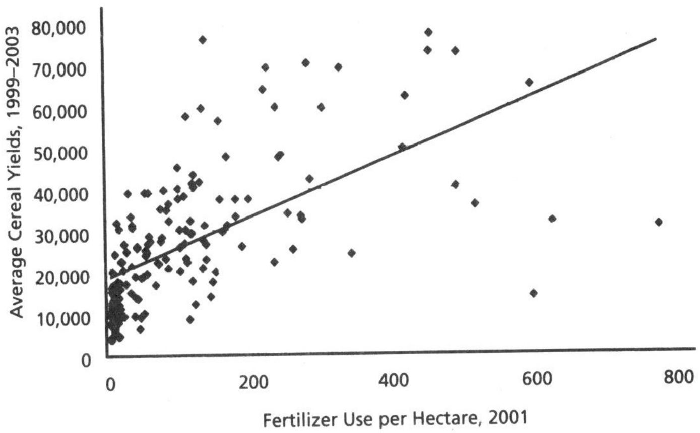

Source: FAOSTAT various years.

Box 13.1 Fertilizer Increases Crop Yields: Experimental Evidence

The best evidence about the impact of fertilizer on crop yields comes not from comparisons between different countries' fertilizer rates and yields, nor from analyses of changing use and yield patterns over time within a country or region. The best evidence is gathered from individual plots on which carefully controlled experiments are carried out, holding all variables constant except the uses of fertilizers.

In the mid-1800s, the British Experimental Station in Rothamsted carried out experiments that first demonstrated that applying nitrogen, potassium, and phosphate to crops could increase crop production. In a series of experiments there between 1848 and 1919, average barley yields were 1,210 pounds per acre when grown without fertilizer, and 2,061 pounds per acre when grown with fertilizer; average wheat yields increased from 1,434 pounds per acre to 1,994 pounds per acre when fertilizer was applied (Cooke 1967:191). The Rothamsted experiments "have often been interpreted as demonstrating that high levels of inputs of inorganic fertilizers . . . can maintain and increase yields for more than a century and a half. While this is true of wheat grown on the soil of the Broadbalk field, it is not always true" (Greenland, Gregory, and Nye 1998:45).

Alan Wild (2003) reports on two fertilizer experiments in Africa. In one experiment in Tanzania, cotton yields were 1,000 kilograms per hectare when grown with compost and phosphate, but no nitrogen. When nitrogen was applied at a rate of 37 kilograms per hectare, yields increased to 1,500 kilograms; when nitrogen application was increased to 112 kilograms per hectare, yields increased further, to 2,000 kilograms per hectare. (Notice how this path of increased input use traces out a curve of the same general shape as that in Figure 13.3.) In a second experiment in Nigeria, fertilized maize fields had yields of 4,369 kilograms per hectare, compared to yields of 56 kilograms per hectare in unfertilized fields.

Despite the dramatic growth of fertilizer use, there appears to be potential for additional fertilizer use in many parts of the developing world, especially sub-Saharan Africa. This can be seen by comparing fertilizer use (per hectare of permanent crop and arable land) in a developing country, to fertilizer use in a more developed country in the same region. For example, average fertilizer application rates in the Democratic Republic of Congo are less than 1 percent of application rates in South Africa; in Nigeria, application rates are 6–7 percent of application rates in South Africa. The experience in Asia is instructive. In 1961, application rates in China, India, and Indonesia were 1–3 percent of application rates in Japan. Today, application rates in China are nearly as high as those in Japan, and those in Indonesia and India have risen to 25–30 percent of the Japanese level. If sub-Saharan Africa can follow in the next thirty to forty years the path taken by Asian countries during the past thirty to forty years, fertilizer use in sub-Saharan Africa will grow substantially.

Table 13.4 shows two different projections about future fertilizer use, one from a paper for the FAO by Daberkow and colleagues (1999), and the other from a paper for the International Food Policy Research Institute (IFPRI) by Bumb and Baanante (1996). Both papers estimate that fertilizer use will continue to increase, but that the rates of increase will slow substantially. The highest growth rate is expected in sub-Saharan Africa.

Fertilizer use may be less than the economic optimal rate in the poorest countries because farmers are unable to borrow money and are unable to save enough money to buy as much fertilizer as they would like. In addition, states the FAO, "In sub-Saharan Africa, where fertilizer use is still very low, consumption is hampered by high distribution costs, the lack of markets for output, lack of a domestic fertilizer industry and poor yield response, and the high

Table 13.4 Two Projections of Future Fertilizer Use

	Actual Use, 2000 (million metric tons)	Annual Growth, 1961–2000 (%)	Projected Use, 2015 (million metric tons)	Annual Growth, 2000–2015 (%)	Projected Use, 2020 (million metric tons)	Annual Growth, 2000–2020 (%)
East Asia	43.7	8.7	49.2	0.7	55.7	1.2
South Asia	21.3	9.9	22.2	0.3	33.8	2.2
Latin America	12.3	6.5	12.3	0.0	16.2	1.3
Near East and North Africa	6.1	7.0	6.0	–0.1	11.7	3.1
Sub-Saharan Africa	1.2	5.2	1.7	2.0	4.2	6.0
Developing countries	85.2	8.2	91.4	0.4	121.6	1.7
Developed countries	50.0	1.5	60.8	1.2	86.4	2.6
World	135.2	3.7	152.2	0.7	208.0	2.1

Sources: Actual use: FAOSTAT. Projections for 2015: Daberkow et al. 1999. Projections for 2020: Bumb and Baanante 1996.

risk of using fertilizer in traditional agricultural settings" (FAO 1996e, paper no. 10). (Box 22.1 discusses policies intended to encourage fertilizer use in sub-Saharan Africa.)

Animal Traction

Another important input in agricultural production worldwide is "animal traction"—the use of draft animals to pull equipment or transport goods. There are an estimated 400 million draft animals worldwide. About half of the land in agricultural production in the world is farmed with draft animals; another quarter is farmed with hand tools only; and the remaining quarter is farmed with mechanized equipment (Gifford 1992).

A draft animal can do the work of three or four adult humans (Stout 1998: 76). By using animal traction rather than hand tools, a farmer is able to farm a greater area of land. This increases the self-produced food supply for the farmer's family and increases the likelihood that the farmer will produce a marketable surplus. In addition, use of animal traction may increase crop yields for two reasons: first, plowing with animals breaks up the soil more deeply and more completely, and thereby aids in plant growth; and second, animal manure is a source of fertilizer for the soil. In addition, animals provide other benefits to subsistence farmers: milk from cows, goats, and sheep is an additional food source; animal hides are used for clothing and shelter; animal manure is used as fuel for cooking and heating (see Box 13.2).

Box 13.2 Sacred Cows

Why is the cow venerated in India? Wouldn't it be better nutritionally to devote scarce agricultural resources to feeding humans rather than to feeding cows—especially because cows will not even be slaughtered for human food? Anthropologist Marvin Harris has an intriguing answer to this question. Cattle contribute in three important ways to improved life in rural India. First, they are a source of power for crop cultivation: "The shortage of draft animals is a terrible threat that hangs over most of India's peasant families. . . . The main economic function of the zebu cow is to breed male traction animals." Second, cows provide milk: "Even small amounts of milk products can improve the health of people who are forced to subsist on the edge of starvation." Third, cattle produce manure: "India's cattle annually excrete about 700 million tons of recoverable manure. Approximately half of this is used as fertilizer, while most of the remainder is burned to provide heat for cooking. The annual quantity of heat liberated by this dung . . . is the equivalent of . . . 35 million tons of coal. . . . Cow dung has . . . one other major function. Mixed with water and made into a paste, it is used as a household flooring material" (1974:10–11, 13).

Pingali (1987) conducted a review of twenty-two published studies about the impact of animal traction on agriculture in various parts of the developing world. He found the following:

- Seventeen of the studies examined the effect of animal traction on the area farmed. All seventeen found that farmers who used animals had larger farms than nearby farmers who farmed with hand tools.
- Nineteen of the studies examined the effect of animal traction on the use of land to produce crops that could be sold, rather than crops to be consumed by the farm household. Twelve of the nineteen studies found that farmers who used animals devoted more land to market crops.
- Fourteen of the studies examined the effect of animal traction on yields. Four of these studies found a positive effect on yields; two found a negative effect; eight found no statistically significant effect.
- Sixteen of the studies found that farmers who used animal traction had higher farm incomes than nearby farmers who farmed with hand tools.

Pingali also identifies impediments to the more widespread use of animal traction in the developing world:

- Cattle in sub-Saharan Africa succumb to a disease—trypanosomiasis—spread by the tsetse fly. Trypanosomiasis occurs in thirty-seven sub-Saharan African countries and threatens 50 million head of cattle. About 3 million cattle die from the disease each year, and another 35 million are treated for the disease. (For more information, see FAO n.d.)
- Many farmers in developing countries lack necessary experience and training in care of animals and equipment.
- The initial cost of animals and equipment is high, making animal traction impossible for farmers with limited access to credit.
- Using draft animals only makes sense when the farmer can expand the land under production. In some situations, all available land is being farmed by individuals who cannot or will not transfer the land to another farmer. In other situations, the land is too steep or rocky to be suitable for animal traction.
- Increasing production beyond what is needed for the farmer's family only makes sense if the farmer can sell surplus production. This requires adequate transportation and access to reliable markets for agricultural goods.

Tractors and Machinery

For the poorest farmers of the world, a major turning point may be the move from hand cultivation to animal traction. For a smaller group of slightly less-poor farmers, the replacement of animal traction with mechanized farming equipment represents a major change. Worldwide data on use of tractors and agricultural machinery are shown in Table 13.5.

The arguments presented above in the context of fertilizer may also apply to machinery—poor farmers with limited access to credit may not be able to buy as much machinery as they would like. But we should be cautious in assuming that machinery can be as efficiently used in the developing world as it is in the United States and other "rich" countries. Agricultural production in the United States tends to be done on large-scale farms using a lot of machinery and relatively little labor. Because the United States is a rich country and a large agricultural exporter, there is a tendency to regard this kind of capital-intensive farming as the "modern" and "efficient" method to which all farmers in all countries should aspire. This type of farming *is* efficient in the United States, because of the relative prices of inputs that prevail there. Labor is relatively scarce in the United States, so labor wages are high. Capital markets are well developed, so farm credit is widely available at relatively low interest rates. Gasoline and equipment prices are also relatively low.

Consider the fact that in the United States, thirty days of farm labor costs the same as 1,600 gallons of gas. In some developing countries, thirty days of rural labor costs the same as 30 gallons of gas. Now consider a simple example in which a farmer is trying to decide whether to adopt a method of production that uses machinery more intensively—weeding between crop rows with a tractor rather than by hiring people with hoes. The method will allow the farmer to use less hired labor (suppose it would save thirty days of hired labor per year), but will require the farmer to use more gasoline (suppose it would require 500 more gallons of gasoline each year). For a farmer in the United States, adopting the new method is sensible and efficient. The farmer saves enough money (in reduced labor costs) to buy 1,600 gallons of gas, but the farmer only needs to buy 500 gallons. For a farmer in a developing country such as that described above, adopting the new method is not sensible and is inefficient. That farmer only saves enough money (in reduced labor costs) to buy 30 gallons of gas, but the farmer needs to buy 500 gallons.

Table 13.5 Worldwide Use of Tractors and Machinery, 1961–2006

	Number of Agricultural Tractors	Number of Combine Harvesters-Threshers
1961	11,256,994	2,224,527
1971	16,350,927	2,639,508
1981	21,673,446	3,623,963
1991	25,100,312	3,943,196
2001	26,339,927	3,804,732
2006	26,947,171	3,647,942

Source: FAOSTAT 2008c.

Of course, "machinery" does not *need* to be the giant large-scale tractors and equipment found in the United States. In developing countries, farmers are more likely to use smaller-scale farm machinery, or machinery that relies more on human or animal power.

Population, Labor, and Agricultural Productivity

In Chapter 8, we saw that according to any reasonable assumptions, population will grow substantially over the next fifty years. As population grows, there are more people available to work in agricultural production. Of course, it is possible that future population growth will occur only in the cities. In fact, Mundlak, Larson, and Crego (1996) found that agricultural labor dropped in 40 percent of countries worldwide over the period 1967–1992. However, it appears likely that population growth will lead to increases in average labor per hectare in many parts of the developing world. Table 13.6 shows trends in the agricultural labor force. The number of agricultural workers per unit of agricultural land has increased steadily in Africa, has remained fairly stable in Latin America, and increased in Asia until the early 1990s and has declined since. Reflecting the generally high population density, Asian agriculture uses labor intensively (more than ten times as many workers per hectare in Asia as in Latin America). (Not shown in the table, workers per hectare are decreasing in the developed world. In the United States there are about 0.007 workers per hectare—the average farm worker tends to over 370 acres. Compare that to Asia with 4 acres per worker, or Africa with 13 acres per worker.)

Adding labor to each hectare of land will increase yield as long as there is productive work for the additional workers to do. Economists describe this situation as one of "positive marginal productivity." One can imagine the opposite situation—"negative marginal productivity"—in which existing workers

Table 13.6 Workers per Hectare of Agricultural Land, Selected Regions, 1961–2005

	Africa	Latin America	Asia	World
1961	0.098	0.065	0.579	0.191
1971	0.114	0.066	0.641	0.207
1981	0.134	0.068	0.713	0.232
1991	0.156	0.064	0.737	0.255
2001	0.182	0.061	0.618	0.267
2005	0.191	0.059	0.637	0.274

Source: FAOSTAT 2008c.

Note: Workers per hectare is based on the size of the economically active population in agriculture (2004 revision).

are already doing all that is possible, and in which adding another worker to a plot of land causes a decline in production as workers begin to get in one another's way.

In those areas that are already under cultivation, what is the marginal productivity of labor in agriculture? That is, by how much would the addition or subtraction of one worker change farm production? In the years before World War II, a considerable literature developed that assumed that such a large pool of unemployed and underemployed labor languished in third world agriculture that substantial amounts could be withdrawn for the industrial labor force with no diminution in agricultural production—in other words, that the marginal product of labor in agriculture was very close to zero (Lewis 1954; Fei and Ranis 1964).

Gary Becker (1975) called that thesis into question with the powerful argument that people attach at least some value to their leisure time. If this is the case, then they will not work their fields up to the point that another minute spent farming yields no product at all. They would rather spend those few minutes at leisure.

Still, we see considerable evidence that the marginal product of labor may be low on small subsistence farms in developing countries. A number of studies have shown that yields on smallholdings in India (so small that all labor is supplied by the farm family) are significantly higher than yields on large farms where a substantial proportion of the labor force is hired (Berry and Cline 1979). In terms of a production function such as that shown in Figure 13.3, the smaller farms have moved up the production function to a point where output per hectare is larger, but the production function is flatter—so marginal productivity of the input is lower. Stevens and Jabara (1988) find similar results for farms in India, Taiwan, and Brazil: yields decline as farm size increases.

Farmers who hire labor are unwilling to hire so much that the product for the last hour worked by the laborer is less than the cost of hiring the laborer for that hour. But when all labor comes from the family, family members may be willing to work for something less than the prevailing wage for those last few hours, because the family will benefit from those last few hours and because they may have no higher-valued use for their time (Mazumdar 1965, 1975; Sen 1964, 1966). All this suggests that the marginal productivity of labor in agriculture is low, and comparisons of wages in agriculture versus nonagricultural activities in the third world support this thesis. The International Labour Organization (ILO 1987) lists the daily wage rate for agricultural activities in the Philippines in 1985 as 23.7 pesos per day (just over one US dollar), but the daily rate for nonagricultural activities as 56.4 pesos per day, or 2.4 times the agricultural rate. The ILO lists the daily rate for farm labor in India at about half a dollar and shows the manufacturing wage to be over five times that amount. More recent data in Table 13.7 show that this pattern has persisted in recent years.

Table 13.7 Ratio of Nonagricultural to Agricultural Wages, Various Countries

	Wage Measure	Gender of Workers	Ratio of Wage for Construction Labor to Wage for Agricultural Farm Labor
Indonesia (2006)	Per month	Male only	1.60
Zambia (2006)	Per month	Male only	1.34
Mexico (2007)	Per month	Both male and female	1.38
India (2000)	Per day, minimum	Male only	2.38
India (2000)	Per day, maximum	Male only	1.00

Source: International Labour Organization, LABORSTAT, 2008.

AIDS and Agricultural Productivity

Chapter 8 discussed the problem of AIDS and the high incidence of HIV infection in some sub-Saharan countries. HIV infection and AIDS reduce labor productivity in the following ways:

- As people fall ill to AIDS, their capacity for work is reduced.
- The labor of healthy family members is diverted from agricultural production to caretaking.
- Premature deaths of AIDS victims remove a source of expertise and experience from the houschold.
- Fear of AIDS and its spread may make HIV-infected people unemployable even when they are capable of working.
- Orphaned children receive less education and care than children who have living parents; thus these orphans grow up to be less productive as adults.

Several reports verify the effects of AIDS on agricultural production. In Thailand, one-third of rural families affected by AIDS saw their agricultural output drop to less than one-half of earlier levels (ILO n.d.). A study of Tanzanian households found that a woman married to an AIDS patient spent 50 percent less time on agricultural activities (UNAIDS 1994). A study in Uganda found that two-thirds of households who had lost a family member to AIDS produced less food than before (FAO 2002).

Population Growth as a Stimulant to Productivity

A number of thinkers have argued that population growth in and of itself is a stimulant to productivity. One of the writers in this school (Clark 1973) capsulized one of its chief arguments as "more people, more dynamism." That is, society is better-off with a large population than with a small one "as a result of there being more knowledge creators" in a large population (Simon 1986:

169). An exploration of the circumstances under which this argument is consistent with economic theory has been done by Kremer (1993).

Critics of this argument note that in today's high-tech world, the presence of large numbers of people does not guarantee creation of a high level of knowledge. If it did, then India and China, with over a third of the world's population between them, should account for a greater share of the world's technological development than do Germany, France, Great Britain, the United States, and Japan, which collectively account for only 10 percent of the world's population. In the third world, many Einsteins may go undiscovered for want of a proper education.

Ester Boserup, in her book *Population and Technological Change,* argues that population growth creates a kind of crisis situation that stimulates the invention of new technology: "Shrinking supplies of land and other natural resources would provide motivation to invent better means of utilizing scarce resources or to discover substitutes for them" (1981:5). Note that in this "necessity is the mother of invention" argument, it is population growth that drives the creation of technology, and not the creation of technology that expands the capacity of the economy to support more people.

Boserup argues that farming is most intense in the densely settled regions of the world (not that people have tended to gather in those regions of the world where soils are most productive). She argues that, because periods of technological innovation and expanding productivity have usually been accompanied by increases in population, growth of population must have caused the increase in technology and production. Critics of this thesis argue that the reality is the other way around—that technological innovation and expanding productivity are what have, in fact, made possible the associated increase in population.

Because productivity is related to income, and income is so closely related to food consumption, those who argue that population growth in itself is a stimulant to productivity imply that population growth would help alleviate the world hunger problem, or at the least impose no threat to a solution. This school of thought must contend with a series of arguments that claim population growth has a detrimental impact on the nutrition of the poor.

Kahkonen and Leathers (1997) suggest an additional reason why increasing population density may be beneficial to agricultural development. If transactions costs—the costs associated with setting up an exchange of goods—decline with the number of transactions, then areas with high population density are more likely to have established markets (or other means of exchange); this in turn gives farmers incentives to produce a marketable surplus.

Small-Scale and Subsistence Agriculture

In the flood of numbers in this chapter, it may be easy to overlook one important fact: many of the poorest and most food-insecure people in the world are subsistence farmers working tiny plots of land to feed themselves. The predominance of small-scale subsistence farming in the developing world was a major theme of the World Bank's 2008 *World Development Report.* It reported, for example, that 85 percent of rural farm households in Nigeria sell less than half their agricultural output on the market. In Ghana, the number is 76 percent (p. 76).

A detailed survey of subsistence and nonsubsistence (or "market-oriented") farmers in Vietnam shows that market-oriented farmers have higher incomes, lower poverty rates, and larger farms; that they are more likely to live in a community that has a market or a commercial enterprise; and that they produce more "high value and industrial crops" (World Bank 2008b:74).

A study of farmers in India shows that, in the early 1990s, 60 percent of India's farmers each had less than 1 hectare of land (accounting for 17 percent of land farmed, and that farms of that size provided no marketable surplus (Singh, Kumar, and Woodhead 2002: tabs. 1–2, 27).

A study of small-scale farmers in Ghana gives some insight into the link between yields and undernutrition in subsistence households. Average maize yields (output per unit land) for the farmers in this study were 0.6–2.7 metric tons per hectare (compared to about 7.9 in the United States). A very poor rural Ghanaian, who relied on maize for all of his calories (say 2,400 per day), would need 0.274 metric tons of maize per year to achieve that diet. In other words, he would need 0.1–0.4 hectares of maize (at a yield of 0.6–2.7 metric tons per hectare) (Aalangdong, Kombiok, and Salifu 1999). We can see how a farm of 1–2 hectares might be just barely enough to provide a subsistence diet for a family, and we can see how precariously the family's food security might rest on the hope of good weather and freedom from pests and diseases.

The study also reports how much labor was used in farming, and shows how labor needs can also be a constraint on food security. The farmers in this study worked between 1,900 and 4,050 days per hectare. One farmer in the study worked 40.5 hours on a 0.01-hectare experimental plot, and produced 10 kilograms of maize, or about 32,000 calories. In order to generate 2,400 calories per capita per day, he would need 27 such plots (0.27 hectares), requiring 1,094 days of labor per year (nearly three times as many days as there are in a year). This gives us a clear indication of how undernutrition can coexist with subsistence agriculture.

Another farmer in the study was considerably better-off. He owned two bullocks and a plow; with only 19 hours of work, he obtained a yield of 25 kilograms of maize on his 0.01-hectare experimental plot (perhaps the animal manure was used as fertilizer to increase this farmer's yields). In order to generate 2,400 calories per capita per day from maize alone, this farmer would need 11 plots (0.11 hectares), or about 200 days of labor.

Partial Productivity Measures

This chapter has focused on yield per unit of land. But this is only one of the possible measures of "input productivity" or "partial productivity"—output per unit of a single input (in this case land). Work by economists Craig, Pardey, and Roseboom (1994, 1997) analyzes differences in partial productivity measures across different countries. Some of their data are shown in Tables 13.8 and 13.9.

Table 13.8 shows two partial productivity measures: the value of agricultural output per hectare and the value of agricultural output per worker. Notice how the numbers here reinforce a concept presented above in the section on machinery use: efficient agricultural practices conserve on use of the scarcest resource. In densely populated parts of the world (such as Japan and Asia in general), returns per hectare are approximately equal to returns per worker; agricultural practices in these parts of the world use relatively high amounts of labor on each plot of land in order to extract the most from the scarce resource—land. In parts of the world where population density is low (such as Australasia and North America), returns per worker are 150 times returns per hectare; agricultural practices in these parts of the world use relatively low amounts of labor on each plot of land in order to extract the most from the scarce resource—labor.

Table 13.9 illustrates how increased output per unit of land is associated with more intensive use of other inputs. In this table, the countries and areas have been sorted according to value of output per hectare. The other columns of the table show per hectare use of labor, fertilizer, and horsepower (both

Table 13.8 Value of Agricultural Output per Agricultural Worker and per Hectare of Agricultural Land, 1986–1990

	Average Value of Output per Worker ($)	Average Value of Output per Hectare ($)
Sub-Saharan Africa	412	123
China	324	300
Asia and Pacific	817	955
Latin America and Caribbean	3,260	239
West Asia and North Africa	2,608	534
Australasia	38,580	216
Western Europe	18,088	1,231
Southern Europe	6,662	652
Eastern Europe	5,324	703
Former Soviet Union	4,432	150
North America	32,948	224
Japan	3,103	2,589
South Africa	3,812	72

Source: Craig, Pardey, and Roseboom 1994.

Table 13.9 Output and Input Use per Hectare, Selected Countries and Regions, 1986–1990

	Output (dollars per hectare)	Labor (workers per 1,000 hectares)	Fertilizer (kilograms per hectare)	Tractors (horsepower per hectare)	Animal Traction (horsepower per hectare)
Japan	2,589	834	373	13,095	4
Western Europe	1,231	70	186	4,222	17
Asia and Pacific	955	1289	80	173	230
Eastern Europe	703	147	171	1,632	32
Southern Europe	652	118	82	1,955	23
West Asia and North Africa	534	292	63	519	116
China	300	925	48	177	70
Latin America and Caribbean	239	179	21	121	71
North America	224	7	35	629	9
Australasia	216	5	13	188	4
Former Soviet Union	150	34	43	274	11
Sub-Saharan Africa	123	354	2	14	18
South Africa	72	19	8	107	3

Source: Craig, Pardey, and Roseboom 1994.

mechanized and animal traction). The countries at the top of the list attain high outputs per acre by using a lot of inputs per acre. The countries at the bottom of the list have low output per acre, but also low inputs per acre. (The data shown are input and output per unit of agricultural land. The fertilizer application rates discussed above for countries such as China and Japan were fertilizer use per hectare of arable and permanent crop land. This accounts for the difference in numbers presented earlier in the discussion and the numbers in Table 13.9.)

14

Increasing Yields Through New Technology

The previous chapter showed that yields per hectare can be increased by using productive inputs more intensively on each hectare of land. We now turn to an alternative way of producing higher yields—improved technology. Technological improvement allows us to gain more output from the same quantity of inputs (or the same output using fewer of the inputs).

The Green Revolution

There can be no doubt that the centuries-long climb in yields reported in Table 13.1 is the result of better agricultural techniques: improved knowledge about crop rotations, timing and levels of fertilizer applications, and selective animal breeding. Much of the new knowledge came from trial and error by farmers themselves, but scientific experiments (such as those reported in Box 13.1) also played a role. Since the 1960s, the most highly publicized new technology has been the new seed varieties developed in the so-called green revolution (described in Box 14.1).

The green revolution began with the work of crop scientist Norman Borlaug in Mexico in the 1940s (for which he was awarded the Nobel Peace Prize in 1970). At that time, wheat yields in Mexico were severely depressed by a fungus disease known as "wheat rust," which shrivels the wheat grain. In his search for a rust-resistant variety of wheat, Borlaug collected 8,500 varieties being grown in different parts of Mexico; two of these proved to be rust-resistant. Borlaug and his associates, using tweezers and a magnifying glass, cross-bred these rust-resistant varieties with other high-yielding varieties. Year after year, Borlaug experimented with new crosses, searching for varieties that gave higher yields. By 1957, the wheat rust problem in Mexico had been solved, and average wheat yield had nearly doubled, from 11 to 20 bushels per acre. In the meantime, Borlaug was working to solve another problem: as yields increased, the now heavier grain caused plants to tip over ("lodging," as illustrated in Figure B of Box 14.1). Crossing with "dwarf" varieties of wheat—wheat with

Box 14.1 What Is the Green Revolution?

Dana Dalrymple

In October 1944, about a year before the close of World War II, the Rockefeller Foundation brought to Mexico a young plant scientist to join a team of agriculturalists that had recently started work to assist in the agricultural development of that country. In a few months, the new man, Norman Borlaug (who was later to be awarded a Nobel Prize for his work in Mexico), was put in charge of the wheat program. He and his team set out to develop new varieties of wheat that would do better than the local varieties. Disease resistance (e.g., resistance to the fungus causing the disease rust) was particularly important at first. In the mid-1950s increased emphasis was given to increasing yields and within a few years varieties had been developed which could produce much more than the traditional ones.

Encouraged by this success, in the 1960s the Rockefeller Foundation joined with the Ford Foundation to establish two permanent research stations for the development of high-yielding cereals, the International Rice Research Institute (IRRI) in the Philippines and the International Maize and Wheat Improvement Center (CIMMYT) in Mexico. The success of these centers in developing high-yielding varieties led to such enthusiasm for the idea of international agricultural research centers that by the late 1980s, thirteen centers, treating various aspects of improving third world agriculture, had been set up worldwide, and sponsorship had spread to a consortium of donors worldwide including both foundations and government agencies.

The high-yielding varieties of wheat and rice have spread more widely, more quickly, than any other technological innovation in the history of agriculture in the developing countries. First introduced in the mid-1960s, they occupied about half of these countries' total wheat and rice area by 1982–1983 (Dalrymple 1985). Figure A shows the remarkable growth in adoption of high-yielding varieties in South and Southeast Asia.

Figure A

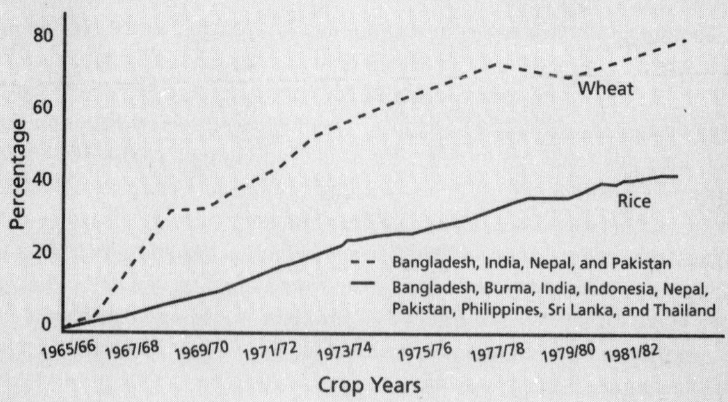

(continues)

Box 14.1 continued

Struck by the remarkable speed with which high-yielding varieties of wheat and rice were being developed for the third world and by their potential to alleviate world hunger, William S. Gaud, who was then administrator of the US Agency for International Development, referred, in a 1968 speech, to the phenomenon of their development and spread as "the green revolution" (Dalrymple 1979:724).

Actually, the green revolution's wheat and rice varieties, also known as high-yielding varieties (HYVs) or modern varieties (MVs), do not do much better than the traditional varieties (they can even do worse) unless they have appropriate amounts of water and fertilizer. In fact, it is largely tolerance of and response to substantial amounts of fertilizer that makes them so successful.

The traditional varieties of wheat and rice were not tolerant of significant amounts of fertilizer. When third world farmers attempted to increase rice or wheat yields by adding fertilizer—especially nitrogen fertilizer—to their fields, their plants would grow so tall that they would fall over. The technical term for this is lodging. What the plant scientists at the international institutes did was to locate plant varieties with genes for shortness and breed these genes into plants that had other characteristics desirable for the third world. The new plants, called semidwarfs, borrowed dwarfism genes from Japan (for wheat) and China (for rice). When used with fertilizer, they grew taller than without the fertilizer, but not excessively so. Thus they were much more resistant to lodging (Figure B).

Figure B

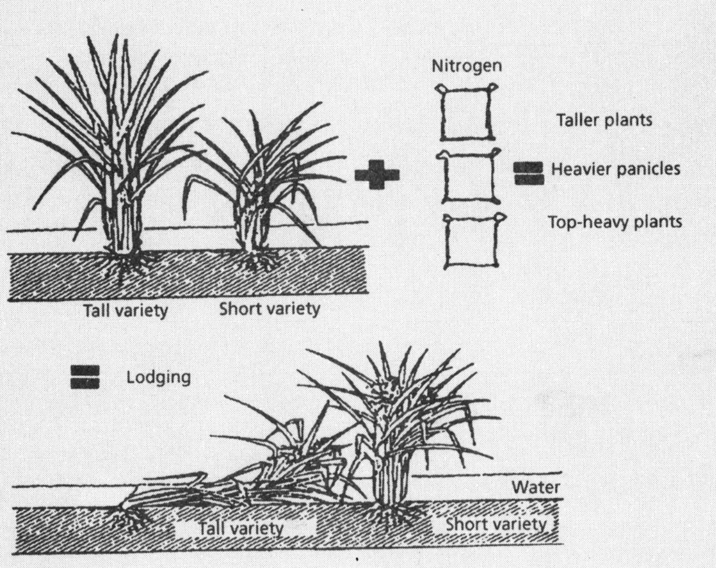

(continues)

Box 14.1 continued

Plant scientists did not stop merely with the development of nonlodging plants. They bred into their new varieties a host of other characteristics such as disease resistance.

One of the more intriguing changes in plant design that they accomplished involved rearranging the location of the seed cluster on the plant. Traditional rice plants sent their cluster of seeds (the panicle) high into the air. The seeds themselves store energy, but they do not make it. Photosynthesis is concentrated in the leaves. It did not make sense for the seed cluster to shade the highest leaves on the plant, so the scientists bred rice plants whose topmost leaf, the flag leaf, extended well above the panicle, thus taking maximum advantage of the available sunshine (Figure C).

Figure C

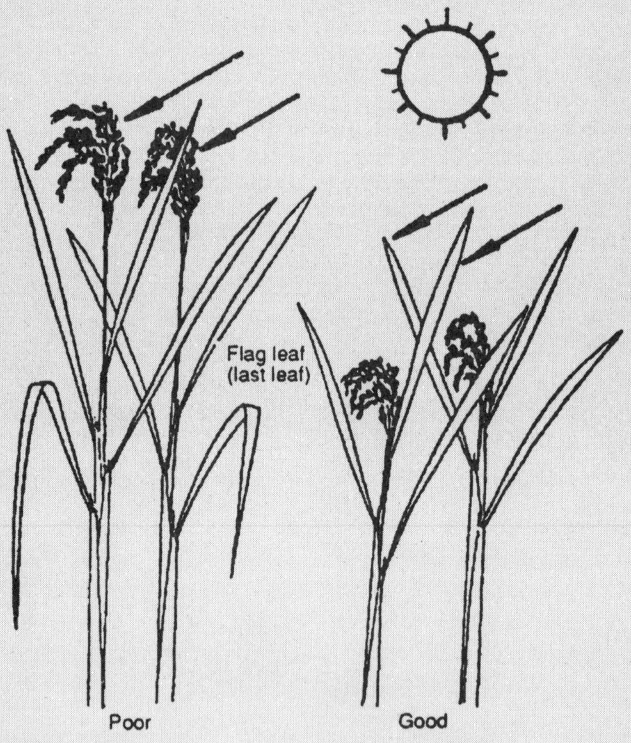

The green revolution, then, is a whole complex of innovations, such as those described above, that combine to make up new plant varieties. The modern plant varieties, when used with a package of appropriate inputs such as fertilizer and water and good management, are dramatically raising crop yields in the third world.

(For more about the green revolution, see Brown 1970; US Department of State 1986a, 1986b; Stackman, Bradfield, and Mangelsdorf 1967.)

shorter and stronger stems—solved this problem. By 1963, Mexican wheat yields had grown to 30 bushels per acre, and experimental plots showed yields of 105 bushels per acre (Paarlberg 1988:102–105).

The shorter and stronger stems of the dwarf varieties meant that the plants could support a heavier crop of grain. The new seed varieties achieved this improved yield in two ways: they were more responsive to fertilizer, and they used the photosynthetic energy of the plant more efficiently for grain growth. The traditional seed varieties responded to fertilizer at a rate of 10 kilograms of increased grain output for each kilogram of increased fertilizer use. The new varieties responded at a rate of 25 kilograms of grain for each kilogram of fertilizer. This made use of commercial fertilizers much more attractive for farmers who used the new seed varieties. The old seed varieties allocated about 20 percent of the photosynthetic energy of the plant to grain production and the remaining 80 percent to growth of roots, stems, and leaves. The new seed varieties improved this ratio to about 50-50.

Borlaug's methods were adapted to developing new varieties of other crops, most notably rice for Asia. The International Rice Research Institute (IRRI) has been the leading institutional support for this research. In the early 1960s, average rice yields in Asia were 1–2 metric tons per hectare. By 2002, IRRI could claim in its project report: "The adoption of improved varieties that have a yield potential of 10 tons per hectare is almost complete."

As Table 14.1 shows, adoption of modern varieties has been most widespread for wheat. By the late 1990s, dwarf wheat varieties accounted for about 80 percent of the wheat planted in the developing world as a whole, about 90 percent in Asia and Latin America, and about 50 percent in sub-Saharan Africa. Modern varieties of rice have been widely adopted in Asia. For all crops, adoption in sub-Saharan Africa has lagged behind adoption in other regions.

Table 14.1 Percentage of Land Planted with Modern Crop Varieties

	1970	1980	1990	1998
Latin America				
Wheat	11	46	83	90
Maize	10	20	30	46
Asia				
Wheat	19	49	74	86
Rice	10	35	55	65
Middle East and North Africa				
Wheat	5	16	38	66
Sub-Saharan Africa				
Wheat	5	22	32	52
Maize	1	4	15	17
Cassava	0	0	2	18

Source: Evenson and Gollen 2003b.

Criticisms of the Green Revolution

The green revolution is not without its critics. As already described, the new seed varieties encourage farmers to use more fertilizer; the green revolution has also brought increased use of irrigation and pesticides. For these reasons, the green revolution has been criticized as potentially damaging to the environment. A second criticism centers on whether green revolution varieties have crowded out traditional varieties and led to a loss of species diversity. But perhaps the most widespread criticism of the green revolution is the fear that it might increase inequality of income; indeed, it might make the poor absolutely worse-off in regions where its practices have been adopted. After all, it has been common for new agricultural technologies to be adopted earlier by the leading farmers in a region, who are often those with the biggest or best farms, the best education, and the greatest willingness to take a risk by trying something new. Furthermore, a tendency exists for public services to be more available to the big farmers than to the small, for technology to carry with it a labor-saving bias that reduces labor's share of the product, and for technological innovations to be more appropriate to some geographical areas than to others. Would not wealthy landlords use the benefits from green revolution technology as a steppingstone to increasing the size of their holdings at the expense of small farmers, thus increasing income inequality?

Some of the above fears turn out to have been justified, but most have not. Early adopters did tend to be the bigger, better farmers, but this has not prevented smaller farmers, who are often slower to change, from adopting green revolution technology.

Still, the benefits of the green revolution have accrued more to farmers in regions where water, especially irrigation water, is plentiful. Growers of lowland rice (rice that spends much of its early growing days with its stalks standing in a few inches of water) benefit more than growers of upland rice. Wheat and rice farmers, who predominate in tropical regions with 40 inches or more of annual rainfall, benefit more than sorghum and millet farmers, who farm the semiarid tropics, where rainfall is usually less than 40 inches per year.

The high-yielding fine grains (wheat and rice) grown with adequate water are far more responsive to fertilizer than are even the best varieties of the coarse grains (sorghum and millet) grown in the dry regions where irrigation water is scarce to nonexistent. The fertilizer subsidies that third world governments frequently provide to their farmers therefore benefit the fine-grain farmers more than the coarse-grain farmers. Thus, the benefits of the green revolution have concentrated mainly in the wetter tropics and especially on the flatter lands where irrigation and water management are easier. This phenomenon may well have increased income inequality between regions.

Nevertheless, within regions, "the benefits from adopting modern varieties have been remarkably evenly distributed among farmers differing in size of holding and tenure status." This is the conclusion of a major study commissioned to examine the impact of the international agricultural research centers (Anderson et al. 1985:4).

In a detailed study of the economy of a northern Indian village where modern varieties had been widely adopted, Bliss and Stern found no strong association between size of holding and the intensity of use of inputs associated with the adoption of green revolution technology. In fact, "the adoption of various newer varieties and intensive practices seemed to be particularly associated with the younger educated farmers" (1982:291).

How is it that the benefits could be so evenly distributed among farms of different sizes and among farmers of different tenure status? For one thing, the capital requirements of the green revolution are minimal. Unlike hybrid corn, which has done so much to increase corn yields in the United States and Europe, the green revolution's fine-grain seeds breed true (i.e., produce offspring almost identical to the parent plant). Hybrid seed corn must be produced annually under technically demanding conditions on specialized farms and sold to farmers each year. A farmer who planted the grain from his hybrid corn crop as seed would be most disappointed in the yield results. The green revolution's rice and wheat, although the result of complicated crosses, are not true hybrids. A farmer can plant a little one year, take the resulting grain as seed for next year's crop, and rapidly and cheaply multiply her seed stock. Thus, a handful of green revolution seed is all a farmer needs to begin farming—and it usually does not cost any more than traditional seed.

On the one hand, green revolution varieties do require expenditures on commercial fertilizer. On the other hand, fertilizer, like seed, is almost infinitely divisible, and a farmer need purchase only as much as he needs for his particular plot. Unlike capital investment in a tractor, the capital investment in the green revolution is not "lumpy"—that is, it does not come only in large, indivisible units.

A second reason for the evenly distributed benefits is that green revolution technology is not labor-saving. To the contrary, it turns out to be labor-using (Hossain 1988b:12; Ranade and Herdt 1978:103; Pinstrup-Andersen and Hazell 1985:11). Much of the third world grain crop is harvested and threshed by hand, and the increased crop yield requires more labor for harvest. More important, though, the fertilizer applied to the crop turns out to stimulate the growth of weeds as well as grain. Farmers are finding it profitable to remove these weeds, and weeding a field, in the third world, is labor-intensive (weeds are pulled by hand or chopped with a blade, for example).

Because the poor spend a greater proportion of their income on food than do the rich, the benefits from price declines associated with increased production following the adoption of green revolution technology favor the poor over the rich. Because of the very low overall elasticity of demand for cereals, consumers benefit substantially from the price drop caused by increased production, but farmers as a whole experience a decline in total revenue.

In a study of the social returns to rice research, Evenson and Flores found that in the Philippines, during the period from 1972 to 1975, the annual loss to producers from the adoption of the high-yielding rice varieties was $61 million, while the annual gain to consumers was $142 million, yielding a net annual gain to society of $81 million (1978:255). Similarly, working with Colombian data, Scobie and Posada concluded that, during 1970–1974, Colombian rice producers lost $796 million, while consumers benefited by over $1.349 billion, for a net gain to society of $553 million. Furthermore, they concluded that "while the lower 50 percent of Colombian households received about 15 percent of household income, they captured nearly 70 percent of the net benefits of the [rice] research program" (1984:383).

The cereal production increases that, because of demand elasticities of well below 1.0, result in the loss of revenue for commercial rice growers, may at the same time result in an increase in total revenue to the near-subsistence farmer. A near-subsistence farmer is defined as one who purchases less than 50 percent of his food supply in the marketplace. (Almost no one is a pure subsistence farmer, purchasing no food supplies, not even spices, in the marketplace.)

Consider a near-subsistence farmer who consumes, say, 90 percent of his rice crop. He has a small amount left over for sale in the market. He sees his neighbors on larger farms, who sell most of what they grow, practicing green revolution technology, and copies them, purchasing some seed and fertilizer and irrigating his crop as usual. His yield increases by the same percentage as that of the larger farmers, but if he consumes the same amount of rice as he did before, the size of the surplus that he has left to sell grows by a much greater percentage than does the size of total production. Thus, even in the face of sharply declining rice prices, he may experience a gain in total revenue, whereas his more commercially oriented neighbors are experiencing losses.

Impact of the Green Revolution on Yields

Whatever one makes of the criticisms of the green revolution, there can be no doubt that it has led to increased yields. How much of the yield increases are attributable to increases in inputs described in the previous section, and how much to new technology? This question is hard to answer because of the nature of the new seed varieties. As we have seen, the new varieties have achieved their higher yields in part by being more responsive to fertilizer; in addition, the new varieties are often more sensitive to drought, and thus have been frequently grown on irrigated land. Therefore, the higher observed yields are the result of a complex interaction between technological improvement and additional levels of inputs.

Several researchers have attempted to unravel this complexity. The Consultative Group on International Agricultural Research (CGIAR 1997) reports that "22 percent of the developing world's [wheat] production increase resulting from higher yields is attributed to the gradual spreading of modern or high yielding varieties." A study of Syrian wheat found that the increase in wheat yields (which more than quadrupled from the 1950s to the early 1990s) was due to several factors: new varieties accounted for about 35 percent of the increase; better management and increased fertilizer use accounted for 23 percent each; and irrigation accounted for 19 percent of the increase (CGIAR 1995). Peter Oram (1995), in an IFPRI paper, writes: "It is estimated that about 50 percent of the gains in farm yields have resulted from plant breeding and the balance from the application of other improved practices." These findings suggest that, of the average yield growth of 2 percent per year from 1968 to 1997, new technology accounted for between 0.5 and 1 percent per year, and input growth accounted for the remaining 1 to 1.5 percent per year.

Evenson and Gollen (2003a) completed a comprehensive review of the effects of increased input use and new seed varieties on yield growth, summarized in Table 14.2. For all developing countries, total yield growth was faster in the twenty years between 1961 and 1980 than in the twenty years between 1981 and 2000. In the earlier period, increased input use was responsible for about 80 percent of the yield growth; in the later period, input use and new technology were about equally responsible for yield growth. Yield growth due to new technology was actually higher in the second twenty-year period. The table also shows that sub-Saharan Africa has lagged behind other geographical areas, both in technological change and in growth of input use.

Technological improvements increase the aggregate supply of food and cause food prices to decrease. In developing countries, prices for crops targeted by the CGIAR have dropped more than the prices for non-CGIAR crops (CGIAR 1997). The US Department of Agriculture (USDA 1991) estimates that research into improvements in agricultural production methods has reduced the

Table 14.2 Effects of Increased Input Use and New Seed Varieties on Yield Growth (percentages)

	1961–1980			1981–2000		
	Average Annual Yield Growth	Contribution of Increased Input Use	Contribution of New Technology	Average Annual Yield Growth	Contribution of Increased Input Use	Contribution of New Technology
Latin America	1.587	1.124	0.463	2.154	1.382	0.772
Asia	3.120	2.439	0.682	2.087	1.119	0.968
Middle East and North Africa	1.561	1.389	0.173	1.505	0.722	0.783
Sub-Saharan Africa	1.166	1.069	0.097	0.361	–0.110	0.471
All developing countries	2.502	1.979	0.523	1.805	0.948	0.857

Source: Evenson and Gollen 2003a.

average yearly food bill in the United States by $400 per person. A model by Evenson and Rosegrant (2003) estimates that the development of new crop varieties has kept food prices 18–21 percent lower than they would have otherwise been.

Federico, following a detailed analysis of relevant studies, concludes that technical change (as measured by "total factor productivity"—discussed below) "accounted for almost all of the production growth in [developed countries], while it accounted for between a third and a half in the [less developed countries]" (2005:81).

Box 14.2 shows how biological and mechanical innovations contributed to productivity growth in US wheat production.

Box 14.2 Technical Change and Productivity in US Wheat Production

The experience of wheat production in the United States gives some insight into the ways that new production techniques contribute to improved productivity.

During the period 1839–1909, US wheat yields were nearly constant, increasing only 24 percent—from 11.3 to 14.0 bushels per acre—over the seventy-year period. Over the same period, wheat output per hour of labor grew fourfold, from 0.3 to 1.3 bushels per hour. The prevailing wisdom attributed this productivity improvement to mechanization. New equipment—the mechanical reaper and thresher and the self-binder—allowed a farmer to cultivate more and more acres of wheat. This mechanization impact was strengthened by the westward movement of the US farm population to flat expanses of the Great Plains.

Olmsted and Rhode (2002) challenge this prevailing wisdom. They argue that biological innovation in the form of new seed varieties was an important component of the productivity increase. The contribution of new seed varieties was twofold. First, the westward expansion of wheat production was made possible by the development of wheat varieties that were suited to the agroclimatic conditions of the Great Plains. One expert in wheat production, writing in 1860, opined that with existing wheat varieties, commercial wheat production was viable only in Ohio, Pennsylvania, and New York; however, new varieties of wheat suited to the growing conditions of states further west led to expansion of commercial wheat production far beyond that limited area. Second, new varieties were introduced to "solve" an increasing number of disease and pest problems that would have caused wheat yields to decline substantially in the absence of the new varieties. In the absence of these biological innovations, Olmsted and Rhode estimate, wheat yields would have fallen from 11.3 to 7.5 bushels per acre (rather than the increase to 14.0 bushels per acre actually observed in 1909), and labor productivity would have increased from 0.3 to only 0.8 bushels per hour (rather than the increase to 1.3 bushels in 1909).

Total Factor Productivity

One systematic way to quantify the impact of all the different kinds of technical change is to calculate *total factor productivity* (TFP). Recall our previous discussion of partial factor productivity, such as growth in output per unit of land or per unit of labor. When looking at growth in partial factor productivity, it was not readily apparent what part of that growth was attributable to technical change and what part was attributable to growth in use of other inputs. Total factor productivity is calculated by first constructing two indexes: an index of output that reflects changes in levels of all outputs, and an index of input use that reflects changes in levels of all inputs (or at least all inputs on which data are available). Total factor productivity is the ratio of these two indexes: outputs divided by inputs. (See Box 14.3 for an example of how index numbers can be constructed.) Therefore (referring again to Figure 13.3), an increase in TFP means that output can be increased without increasing input, either because inefficiency has been eliminated (moving from A to B) or because a technological improvement has caused an upward shift in the production function (moving from B to C).

Pingali and Heisey (1999) review TFP estimates from fifteen separate studies, covering thirty-two different countries, for a variety of time periods. They report seventy-two total measurements of annual percentage growth in TFP; sixty-six are positive and six negative; seventeen of the estimates are greater than 2 percent per year. To put this into perspective, worldwide production of all cereal crops grew at an annual rate of 2.16 percent from 1961 to 2001; total calories worldwide increased by 2.24 percent per year over the same period.

Federico does an even more comprehensive review of TFP studies. He finds more variation in estimates than do Pingali and Heisey, and finds a higher prevalence of negative estimates of average TFP in developing countries. He concludes (2005:77–79):

> 1. Agricultural productivity has grown remarkably in the large majority of countries and periods. Almost 70 percent of estimates (438 out of 636) are positive; and the average TFP growth for the whole database is 0.58 percent per annum. . . . Out of 175 country/period observations, there are 109 "good performers" [a majority of estimates show positive TFP growth] [and] 53 "poor performers" [a majority of estimates show negative TFP growth]. . . .
> 2. The performance of OECD countries has been quite good. TFP has grown in almost all cases. . . . Furthermore, the growth in TFP has been accelerating in almost all cases. . . .
> 3. The performance of LDCs has been decidedly mixed. . . . The main feature is the large number of "poor performers"—forty-five out of a total of ninety-four LDC countries.

Box 14.3 Index Numbers

An index number is essentially a weighted average of quantities, scaled to reflect the level of that average compared to some base period. Suppose a farm produces wheat and corn. In 2002 the farm produces $y^1{}_{corn}$ units of corn and $y^1{}_{wheat}$ units of wheat. In 2003 the farm produces $y^2{}_{corn}$ units of corn and $y^2{}_{wheat}$ units of wheat. Let p_{corn} be the price of corn and p_{wheat} be the price of wheat. To construct an index of output, first choose a "base year." This is arbitrary, so we could choose either 2002 or 2003; for this example we will choose 2002. The index number for each year is:

$$\text{Output index year } t = 100 \times \frac{p_{corn}\, y^t_{corn} + p_{wheat}\, y^t_{wheat}}{p_{corn}\, y^{2002}_{corn} + p_{wheat}\, y^{2002}_{wheat}}$$

where t can be 2002 or 2003.

The sample outputs and prices in the following table show how the index is constructed:

Year (t)	Corn Output ($y^t{}_{corn}$)	Wheat Output ($y^t{}_{wheat}$)	Corn Price (p_{corn})	Wheat Price (p_{wheat})
2002	125	50	3	5
2003	150	45	3	5

For 2002, the sum of price times quantity is (125 × 3) + (50 × 5) = 375 + 250 = 625. For 2003, the sum is (150 × 3) + (45 × 5) = 450 + 225 = 675. The index numbers are therefore:

$$2002\text{: } 100 \times \frac{625}{625} = 100$$

$$2003\text{: } 100 \times \frac{675}{625} = 108$$

A number of points should be obvious:

- For the base year, the value of the index is 100 by definition.
- If all outputs increased by exactly 10 percent from 2002 to 2003, the index number for 2003 would be 110. If all outputs decrease by exactly 10 percent, the index number would be 90.

(continues)

Box 14.3 continued

- In the example, the increase from 100 to 108 reflects an average of the 20 percent increase in corn production and the 10 percent decrease in wheat production. The weight assigned to the different commodities depends on the price of those commodities.
- This concept is easy to extend to more than two outputs, or more than two years. It can be applied to output of a single farm, or a state, or a nation, or a combination of nations.

If the farm used three inputs—land, seed, and labor—we could construct an index of input use in a similar fashion:

Year (t)	Land Input (x^t_{land})	Seed Input (x^t_{seed})	Labor Input (x^t_{labor})	Land Price (r_{land})	Seed Price (r_{seed})	Labor Price (r_{labor})
2002	2	50	10	30	2	40
2003	2	55	12	30	2	40

For 2002, the sum of price times quantity is (2 × 30) + (50 × 2) + (10 × 40) = 560. For 2003, the sum is (2 × 30) + (55 × 2) + (12 × 40) = 650. The input index numbers are therefore:

$$2002\text{: } 100 \times \frac{560}{560} = 100$$

$$2003\text{: } 100 \times \frac{650}{560} = 116$$

Total factor productivity is the ratio of the output index to the input index. The total factor productivity measures are therefore:

$$2002\text{: } 100 \times \frac{100}{100} = 100$$

$$2003\text{: } 100 \times \frac{108}{116} = 93$$

In this example, even though output increased (as indicated by the increase in the output index from 100 to 108), input use increased faster (as indicated by the increase in the input index from 100 to 116); so total factor productivity declined by 7 percent.

> 4. . . . [T]he agricultural sector moved from laggard to leader, at least from a purely statistical point of view.

In a 2003 article, Nin and colleagues used FAO data on 115 countries and country-groups for the period 1965–1994 to estimate TFP separately for crops and animal products. The results of that study are summarized in Table 14.3. TFP for all countries is estimated at 0.51 percent per year for livestock and 0.63 percent per year for crops. This is one-quarter to one-third of the growth rates in worldwide output reported by Pingali and Heisey (1999). This is consistent, at the low end, with our previous discussion about the extent to which yield growth is attributable to new technology and the extent to which it is attributable to increased input use.

Prospects for Future Yield Growth

Can crop yields continue to grow? And if so, at what rate? The answers to these questions go a long way to determining whether the future food supply will keep pace with demand. Above, we reviewed the different opinions about the prospects for adding irrigation, fertilizer, and labor to agricultural production. Even the most optimistic projections see input usage growing more slowly than in the past. But what about technology? Can technological advances pick up the slack caused by slower growth in input use? We find optimists and pessimists on this question.

Reasons for Concern

The pessimists point out that growth in yields has been slowing (see Figures 13.1 and 13.2). Imagine (say the pessimists) that crop yields are growing according to

Table 14.3 Annual Percentage Growth in Total Factor Productivity, 1965–1994

	Livestock	Crops
Middle East and North Africa	0.01	0.20
Sub-Saharan Africa	–0.01	–0.32
South America	0.52	0.98
Central America	0.83	0.03
Eastern Europe	0.63	1.55
China	1.80	0.69
India	0.83	–1.74
Western Europe	1.19	2.50
All	0.51	0.63

Source: Nin et al. 2003.

an S-shaped curve; we have gone through a period of rapidly rising yields, but we are now reaching the top, where yields become flat: "Countries that have doubled or tripled the productivity of their cropland since mid-century are the rule, not the exception. But with many of the world's farmers already using advanced yield-raising technologies, further gains in land productivity will not come easily" (Brown and Kane 1994:132). A 2002 IRRI project report warns: "Yield at the farm level is approaching a plateau."

Some pessimism about yields is based on a belief that scientists cannot discover new ways to increase yields. "Rising grain yield per hectare . . . must eventually give way to physical constraints. . . . Yields may now be pushing against various physiological limits such as nutrient absorption capacity or photosynthetic efficiency" (Brown and Kane 1994:138). Thomas Sinclair, a horticulturist at the University of Florida, Gainesville, explained to *Science* magazine: "To grow corn, . . . you have to have leaves, stalks, and roots, so there's got to be mass committed to what you don't harvest. . . . At the beginning of this century, . . . many crops had harvest indexes on the order of 0.25 of their weight in grain, and now many crops are approaching 0.5. . . . Maybe you could go up to 0.6 or 0.65, but beyond that you can't have a viable plant" (quoted in Mann 1997:1042).

Another source of pessimism about yields is concern about whether they are "environmentally sustainable." According to this school of thought, the current high yields have been obtained by putting extreme pressure on the natural environment. As discussed in Chapter 13, the environment can stand this pressure for only a short time; yields then naturally begin to decline.

A Second Green Revolution: Research Efforts to Reduce Environmental Degradation

Perhaps in response to these legitimate concerns about the environmental impact of agricultural production, the emphasis of international agricultural research shifted during the 1990s, from increasing yields to reducing environmental impact. Some of the new production techniques that have been developed include low-till and no-till cultivation methods that reduce soil erosion, and drip irrigation techniques that improve the efficiency of delivery of irrigation water to the plants.

Not all of the innovations rely on high-tech solutions. One example comes from China, where a fungus was destroying rice fields. Wherever the fungus emerged, windblown spores would spread it from row to row until the entire field was affected. Scientists discovered a lower-yielding wild variety of rice that was resistant to the fungus. By alternating rows of the higher-yielding but fungus-susceptible rice with rows of the lower-yielding but fungus-resistant rice, scientists were able to halt the spread of the fungus. Spores would be blown from the infected rows to the neighboring resistant rows, which they could no longer infect. This research simultaneously pursued the objectives of improved yields and reduced stress on the environment, since it found an alternative to chemical fungicide.

Some research into indigenous farming practices also has found ways to increase agricultural production in environmentally friendly ways. For example, some indigenous populations use fish or anthill waste as a source of fertilizer. In Sudan, planting millet under the acacia albida tree improved millet yield. The tree roots drew nutrients from deep in the soil and the dropping tree leaves transferred those nutrients to the top of the soil. Furthermore, the timing of the leaf growth and decay allowed the millet plants to receive sun and shade at crucial times in their life cycle. A possible application of biotechnology to sustainable aquaculture is described in Box 14.4.

Reasons for Hope

The change in focus of agricultural research from increasing yields to reducing environmental impact may help explain why yields in experiment stations

Box 14.4 Genetic Engineering to Save Fish Species

The original problem was overfishing the ocean's supply of salmon. For economists this is a textbook example of the "tragedy of the commons." Because the fish population does not belong to any individual or group of individuals (property rights are not defined), and because anyone is free to catch fish (there is no way to exclude users), no one has the appropriate incentive to watch out for the long-run health of the fish population, and to ensure that enough fish are left this year to replenish the fish stocks for next year. (The same argument explains why wild species may be hunted to extinction, and why communally owned pastures may be overgrazed.)

The introduction of aquaculture, or fish farming, provided a way to grow salmon in cages or in ponds. Under these conditions, the fish population was no longer a common resource: the fish in the cages or ponds were owned and no one could remove them without permission of the owner. The owner's future profitability depended on maintaining a healthy fish population.

But a new problem emerged. Fish farms feed their salmon a diet of fish meal made of ground-up herring and anchovies. This fish meal is high in omega-3 fatty acids, which is an important nutrient in the diets of salmon. As fish farming grew, it put more and more pressure on wild populations of herring and anchovies, which were being overfished to provide the fish meal.

Scientists at Montana State University discovered that a plant called carpetweed produces omega-3 fatty acids. Unfortunately, the salmon had trouble digesting carpetweed meal; fish require low-fiber diets, such as a meal made of hulled sunflower seeds. Scientists are now attempting to modify sunflower genes to incorporate the omega-3 gene from carpetweed. If they are successful, the new feed will reduce the pressure on wild fishing populations.

Source: Baden 2003.

(the laboratories of crop production) have leveled. A related explanation is that international crop research has been underfunded (Pardey and Alston 1995). Alexandratos (1995) argues that the slowdown in yield growth is less a result of technological feasibility and more a result of low farm-level prices. Finally, worldwide average crop yields were brought down during the 1990s by a large drop in yields in the former Soviet Union, probably a temporary phenomenon.

Given these explanations, many analysts are more optimistic about future yields. The FAO projects that cereal yields will grow at 1.4 percent per year between 1990 and 2010, compared with a 2.2 percent growth rate between 1970 and 1990 (Alexandratos 1995). A World Bank study estimates that grain yields will continue to grow at a 1.5–1.7 percent annual rate (Mitchell, Ingco, and Duncan 1997). An IFPRI report on food supply and demand in 2020 presents a base scenario in which yields continue to grow at the rate observed in the late 1980s and early 1990s (Pinstrup-Andersen, Pandya-Lorch, and Rosegrant 1998). Rejesus, Heisey, and Smale (1999) review a number of studies that project annual growth rates for wheat yields of between 1.4 percent and 1.9 percent for developing countries during the first decades of the new millennium, and slightly lower growth rates for developed countries. (For a criticism of these projections, see Worldwatch Institute 1996b. For an FAO defense, see FAO 1996d. For an analysis of the debate about future yields, especially focusing on China, see Crosson 1996b.) To put the debate into perspective, consider that in 1989, the Centro Internacional de Mejoramiento de Maize y Trigo (CIMMYT) projected that wheat yields between 1987 and 2000 would grow at a rate of between 1.5 percent per year (the "realistic" estimate) and 2.3 percent per year (the "optimistic" estimate); the actual growth rate was 1.85 percent (the CIMMYT is the division of the Consultative Group on International Agricultural Research that focuses on wheat and maize research). Box 14.5 presents some optimistic evaluations by crop scientists.

The optimistic beliefs cited above are based on several observations. A European study (see Penning de Vries et al. 1995 for a description) concludes that potential yields of cereal crops are close to 10 metric tons per hectare, compared to current yields of about 3 metric tons per hectare. Some dramatic reports about new technological breakthroughs have been published. A "super rice" developed in 1994 by the IRRI has potential yields of 15 tons per hectare, compared with current yields of about 5 tons per hectare. A new cassava strain promises yields that are ten times higher than current yields.

Biotechnology and Genetically Modified Food

One source of optimism about future yield growth comes from the scientific advances in the field of genetic engineering and biotechnology. In the past few years, the issue of genetically modified (GM) food has become a subject of public debate and discussion. Genetic engineering creates new seed varieties that contain specific genetic characteristics. Genes from one organism can be

Box 14.5 Agronomists' Perspectives on Potential for Future Yield Growth

A 1998 symposium (see Waterlow et al. 1998) heard a number of papers reviewing the scientific literature on specific ways that crop yields could be increased. The tone of the papers was optimistic:

> "The rapid rise in yield potential and response to inputs . . . seems unlikely to be maintained beyond the next two decades . . . unless crop growth rates can be enhanced. Although these may . . . be limited by the photosynthetic rate, other processes . . . may also have a significant effect. . . . Even in the absence of any further rise in genetic yield potential, crop yields could continue to rise with improvements in climatic adaptation, pest and disease resistance, and agronomic support." (Evans)
>
> "I have no hesitation in stating that . . . plant biotechnology can bring major progress to tropical agriculture." (Van Montagu)
>
> "Before writing this article I approached 10 or more leading experts on photosynthesis research throughout the world. . . . All agreed that the efficiency of photosynthesis in the field is far from the theoretical maximum and is restricted by environmental factors. . . . [Research into ways of] enhancing . . . protective and repair mechanisms [of plants] will help us to approach levels of photosynthetic efficiency observed under optimal/nonstressed conditions." (Barber)
>
> "The possible prize for successful engineering of Rubisco [an enzyme that reduces photorespiration in plants] in the world's major . . . crops—an increase of 20% in the potential yield in temperate regions and of 50% in the tropics—must surely justify increased effort toward this single charge." (Long)
>
> "Metabolic engineering of source-sink relationships is a promising approach to increase . . . genetic yield potential of cereal crops. . . . By combining the expression of several different transgenes . . . a significant increase in . . . crop yields may be achieved." (Choi et al.)
>
> "With the availability of biotechnological tools . . . procedures for breeding genetically diverse parental lines and hybrids can be made more efficient." (Khush, Peng, and Virmani)
>
> Other papers heard at the symposium discuss research efforts to promote tolerance to salinity and drought (Verma) and resistance to disease (Lamb).

inserted into the DNA of another organism to create certain characteristics. To date, the most common types of genetic changes are the following:

- Plants are modified so that they contain a bacterium (*Bacillus therogensis,* or Bt) that is a natural pesticide. Thus these plants are protected from pest damage without the need for commercial chemical pesticides.

- Plants are modified so that they are particularly resistant to certain weed-killing chemicals. Therefore, the chemicals can be applied more heavily to kill weeds without killing the crop.
- Plants are modified to thrive in adverse conditions—for example, to be more resistant to disease or pests, or to be more tolerant of frost, drought, or saline soil.
- Plants are modified so that the crop contains certain nutritional characteristics—for example, a potato that is rich in protein.
- Salmon grown in fish farms are genetically modified to change their inbred eating habits and increase weight gain.
- Genetically modified microbes are applied to soil to assist in nitrogen fixation.
- Animals are genetically modified to produce more milk and less manure, and use feed more efficiently (CAST 2003).
- Pharmaceutical drugs are produced in genetically modified plants and animals.

Genetically modified crops have been adopted by farmers in growing numbers since they were commercially introduced in the mid-1990s. Figure 14.1 shows the growth in worldwide acreage planted with GM crops. For example, in 2002, 58.7 million hectares of land were planted with such crops, 73 percent of which was in developed countries (almost all of it in the United States and Canada), and the remainder of which was in developing countries (led by Argentina with 23 percent of the total GM land). Soybeans, corn, and cotton account for most of the GM crops. Table 14.4 shows that more than half

Figure 14.1 Land Area Planted with GM Crops Worldwide, 1996–2007

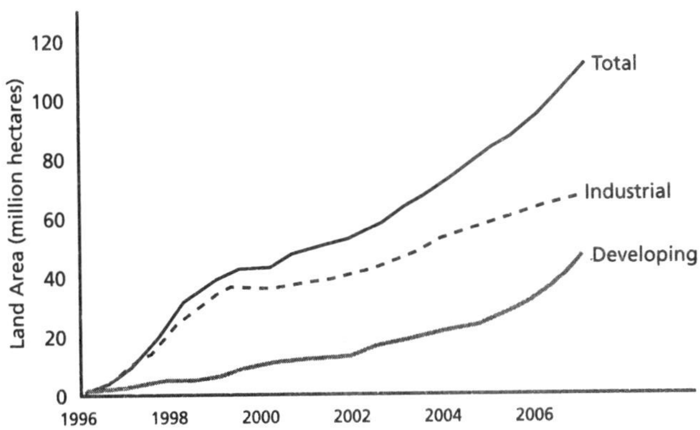

Source: James 2007.

Table 14.4 Land Area Planted with Genetically Modified Crops, 2007

	Million Hectares Planted with GM Varieties	GM Area as Percentage of Total Area in That Crop
Soybeans	58	64
Maize	36	24
Cotton	15	43
Canola	5	20

Source: James 2007.

of the land planted with soybeans worldwide is planted with GM varieties. It is estimated that 70 percent of the processed food sold in the United States contains GM products either directly or indirectly, as when GM crops are fed to animals (General Accounting Office 2008:78).

Adoption of genetically modified varieties is growing faster in developing countries than in developed countries. One explanation for this is found in a 2003 paper by Qaim and Zilberman. They showed that some genetically modified crops that reduce pest damage may be especially effective in boosting yields in developing countries, where farmers are unable to afford commercial pesticides. The study examined cotton yields in India and found that Bt cotton had yields 80 percent higher than non–genetically modified varieties. However, the advantages of Bt to the farmer evaporate if the price of the genetically modified seed is too high. In a study of Bt cotton adoption in Argentina, Qaim and de Janvry (2003) found that the cost of the seed was twice as high as the average expenditures on chemical pesticides by farmers who adopted non-GM cotton.

Concerns about GM crops, especially food crops, have been so strong that many countries, notably those of the European Union, have banned their use. The EU ban on GM imports has been challenged in the World Trade Organization (WTO) as a violation of the EU's commitments to free trade. During the 2002 famine in southern Africa, some countries felt pressure to refuse food aid from countries that permitted GM crops, or refused to distribute GM food to their starving citizens. What are the concerns that drive opposition to GM crops?

Human health concerns about GM foods focus especially on allergies. For example, one kind of GM soybean includes a genetic sequence derived from the brazil nut. Will a person who is allergic to brazil nuts have an allergic reaction if exposed to GM soybeans? How can a person who is assiduously avoiding foods with brazil nuts know that he or she must also avoid this other food containing soybeans? To date, no verified cases of allergic reaction to GM food have been found, but scientists do not fully understand the mechanisms of allergy, so the possibility of hidden allergic reaction cannot be eliminated (see Haslberger 2003 for a recent discussion). The concern about allergies erupted onto front-page headlines when the StarLink brand of GM maize, which had been approved by the Food and Drug Administration (FDA) for animal feed but not for human food, was mistakenly used to make tortilla chips for a fast-food chain. In response to the publicity and a multimillion-dollar recall effort, twenty-eight people reported that they had experienced possible allergic reactions to the products containing StarLink corn. However, further lab tests found that none of these people were allergic to StarLink (CDC 2002).

A second human health concern arose out of experiments with rats in the United Kingdom. Researchers found that, when fed a diet of GM potatoes, the rats developed weakened immune systems (Ewen and Pusztai 1999). The research was criticized for poor experimental design by the Royal Society, which found "no convincing evidence of adverse effects from GM potatoes" (1999:1). A further Japanese study on the effect of GM soybeans found no effect on the immune system (Teshima et al. 2000).

A third human health concern about GM foods is that they may promote the evolution of diseases that are immune to antibiotics. One common method of genetic engineering is to introduce a virus containing the foreign gene into the host organism. The virus is constructed to contain an antibiotic-resistant marker gene to allow scientists to select the transformed cells. Thus, according to some scientists, "The urgent question which needs to be addressed is the extent to which genetic engineering biotechnology, by facilitating horizontal gene transfer and recombination, is contributing to the resurgence of infectious, drug-resistant diseases" (Ho et al. 1998:36). Other research (Jackson et al. 2001) shows that it may be possible to genetically engineer viruses to overcome the natural immune responses. The concern here is that the introduction of human and animal genes into field crops may lead to production of insect viruses bearing these human genes; in turn, these new viruses could be able to overcome natural immune responses in humans. Other human health effects continue to be studied as well (see http://www.biotech-info.net/health_risks.html for up-to-date news on recent research).

Several environmental concerns about GM crops have been raised. The inclusion of the natural insecticide (Bt) gene raises the possibility that nontarget insects could be harmed by GM crops. One of the earliest studies to raise this possibility suggested that monarch butterflies could be exposed to Bt because they eat milkweed plants that grow near cornfields. However, a review of the findings by USDA's Agricultural Research Service (2003) found that "there is no significant risk to monarch butterflies from environmental exposure to Bt corn." However, other studies have shown possible impact of Bt crops on nontarget insect populations (for a review, see Wolfenbarger and Phifer 2000: tab. 2).

A second environmental concern is the possible development of a "superweed" that is resistant to insects and to commercial herbicides. If the genetic material that makes crops resistant to insects or to herbicides "escapes" into the genetic material of weeds, it may make the weeds difficult or impossible to eliminate (Snow 2002).

A third environmental concern is that GM organisms will upset the current ecological balance. For example, herbicide-tolerant crops lead to heavier application of herbicides and more complete elimination of weeds, but animals that feed on those weeds may be adversely affected. Also, genetic modifications can enhance an organism's ability to become an invasive species. GM crops may interbreed with wild relatives, ultimately leading to the extinction of the wild varieties (Wolfenbarger and Phifer 2000). More broadly, there is concern that GM organisms may lead to an erosion in species diversity (for example, see Schaal 2003).

All of these concerns suggest that there is a degree of uncertainty in the scientific community about the effects of biotechnology. The "precautionary principle" has been proposed to urge policymakers to oppose use of GM organisms until the scientific uncertainty can be resolved and the organisms can be proven safe. Of course, this raises the possibility that unscrupulous or ideological scientists could deliberately create or perpetuate uncertainty in order to achieve their own policy goals.

Not all environmental impacts of GM organisms are negative. Bt crops allow farmers to use less insecticide. Herbicide-resistant crops allow application rates and chemicals that may in the aggregate be less stressful on the environment. GM fish may reduce pressure to overfish wild stocks. The Council on Agricultural Science and Technology (CAST) convened a colloquium of agricultural scientists to discuss the overall environmental impact of GM organisms. They concluded that "biotechnology-derived [crops] . . . are consistent with improved environmental stewardship . . . [and] can provide solutions to environmental . . . problems" (CAST 2002:2).

Some opposition to GM organisms is based on political or ethical arguments rather than on scientific evaluations. Some doubt whether any social benefit can be derived from multinational corporations whose primary motivation is to increase their own profits. "The vast majority of scientific research being undertaken today is driven more by the goal of being first in line at the patent office than that of meeting profound social needs" (Dawkins 2003:39). A related argument is that GM crops put small subsistence farmers at a disadvantage and force farmers to deal with giant corporations on terms of unequal power. Other critics question whether laws should permit a genetic sequence to be "owned," patented, and sold. Finally, some are uneasy about biotechnology for religious reasons, questioning whether the scientists are "playing God" with their experiments.

Balanced against the health, environmental, and other concerns about biotechnology are the benefits of increased yields. A panel of experts reporting to the World Bank and the Consultative Group for International Agricultural Research in 1997 concluded that GM crops could boost world crop yields by 25 percent (CGIAR 1997). Nigeria's minister of agriculture, writing on the op-ed page of the *Washington Post,* stated: "We do not want to be denied this [GM] technology because of a misguided notion that we don't understand the dangers or the future consequences. . . . The harsh reality is that, without the help of agricultural biotechnology, many will not live" (Adamu 2000: A23).

Postproduction Food Losses

Before leaving the subject of food production, we should recognize that food consumption theoretically can be increased *without* increasing food production—if we can reduce losses between the field and the consumer (FAO 1996e: papers nos. 4, 8). These losses are estimated to be as high as 30 percent (Erlich and Erlich 1991). In developing countries, postharvest food losses have been attributed to pests (Angé 1993), poor facilities for storage and transportation (James and Schofield 1990), and on-farm handling (FAO 1996e: paper no. 8). One might think that postharvest food losses would decline as countries develop economically, because of improvements in roads, credit, and information. However, a study in the United States (Kantor et al. 1997) estimates that over 25 percent of food is lost in retailing, restaurants, and at-home consumption. So in rich countries there may be less food lost to pests and poor storage facilities, but more food is wasted by consumers or thrown away by restaurants.

15

The Interaction Between Undernutrition and Health

This book is about food supply and demand, and undernutrition caused by insufficient food consumption. But before turning to policy issues, we need to recognize the ways that nutrition and health interact.

The Synergisms Between Nutrition and Health

A healthy person has a good appetite, likely has a good diet, digests their food well, and makes efficient use of it in their body. A well-nourished person can keep their immune system functioning at a high level and is likely to be healthy. In contrast, a sick person is likely to lose their appetite, have a poor diet, digest their food poorly, and use some of their nutrients to fight infection. A poorly nourished person suffers a weakened immune system and is susceptible to infections.

We have already discussed deaths caused by undernutrition. Typically, these deaths occur because a person weakened by undernutrition catches an infectious disease, and the weakened body cannot fight off the disease, so the person dies. A well-nourished person might get the same disease and survive it. An undernourished person who avoided the disease would likewise survive. Death is caused by the combination of undernutrition and disease. The impacts of undernutrition can be reduced by improving the general health of the population.

The interrelationship is so important that the definitive review of the literature concludes: "Where both malnutrition and exposure to infection are serious, as they are in most tropical and developing countries, successful control of these conditions depends upon efforts directed equally against both" (Scrimshaw, Taylor, and Gordon 1968:267).

Infection Exacerbates Malnutrition

Infection increases the potential for and severity of malnutrition. Most common infections have a heavy impact on nutritional status in three important ways: (1) through loss of appetite or intolerance for food (e.g., vomiting); (2) through cultural factors (e.g., relatives of the sick individual substitute less nutritious diets for the regular diet and administer purgatives, antibiotics, or other medicines that reduce absorption of specific nutrients); and (3) through loss of body nitrogen (protein).

This last pathway to malnutrition through infection (loss of nitrogen) is complex enough that it deserves separate discussion. What happens is that protein tissue in the body is used up to fight the infection. To manufacture such disease-fighting materials as interferon, white blood corpuscles, and mucus, the body needs amino acids, which it acquires in part by breaking down previously existing protein—chiefly from the muscles. This borrowing of muscle tissue for fighting infection is one of the reasons a person feels so weak following a serious illness. It might seem reasonable to try to keep up the body's supply of protein during an illness through eating, but this is usually impracticable. Sick people often have little appetite. During convalescence, with an appropriate diet, the lost body protein is usually replaced.

Infection Promotes Dietary Deficiency

A reasonably healthy person who is presently on the borderline of nutritional deficiency may not show clinical signs of nutritional difficulties. But, owing to the above problems associated with infection, an illness can increase a person's nutritional deficiency, and that person can then develop any of a number of conditions caused by micronutrient deficiencies (Scrimshaw, Taylor, and Gordon 1968:265):

- Keratomalacia (a softening and ulceration of the eye's cornea), caused by a shortage of vitamin A. If the shortage continues long enough and is severe enough, xerophthalmia (a dry, thickened, lusterless condition of the eyeball resulting in blindness) may ensue.
- Scurvy (spongy gums, loosening of the teeth, and a bleeding into the skin and mucous membranes), caused by lack of ascorbic acid (vitamin C).
- Beri-beri (inflammatory or degenerative changes of the nerves, digestive system, and heart), caused by lack of thiamin (vitamin B_1).
- Pellagra (a condition marked by dermatitis, gastrointestinal disorders, and disorders of the central nervous system), resulting from insufficient niacin (one of the B-vitamins).
- Macrocytic anemia (anemia associated with exceptionally large red blood cells), caused by a deficiency of vitamin B_{12} or folic acid (one of the B-vitamins).
- Microcytic anemia (anemia associated with exceptionally small red blood cells), caused by a shortage of iron.

Also, cataracts are significantly less likely to develop in elderly people who take vitamin supplements (or beta-carotene, or riboflavin and niacin). There is also evidence that undernutrition can increase the possibility that a person who is HIV-positive will develop full-blown AIDS. A study that followed HIV-positive men for seven years found that those who consumed three to four times the RDA of niacin and vitamin A had a 40–50 percent lower chance of developing AIDS (Tang et al. 1993).

Diarrhea and Nutrition

The most common instance of an illness seriously affecting nutritional status is that of undernutrition induced or exacerbated by a gastrointestinal infection (gastroenteritis) that causes diarrhea or, in its more extreme form, dysentery. An outstanding feature of kwashiorkor, for instance, is the frequency with which it is precipitated by an attack of acute diarrheal disease (Scrimshaw, Taylor, and Gordon 1968:27).

For the world as a whole, diarrhea is not as ubiquitous as the common cold, but in many third world localities it is nearly as frequent. Diarrhea particularly affects children under 5 years old, and childhood fecal matter is a main source of the infective material. Food and water are key transmission routes. Children often make their first contact with diarrheal disease organisms through weaning foods (Martorell, personal communication). In fact, an outstanding feature of diarrheal disease in the third world is the concentration of cases among children during and immediately after weaning (Durand and Pigney 1963), illustrated in Figure 15.1.

In poor countries, the onset of diarrhea at weaning time (so common that it has sometimes been called by the special name of *weanling diarrhea*) is typically acute and rapidly progressive, with liquid or semi-liquid stools, varying from three to as many as twenty a day. About one-fourth of patients have blood or mucus, and frequently pus, in the stools. Fever may be absent, but low-grade fever is usual, along with malaise, toxemia (buildup of toxic substances in the blood), intestinal cramps, and tenesmus (a distressing but ineffectual urge to evacuate the rectum or bladder). The usual clinical course runs four to five days. Repeated episodes can result in a month or more total time spent fighting diarrhea during a year's time. De Zoysa and colleagues (1985:11) report that the average child in Africa suffers from five bouts of diarrhea a year, with each bout lasting five days. By comparison, the average child in the United States suffers one such bout. In malnourished children, a low-grade indisposition often continues for a month or more, sometimes as long as three months, with irregularly recurring loose stools, a progressively depleted nutritional state, and occasional recurrent acute episodes (Scrimshaw, Taylor, and Gordon 1968:220).

Figure 15.1 Third World, Age-Specific Diarrheal Morbidity Rates

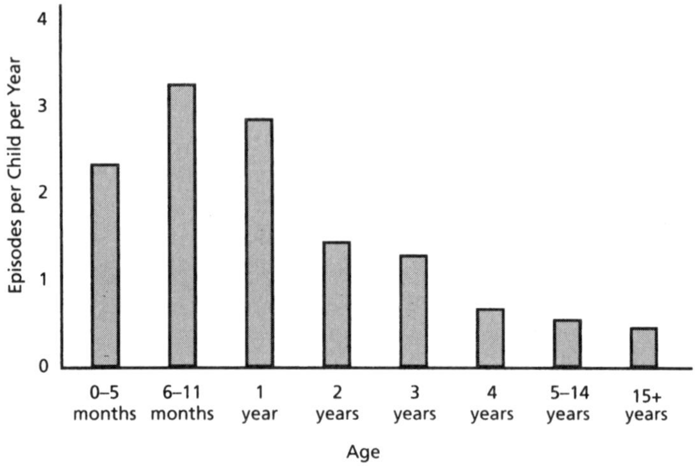

Source: Adapted from de Zoysa et al. 1985:8 using data from Snyder and Merson 1982.

Although there are some twenty-five different organisms (bacteria, viruses, and parasites) that can cause diarrhea, all cases result in a shortage of water and salts (electrolytes) in the body. This dehydration of the body can be the most serious consequence of diarrhea. By the time a weanling child is seriously dehydrated from diarrhea, it is lethargic; its eyes are dulled and when it cries there are no tears; its skin is wrinkled like an old man's; it stops urinating; the fontanel (soft spot at the top of an infant's skull) is sunken. If the child's skin is pinched, it only slowly returns to the normal conformation (Goodall 1984). At best, this dehydration stands in the way of a quick recovery from the diarrhea. At worst (if the child loses more than 15 percent of his or her body fluids) it is fatal. A baby may well die within twenty-four hours of the arrival of these signs of serious dehydration. Some 60 to 70 percent of the 5 million annual diarrheal deaths are caused by this associated dehydration (WHO 1985c:6).

Malnutrition Exacerbates Infection

Malnutrition often amplifies the impact of infection. An example from the Philippines is illustrative. Severely undernourished children admitted to a hospital for acute respiratory infection are found to be thirteen times as likely to die from the disease as children whose nutrition is normal (see Figure 15.2).

Malnutrition is almost always synergistic with intestinal diseases caused by worms or protozoa and with any disease caused by bacteria (Scrimshaw,

Figure 15.2 Acute Respiratory Infection Mortality by Nutritional Status, Philippine Hospital Cases

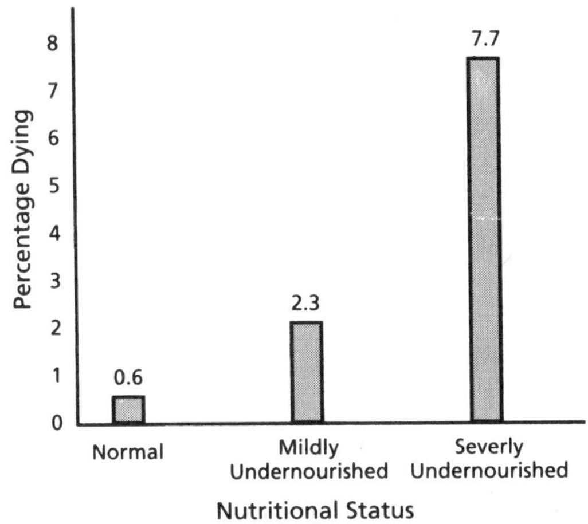

Source: Adapted from Galway et al. 1987:23 using data from Tupasi 1985.

Taylor, and Gordon 1968:263–264). That is, malnutrition aggravates the course of the disease, and the disease, in turn, intensifies the malnutrition.

A wide variety of nutrients have been demonstrated to have an impact on the competency of the body's immune system (Gershwin et al. 1985:2; Phillips and Baetz 1980). The impact of nutrition on infection begins at birth and lasts throughout life. Not only is breast milk loaded with the appropriate nutrients for an infant's diet, but it carries with it a load of substances that help protect the infant against disease: immunoglobulins, macrophages, lymphocytes, neutrophils, components of the complement system, and so on (Rivera and Martorell 1988). (The complement system involves a set of more than eleven proteins normally found in the bloodstream. These proteins act in conjunction with the blood's antibody system in fighting infection, complementing the work of the antibodies.) Undernutrition increases the duration of infections, especially diarrhea. Unequivocal evidence shows that the immune response is reduced in severe undernutrition, and some evidence suggests a diminished immune response in moderate undernutrition, particularly in wasted children (Rivera and Martorell 1988).

Worldwide, malnutrition is the most common cause of deficiencies in the immune system, even among adults. Two examples help to illustrate this point: (1) nutritional supplements given to the elderly have been found to improve their response to the influenza virus vaccine; and (2) among individuals given a vaccine to protect them from tuberculosis, a positive correlation between better nutrition and resistance to the disease has been observed (Chandra 1988). Malnutrition interferes with various bodily mechanisms that attempt to block the multiplication or progress of infectious agents. The list of ways it can do this is long, but significant among them are a decrease in the response and activity of white blood corpuscles; a reduction in the production of interferon; a decrease in the integrity of the skin, the mucous membrane, and other tissues that serve to bar the entrance of infection; and interference with normal tissue replacement and repair (Scrimshaw, Taylor, and Gordon 1968:263–264).

The exploration of the mechanisms by which undernutrition affects the immune system has only just begun, but already we have enough information to provide interesting clues as to what is happening. For instance, the mucous membrane not only provides a physical barrier against the entrance of foreign particles that might cause infection (bacteria, viruses), but provides a chemical barrier as well. Mucus contains a variety of biochemical and immunological disease fighters, one of which is an enzyme called *lysozyme,* which has the capacity to attack the cell walls of invading bacteria. Colombian children suffering from protein-energy malnutrition were found to be producing reduced levels of lysozyme. In a process called cell-mediated immunity, T-lymphocytes play a key role in attacking disease-causing microbes. Children with protein-energy malnutrition are also likely to suffer an atrophied thymus, the organ primarily responsible for the "education" and proliferation of T-lymphocytes, and at the same time are more likely to produce fewer of these lymphocytes than expected when their bodies are challenged with invading disease organisms (Sherman 1986).

A final point we must remember: not only does malnutrition reduce resistance to infection, but it also decreases stamina, which in turn decreases the capacity to cope with life and to perform on the job, making it more difficult to earn money to pay for transportation to healthcare centers, to pay for the services themselves, and to pay for appropriate drugs to combat infection.

Policies to Improve Health

In this section we review policies that can improve the general health of people, especially focusing on developing countries.

Policies to Combat the Source of Disease

For some diseases and conditions, it may be possible to attack the problem at its source. For example, rich countries put a substantial effort into identifying and eradicating Bovine Spongiform Encephalopathy (BSE, or "mad cow" disease), which can cause serious disease in humans who consume the brains of infected cows. Efforts at universal inoculation have resulted in worldwide eradication of smallpox, and raised hopes for worldwide eradication of measles, polio, rubella, and other diseases.

Policies to Impede the Spread of Disease

The spread of infectious diseases can be reduced by interfering with the mechanisms or vectors by which disease moves from one person to another. For example, use of treated malaria nets keeps malaria-infected mosquitoes from biting people as they sleep (Hoffman [2009] investigated whether families who were given mosquito nets used those nets differently than did families who purchased the nets). Use of condoms interferes with the transmission of HIV—the virus that causes AIDS. Moses and colleagues (1991) studied a condom promotion program for Kenyan prostitutes that reduced the spread of HIV at a cost of $8–12 per case prevented. Where a disease is limited geographically, quarantine may reduce the spread of the disease to the general population. Immunization and sanitation measures are discussed below.

Policies to Promote Immunization Against Disease

Vaccination or inoculation exposes an individual to a small dose of a disease-causing agent in order to induce the development of disease-fighting mechanisms inside the person's body. Thus, when the individual is exposed to the disease-causing agent in the real world, he or she either does not contract the disease, or contracts only a mild episode. As noted above, in the case of smallpox, immunization has been so successful that the organism that causes the disease has been completely eradicated (outside the laboratory).

In 1974, less than 5 percent of children in the developing world were immunized against six common vaccine-preventable childhood killers—measles, tetanus, diphtheria, pertussis, tuberculosis, and polio. Today, that number is more than 80 percent. This enormous increase is due in large part to government policies—in particular the World Health Organization's expanded immunization program. UNICEF estimates that over 2 million lives per year are saved by immunization (UNICEF n.d.). However, the work is not finished. About 36 million infants worldwide are not immunized each year (WHO 2003a).

Measles-associated diarrhea is more severe and is more likely to lead to death than other diarrheas. One estimate suggests that up to a quarter of diarrheal deaths among preschoolers could be prevented by an effective measles immunization program (Feacham and Koblinsky 1983).

The World Bank, in its 1993 *World Development Report,* urged public health officials to promote immunization with a new vaccine—for hepatitis B and yellow fever—and recommended that this be combined with supplements of vitamin A and iodine, stating that a combination "would have the highest cost-effectiveness of any health measure available today."

Policies to Promote Clean Drinking Water

In developed countries, governments have been involved in developing and maintaining public systems of clean water for nearly 200 years. Modern engineering principles were applied to water-borne sewage disposal systems in the West during the 1840s. By the early 1900s it was common for local governments in the Western world to consider the supply of drinking water to be properly within their purview. The World Health Organization (WHO 2003b) estimates that 1.1 billion people do not have access to improved water supplies. Improving the local water supply in rural areas can involve something as simple and effective as technical assistance and encouragement for the construction and use of rainwater-gathering vats. Use of such vats provides not only a clean but also a convenient source of household water. When a mother does not have to walk as far to get her household supply of water, she has more time available for other activities, including childcare.

Providing an ample supply of clean water in a third world city can be a political as well as an engineering problem. Port-au-Prince, Haiti, provides an extreme example. During 1976, a shortage of water outlets promoted a substantial and profitable private market for what was, ostensibly, a publicly provided city service (see Box 15.1). The issue of whether water should be provided by public utilities or by private companies continues to be a subject of heated debate (World Bank 2001).

Policies to Require Food Fortification

Micronutrient deficiencies can be reduced or eliminated by government programs that require commonly purchased foods to be fortified with the missing micronutrients. In the twentieth century, public programs promoting the iodization of salt and the fortification of flour with vitamins became routine in the West. (The history of such programs in the United States is outlined in Box 15.2.)

Direct Treatment of Diseases and Symptoms

Preventative efforts notwithstanding, people still get sick, and turn to the health system for treatment. One notable example of a policy effort to treat diarrhea directly is the effort to subsidize the use of oral rehydration therapy to treat the impact of extreme diarrhea. For generations, it was thought that the only way to replace electrolytes lost because of diarrhea was through intravenous injection. Oral replacement using the salts alone simply did not work. In the late 1960s, researchers in India and what is now Bangladesh found that merely adding common cane sugar to the missing salts produced a formula that worked by mouth. This simple technology—oral rehydration therapy—was first used

Box 15.1 The Political Economy of Drinking Water
Simon M. Fass

An in-depth examination of one small segment of the public service sector in Haiti, specifically water distribution management in Port-au-Prince, highlights the severe consequences which deficient administration can bring to bear upon a relatively large number of people.

In Port-au-Prince, in 1976, about 50 percent of the water input leaked out of the municipal water system. For the most part the loss was due to breaks and leaks in pipes, but much of it occurred because connections to reservoirs in homes and establishments were not equipped with automatic shut-off valves, or sometimes any valves at all. When such reservoirs, including swimming pools, were full, the overflow spilled into the streets. Since there were no metering devices or enforced penalties for not having valves, and the water tariff was on a flat-rate basis, subscribers had little incentive to invest in appropriate valves. The United Nations estimated that control mechanisms at private connections could have reduced losses to a more reasonable rate of 30 to 35 percent.

Individuals installed reservoirs because of irregular distribution and pressure of water flows. Heavy demand fluctuations, variations in rainfall, limited public storage capacity, and high distribution losses caused the irregularities. Subscribers had to be prepared for periods of several days, sometimes weeks, between deliveries through the pipes.

The 30,000 legal and clandestine private connections in 1976 provided direct service to only about 150,000 residents. The remaining 490,000 residents were in theory serviced by the 36 officially designated standpipes [public water outlets employing multiple spigots] that were supposed to exist at the time. In reality only 27 standpipes functioned. The others had long since been destroyed. The total outflow from the operating standpipes on any given day did not exceed 1.4 million liters. Thus under the best of circumstances their total supply could not provide more than 2.8 liters per person each day for the 490,000 residents presumably dependent on them. To put this in perspective, a single flush of a modern toilet facility requires 19 liters.

The extreme scarcity caused by the municipal distribution system led the city's population to adapt in a number of ways. There were a number, something on the order of 40,000, who relied on leaks and breaks in the pipes. It was common to knock a hole in a pipe, if necessary digging into the street to find one, plugging it with a wooden spike when not in use, and attaching a short rubber hose from the hole to a bucket when drawing water from it. Since penalties for this practice could be quite severe, this method of obtaining water was not widespread.

A more common practice, providing water to some 95,000 residents, was sharing among neighbors. In a number of areas, high, middle, and low-income homes are located side by side. In such neighborhoods, the proportion of families with connections tend to be relatively high, and so the number of individuals requesting water from a particular subscriber on any given day, usually in the form of a request to a household servant, is low, typically less than 10 or 12.

The majority of nonsubscribers, however, some 300,000 low income residents who lived in downtown areas with relatively few private connections, were

(continues)

Box 15.1 continued

obliged to buy water from fixed and mobile vendors. The shortage of water outlets had in effect given rise to a rather substantial private market for a publicly provided service.

The Private Water Market, 1976. The private water market in 1976 contained three principal sets of vendors. The first set were tanker truck operators who drew water from the fire hydrants and transported it to industrial and commercial establishments and to about 1,200 higher income homes located in areas without piped service or with very irregular service. Charges for truck supply varied between US\$3.00 and US\$6.00 per m^3. The gross revenues of truckers, who paid nothing for water, whether in the form of user charges or in the form of license fees, amounted from these consumers alone to an average of \$2,700 per day, \$980,000 per year, and an annual return of \$39,000 per truck.

There were substantial profits to be made, and often truckers were known to break pipes in order to create demand for transported water over the extended periods required to locate the breaks and repair them.

The second set of vendors consisted of some 2,000 households which had connections to the system and sold water in lower-income areas to neighboring consumers and/or to mobile sellers who would transport it further afield. The common method was to sell water by the bucketful (about 18 liters per bucket).

The minimum price, in effect during the rainy season if water was flowing in the pipes, was two cents a bucket, or about \$1.10 per m^3. In the dry season the typical price would be ten cents, equivalent to \$5.60 per m^3. During drought periods, as happened in 1975 and 1977, unit prices could reach anywhere from \$10.00 to \$20.00 per m^3 for several months at a stretch.

The third set of vendors were mobile vendors who bought water from connected households and transported it to consumers. They numbered 14,000 or about 4.5 percent of the urban labor force.

The margin charged by the vendors in ordinary circumstances was one cent per bucket, or two if the transport distance was long. These were margins on top of whatever the vendors themselves paid to connected families. For 1976, the aggregate net earnings of mobile vendors came to \$930,000, or approximately \$5.50 a month for each vendor.

Total expenditures by consumers in the private water market thus amounted to about \$3.8 million a year, a quarter being paid largely by a very small group of high-income residents and the balance by some 300,000 low-income families. By contrast, [the municipal water authority's] total annual revenue from the sale of water was \$650,000 during the same period.

Impacts on Low-Income Families. At a price of \$2.30/m^3 a typical family of five would have to spend about \$4.00 a month in order to consume 11 liters of water per day. In 1976 about 40 percent of urban families had incomes of \$20 per month or less.

Given all the various daily demands placed on the use of money, many of these families found it impossible to spend a fifth of their income on water. They responded in a number of ways.

(continues)

Box 15.1 continued

The poorest of them, with incomes of less than $10 per month, used purchased water only for cooking and drinking. They used surface runoff for cleaning themselves. They might also launder and wash less often. A major hazard was that of illness caused by contaminated water or by residence in areas with disastrous sanitary conditions, susceptibility to which was aggravated by the extra energy expended by already malnourished bodies to trudge 20-kilogram buckets of water several kilometers each day. With the risk of illness came the possibility of seriously compromising the capacity to generate income streams. Curative medical services would require such families to curtail other expenditures further, to dig into savings, or to incur heavy debts.

Source: Extracted from Fass 1982.

Box 15.2 A Brief History of Food Fortification in the United States
Richard Ahrens

The leading cause of draft deferment in the United States during World War I was the swelling of the thyroid gland called goiter. . . . [B]oys who had the condition could not fit into the tight collars of the military uniforms. In 1923 the Harding Commission, appointed by President Warren Harding, recommended a voluntary program for the iodization of salt to combat goiter. A gentlemen's agreement was worked out between the salt companies and the executive branch of the government that there would be no price difference charged between the iodized and uniodized product.

During Word War II a bill was introduced into Congress that would have made it mandatory that all table salt be iodized, but the bill was defeated in committee when a number of medical doctors testified that there were probably some people in the United States who were sensitive to iodine and who would have skin problems as a result of being unable to obtain iodine-free salt.

Fortification of flour arose at the start of World War II, after President Franklin Roosevelt asked the National Academy of Sciences (NAS) to evaluate the nation's readiness for war. As one of their recommendations, the NAS came up with a proposal to ask flour millers to fortify wheat flour with iron, thiamin, riboflavin, and niacin. The flour fortification program became policy in 1940 . . . and now it is the province of the US Food and Drug Administration.

Source: Richard Ahrens is professor emeritus of nutrition at the University of Maryland.

to fight cholera (the most virulent form of dysentery) in an epidemic in India in 1971. Since then it has become a third world public health mainstay and has been vigorously promoted by both the UN Children's Fund and the World Health Organization. UNICEF (1987:8) has called oral rehydration therapy the cheapest and most effective health intervention that can be implemented in the home to decrease childhood mortality. UNICEF estimates that 1 million deaths per year are prevented by this intervention.

Direct "Medical" Interventions to Reduce Undernutrition

Perhaps the most direct "health policies" that address malnutrition are programs deliberately designed to treat patients who are suffering from (or who have a high probability of suffering from) a particular kind of malnutrition. For example, an interesting controlled experiment was conducted in Ache province in northern Sumatra, where vitamin A deficiency may well be the most severe in the world. During a one-year period, some preschoolers in Ache were given one capsule containing 200,000 international units of vitamin A every six months. Others were given no vitamin A supplement. The vitamin A supplement was shown to reduce dramatically both the risk of xerophthalmia and the death rate (Sommer et al. 1986; Gopalan 1986). Vitamin A pills such as those used in this experiment can be manufactured for less than 5 cents each. The cost of distribution far exceeds the cost of the pills. The presence of an ongoing maternal and child healthcare center permits the cost of a vitamin A supplementation program to be shared among the costs of other health delivery programs.

Recently, a similar "prescription" has been introduced to treat calorie deficiency. The group Doctors Without Borders (2008a) has developed a high-calorie food supplement (Ready-to-Use Food [RUF], sometimes called Ready-to-Use Therapeutic Food, or Plumpynut). This is a paste made from peanuts and milk that comes in airtight foil packets. The packets can be prescribed to individuals who are suffering from acute calorie deficiency. Some research about the use of these RUF supplements in developing countries was delivered at a recent symposium (Doctors Without Borders 2008b).

One glimpse into how calorie malnutrition is dealt with in developed countries can be found in a pamphlet from the United Kingdom's National Health Service (2006). This pamphlet identifies five underlying causes of malnutrition among the elderly—loss of appetite, difficulty eating, inability to absorb food, other disease, social situation—and recommends actions to deal with each of the underlying causes, including (in some circumstances) the prescription of a high-calorie food supplement.

General Policies to Improve Medical Care

The final issue we consider here is policies to improve general medical care. Questions arise, such as: Should healthcare professionals work for the government, or as private entrepreneurs? Should government promote the building and use of small community clinics, or large hospitals?

Over the years, the view became widespread that, generally, it is cheaper to maintain good health than it is to make people well after they get sick. This idea is coloring many of the current healthcare developments in the third world, which are emphasizing low-cost delivery of health services to the poor. We see "barefoot doctors" in China, "nutrition huts" in the Philippines, and "health huts" in Haiti. Most common third world illnesses can be successfully treated in the field by paramedical workers using simple equipment and a limited range of medicines. Thus a strong argument develops for placing increased emphasis on preventive medicine carried out by lower-level technicians in clinics close to people's homes, contrasted with curative medicine carried out by highly trained physicians surrounded by a hierarchy of staff in expensive urban hospitals.

In a review of the literature, Filmer, Hammer, and Pritchett found that there was little empirical evidence to back up the prejudice in favor of publicly financed low-cost "primary" healthcare delivery:

> A combination of experiences has led to a strong consensus among public health specialists who focus on developing countries. They argue that the existing allocation of health expenditures toward curative care in secondary and tertiary facilities, such as hospitals and clinics to which patients are referred, is inappropriate and that a reorientation of government efforts toward primary health care would bring both health gains and cost savings. . . . Although the images and statistics that motivate primary health care appear compelling, the gains have rarely been demonstrated in practice. . . . Has public spending on health and, more particularly, on primary health care promoted good health? If so, there should be empirical regularities at both the national and local levels. First, given a level of total health expenditures, more spending on primary healthcare activities and greater access to primary healthcare services should be associated with lower aggregate mortality. Second, at the local level (household, village) greater access to primary healthcare facilities should reduce mortality. Third, projects that develop primary care facilities should reduce mortality. None of these regularities finds much support in the data. (2000:199–201)

Rather, they found that almost all of the observed differences in healthcare outcomes can be explained by socioeconomic variables, especially income per capita. In other words, people in higher per capita–income settings have better health outcomes than people in lower per capita–income settings, regardless of the level of government spending, or the level of access to primary healthcare. Descriptions of how healthcare is provided in developing countries are found in Box 15.3.

Development programs that lean toward emphasizing human capital development (which would include primary healthcare and public health programs) not only serve to improve people's productivity, especially at the bottom end of the income distribution, but may also improve health outcomes by raising per capita income. Policies to promote economic growth will be discussed in Chapter 17.

Box 15.3 Healthcare Quality in Developing Countries

Ken Leonard and his colleagues report on their experiences observing how healthcare is delivered in developing countries. Their stories paint a picture of healthcare systems that are very different from the healthcare system of the United States. Below are two stories from their report (Das, Hammer, and Leonard 2008: 94).

"Dr. SM and his wife are the most popular medical care providers in the neighborhood, with more than 200 patients every day. The doctor spends an average of 3.5 minutes with each patient, asks 3.2 questions, and performs an average of 2.5 examinations. Following the diagnosis, the doctor takes two or three different pills, crushes them using a mortar and pestle, and makes small paper packets from the resulting powder, which he gives to Ms. Sundar [the patient] and asks her to take for two or three days. These medicines usually include one antibiotic and one analgesic and anti-inflammatory drug."

"In rural Tanzania, Ms. M brings her nine-month-old to the local health clinic, carrying the child on her back. When she enters, Dr. K (an Assistant Medical Officer with O-level education and four years of medical training) asks her what the problem is. Still standing in front of his desk, she replies that her daughter has a fever. Dr. K fills a prescription for malaria based on this statement, even though he cannot see the child, much less observe her condition. The consultation and medicine are both free and Ms. M leaves the facility with the prescribed medicine. During the exit interview, a nurse on our team notes that the child is suffering from severe pneumonia. The health facility has the medicine to treat both malaria and pneumonia. Dr. K is trained in the diagnosis and treatment for these diseases and saw only 25 patients that day. Yet, but for the intervention of the nurse on our research team, the child would have died."

Maternal and Newborn Childcare

Healthcare delivery can be particularly effective in reducing impacts of malnutrition when it is aimed at women in the months before and after childbirth. A number of studies show that low-birth-weight babies suffer more health problems as adults than do normal-birth-weight babies. These problems include high blood pressure, too much cholesterol and sugar in the blood, cardiovascular disease, and diabetes. So better nutritional information and healthcare for pregnant women can improve the health outcomes of their unborn children.

The health and nutritional advantages of breast-feeding were described in Chapter 5. The vitamin A distribution program described previously in the present chapter depended on an existing maternal and infant health program to make distribution of the supplement cost-effective. Maternal and infant health clinics also improve infant health by promoting breast-feeding. Not only do breast-fed infants have a lower morbidity rate from diarrhea, but breast-feeding also protects against death from diarrhea. Infants who receive no breast milk are about twenty-five times more likely to die of diarrhea than those who are exclusively breast-fed (Feacham and Koblinsky 1984).

The alternative to breast-feeding—feeding of infant formula—is problematic. Commercial infant bottle-feeding formula can seem exorbitantly expensive; this can lead mothers to overdilute the mixture with water or even to mix it with white flour or sugar, which can in turn lead to marasmus (emaciation). Preparation of formula requires mixing the powder with water, and this water may be unsanitary, exacerbating diarrhea in the infant (see Latham 1984).

Even in relatively well-off third world households, knowledge regarding nutrition may be lacking, such that substituting the bottle for the breast can still lead to marasmus (see Box 15.4).

Box 15.4 Marasmus in a Newly Rich Urbanized Society

Peter Pellett

I was long of the opinion that infantile marasmus would be essentially eliminated when social and political change were accomplished such that abject poverty no longer existed. However, . . . experience in Libya, a rich but still developing nation, has caused me to reconsider somewhat this view.

The Libyan Arab Republic was formed in 1969 by a coup d'état. . . . During the eight years between then and 1977, real incomes for ordinary workers in Libya increased fourfold. In 1977 the major food items (flour, rice, tomato paste, meat, olive oil, coffee, tea, and sugar) were subsidized by the government, and baby foods were tax-free. Both gross poverty and inadequate housing were largely eliminated and phenomenal social progress was accomplished.

Despite this, in 1977 infantile marasmus in Libya remained a widespread problem. As elsewhere in the developing world, breast-feeding had declined.

In a study in Tripoli, the capital city of Libya, we compared the family backgrounds of 50 marasmic infants with the backgrounds of 50 essentially healthy infants of similar age. Total income was similar in both sets of families, and major consumer items such as TV sets, cars, and refrigerators were widely present in both groups. However, families with marasmic infants had less-literate mothers who tended to breast-feed for shorter periods and to feed purchased pureed baby foods more frequently. We concluded that the causal factor for marasmus in most of these instances was probably unhygienic infant feeding, despite the availability of clean water and modern kitchen facilities.

Source: Extracted from Pellett 1977:53–56. For more information on the topic, see Mamarbachi et al. 1980.

Part 3

Policy Approaches to Undernutrition

IN PART 2, WE LOOKED at the causes of undernutrition—vehicles by which undernutrition is delivered to families—and identified economic, demographic, agricultural, environmental, and health factors. The central focus of Part 3 is to explore public policy alternatives of interest to nutrition planners.

16

Philosophical Approaches to Undernutrition

In this chapter we explore philosophical approaches to reducing worldwide hunger. What motivates governments, societies, individuals, or groups of individuals to concern themselves with the issue of undernutrition? How does the motivation influence the policy decisions?

From the Standpoint of a Moral Philosopher

Charity or Concern for the Poor and Hungry?

Our first chapter opened with a reference to starving Ethiopian babies—babies with bloated bellies, spindly arms and legs, and bodies too weak to sit up. The device is a standard technique for grabbing the attention of people attuned to Western culture and making them stop to think about the world food problem.

Those who live in the Western world are exposed to repeated appeals to conscience, asking them to join the battle to end hunger. In 1980, the Presidential Commission on World Hunger urged that the United States "make the elimination of hunger the primary focus of its relations with the developing world." Commenting on this in a paper written for a religious audience, McLaughlin stated that "the moral and humanitarian reasons for such a policy seem self-evident" (1984:3).

A common argument in favor of studying and solving the world hunger problem is the moral dictate that each of us should help individuals who are less fortunate than ourselves. The pope's statement to the World Food Summit, for example, contains the following admonition:

> In the analyses which have accompanied the preparatory work for your meeting, it is recalled that more than 800 million people still suffer from malnutrition and that it is often difficult to find immediate solutions for improving these tragic situations. Nevertheless, we must seek them together so that we will no longer have, side by side, the starving and the wealthy, the very poor and the very rich, those who lack the necessary means and others who lavishly

> waste them. Such contrasts between poverty and wealth are intolerable for humanity.
>
> It is the task of nations, their leaders, their economic powers and all people of goodwill to seek every opportunity for a more equitable sharing of resources, which are not lacking, and of consumer goods; by this sharing, all will express their sense of brotherhood. It requires "firm and persevering determination to commit oneself to the common good; that is to say, to the good of all and of each individual, because we are all really responsible for all" (Sollicitudo rei socialis, no. 38). This spirit calls for a change of attitude and habits with regard to life-styles and the relationship between resources and goods, as well as for an increased awareness of one's neighbour and his legitimate needs. (Pope John Paul II 1996)

Philosopher Peter Singer approaches the issue from a distinctly nonreligious point of view. He concludes: "I begin with the assumption that suffering from lack of food, shelter, and medical care are bad. . . . My next point is this: if it is within our power to prevent something bad from happening, without thereby sacrificing anything of comparable moral importance, we ought, morally, to do it" (1972).

Garrett Hardin (1974) reached a completely opposite conclusion. To Hardin, the most ethical action was staunchly to *refuse* to help the poorest of the poor. His reasoning was this: our planet is like a lifeboat with a limited capacity; the altruistic impulse to save people will overcrowd that lifeboat and thereby doom everyone, even those who could have been saved. Of course, the underlying logic here requires that the lifeboat actually be full, and the experience since 1974, when Hardin put forth his analogy, contradicts that assumption—the world's population has increased more than 60 percent and the number of undernourished people has declined.

The arguments presented in the past few paragraphs describe the issue of helping the world's undernourished as a matter of religious conviction or personal ethics. The motivation to help comes from within the individual. I may believe that I "owe" compassion to the hungry, but that debt is a product of my beliefs, something generated from within myself. In that regard, it is quite different from the debt I owe to the government as taxes.

Food as a Right

This distinction (between being motivated by personal ethics and being motivated by an obligation to society) is important as we consider a second type of moral argument about why we should be interested in the problem of world hunger: the issue of "food as a right." Some eighty-five countries have endorsed the International Covenant on Economic, Social, and Cultural Rights (adopted by the United Nations General Assembly in 1966), which defined and formalized the right to food as a basic human right. The right to food was widely discussed in preparation for and during the World Food Summit of 1996 (Pinstrup-Andersen, Nygaard, and Ratta 1995; Alston 1997).

If food is a right, then hunger is a violation of that right, and we have a second ethical motivation to be concerned about hunger—the moral requirement that we seek justice and oppose violations of rights. There is a difference between the "charity" motivation (we have an ethical obligation to help the hungry) and the "justice" motivation (food is a right). That difference is illustrated by the following:

- If you accept the view that there is a fundamental right to food, then you are motivated to address the hunger problem even if you do not believe that you have a moral duty to help the poor and hungry. Your motivation here is simply to ensure the protection of that fundamental right.
- If you believe that you have a moral duty to help the poor, then you are motivated to address the hunger problem even if you do not believe that there is a fundamental right to food. Your motivation here is your duty to be charitable.

To illustrate this difference, consider the right to religious freedom. A Christian who embraces the concept of this right could simultaneously believe (1) that people have a right to worship as Jews or Muslims, and (2) that nobody ought to exercise that right, because those religions deny the divinity of Christ. Or consider the right to "free speech." A person might simultaneously believe (1) that people have the right to read pornographic books, and (2) that nobody ought to exercise that right.

The assertion that people have a right to food is in this sense stronger than the assertion that people have a moral responsibility to help the poor and hungry. The latter is an assertion of a principle that will guide the speaker's behavior, and a plea to others to adopt the same principle. The former is an assertion that other people have a responsibility to help the poor and hungry even if those people do not choose to adopt the moral principle that would motivate this behavior. In other words, a coercive element is embedded in the "rights" assertion that is absent from the "moral principle" assertion.

The assertion that food is a right (or that people have a fundamental right to food and other necessities or "basic needs") is highly controversial. Let us consider some of the sources of controversy by means of analogies.

Consider a right that we accept as fundamental in the United States: the "right to remain silent" or the right not to incriminate oneself. We accept the existence of this right even when we disapprove of its exercise. For example, if a kidnapper refuses to tell where he has hidden his victim, we may doubly abhor the kidnapper for his silence as well as his violence. But we do not argue that laws should permit police to torture suspects. The widespread acceptance of the right to remain silent sets this issue beyond the reach of political debate. It simplifies decisionmaking; we don't need to consider the pros and cons of any action, we need only to answer the question: "Does the action violate the right?"

This may explain why activists have pushed to have the right to food accepted as a fundamental right. They may hope to eliminate debate over the costs and benefits of various programs; the existence of the right trumps all other arguments. The FoodFirst Information and Action Network (FIAN 1997), in its fact sheet "Twelve Misconceptions About the Right to Food," states that "governance is negotiable; rights are not." A panel of constitutional experts supporting the concept of economic rights stated: "Fundamental needs such as social welfare rights should not be at the mercy of changing governmental policies and programmes, but must be defined as entitlements." The intention of advancing the "right to food" concept is to force acceptance of more active government programs to combat world hunger without having to justify those programs economically.

The strongest objection to the concept of food as a right is that unlike traditional civil rights, which require government *not to act* in certain ways, economic rights appear to require the state *to act* in certain ways. Traditional civil or political rights do not require government to act. Consider the right to religious freedom, or the right to worship as one chooses. This right imposes on the state the restriction that it cannot pass laws or take actions that interfere with an individual's right to worship. Suppose you want to attend a Zoroastrian temple for weekly worship; but suppose the nearest such temple is in Chicago, and suppose further that you cannot afford to travel to and from Chicago each week. Does the government have any obligation to buy a weekly plane ticket for you? No, at least not as the right to religious freedom is interpreted in the United States.

Economic rights do not just require the government to avoid actions that would interfere with any individual's ability to obtain food; they additionally require the government to take actions to increase the ability of hungry people to obtain food. Legal scholars refer to this as the difference between "positive" and "negative" rights.

The most extreme objections to the concept of economic rights assert that these rights are immoral themselves because they require government to limit the freedom of some members of the society. It is hard to conceive of any effective assertion of economic rights that does not require extensive redistribution of income from rich to poor. If we accept the argument that governmental limits on freedom are immoral, then any taxation is immoral, since the taxation itself is coercive, restricting individual liberties. But isn't taxation required to guarantee other civil rights? To ensure the right to be free of "cruel and unusual punishment," the government must use tax revenue to build new prisons (or free some prisoners). As long as it is costly to guarantee individuals their civil rights, some element of government coercion through the taxation system is necessary (see Holmes and Sunstein 1999).

The word *rights* in the traditional sense refers to entitlements that are in most applications absolute. The US government cannot censor a newspaper, or ban a religion, because those rights are absolute. Of course, it is easy to find examples of ways in which "absolute rights" are not absolute. The right of free speech does not extend to cover the right to yell "FIRE" in a crowded theater. The right of freedom of association (the right to choose your own friends) does not mean that an employer has the right to hire individuals of only one race. Absolute rights become limited only when the exercise of the right interferes with another person's exercise of his or her rights. The false yell of "FIRE" interferes with other people's right to congregate safely in a theater. Racial discrimination interferes with employees' rights to be free of discrimination. When one right conflicts with another, as in these cases, it is impossible to guarantee both rights absolutely.

On the one hand, what makes the concept of economic rights so controversial is that economic rights, because they require government expenditures, inevitably conflict with other rights. On the other hand, the assertion that food is a right gives those who favor government intervention an important argument to use against libertarians. The civil libertarian argues: "The government cannot take my money (through taxes) to buy food for a poor person, because I have a right to control my own property" (notice how the assertion of a right is used to trump other arguments about whether a policy is a good or bad idea). The hunger activist can respond: "You have a right to property, but the poor person has a right to food. This is a conflict of rights and the government has an appropriate role in settling that conflict."

Further, because economic rights are in inevitable conflict with rights to property, economic rights can never be absolute. So a right to food does not mean that as long as a single hungry person exists in the world, the United States cannot devote any governmental expenditures to defense, or student loans, or drug interdiction, or civil rights enforcement. We have competing social goals that must be pursued with limited resources.

Who decides the priorities for these competing social goals? In the United States, conflicts between rights are typically resolved in the court system, not by democratically elected representatives. This raises the additional question of whether we have a fundamental right to control the level of taxation through a political process. If we have no such right, then courts could require higher and higher taxes to ensure economic rights. If there *is* such a right (to a social contract on taxes), then this right must be balanced against economic and other civil rights.

If we maintain the current system of establishing priorities through a political process, we impose a severe limit on economic rights. The process of simultaneously "guaranteeing" economic rights and property rights is really no different than the process of setting policy goals and balancing competing interests.

From the Standpoint of an Economist

The preceding section started as a discussion of moral imperatives, moved on to the notion that the assertion of a right makes economic policy analysis unnecessary, and ended by raising the question: How should we allocate scarce resources to accomplish competing objectives? This question covers familiar ground for economists. Whether the trade-offs are made by courts, legislatures, or administrators, the economic rule for policymaking is to *maximize total benefits minus total costs* (see Posner 1986). Much of the remainder of the book will look at policy from the standpoint of an economist; before proceeding, we lay out some of the basic doctrines of economic policy analysis, and critiques of those doctrines.

Perhaps the questions they ask say more about economists than the way they answer those questions. The two questions that identify the asker as an economist are: (1) What is the appropriate ("optimal") policy for the society as a whole? (2) How can government best manipulate human greed to achieve its policy objectives?

What Is the Best Policy?

A political scientist wants to know: What do different people or different groups care about? Then: How will those differences in those objectives be resolved? What political processes will be involved in resolving those conflicts? What are the levers of power, and who controls those levers?

Economists start at the same place: What do different people want? And in one sense, economics is inherently about resolving conflict; after all, the buyer wants to pay a low price and the seller wants to receive a high price. But from the economists' perspective, there is some ideal way of balancing the conflicts. Any decision or choice imposes costs on some people and provides benefits to some people. The economists' ideal (at least in its simplest and purest form) says that the best choice is the one that maximizes the extent to which benefits exceed costs.

Policymaking as a rational process. Economists implicitly view policymaking as a rational, orderly process managed by benevolent, well-informed, rational, analytical policymakers. This view tends to be so ingrained in the economics literature that many professional economists may not have even considered that they have adopted this view. If pushed on the subject, few even among economists would say that this is a realistic view—that policies are actually made according to this idealized process. Nor do economists really propound this as the way policy should be made; certainly it would be nice (in the minds of economists) if policy were made like this, and the policy outcomes are likely to be improved if made this way, but no one can conceive of a practical way of implementing such a process, except perhaps through the dictatorship of an enlightened, well-trained economist, and no one is striving to have such a process implemented.

Economists do their analytical work in the hopes of tweaking the consciences of the actual policymakers, saying in effect, "Of course you can do whatever you want, but a benevolent, well-informed, rational, analytical policymaker would do the following . . ." The hope here is that actual policymakers who like to think of themselves as benevolent, well informed, and so forth, will adopt the recommended policy to avoid the shame of doing otherwise.

This idealized view of the policymaking process is that policy debates are more like scientific inquiries than like forensic debates. The scientific method presumes that truth is discovered through a series of interchanges among scientists who share the same objective—uncovering the truth. In jurisprudence, the truth is presumed to be arrived at through an adversarial contest of advocates. The prosecutor presents the very best case as to why the defendant should be found guilty, and the defense presents the very best case as to why the defendant should be found not guilty; the judge or jury, balancing those two cases, comes to a decision (in most cases) in which one side wins and the other side loses; the defendant is found guilty of the charge or not guilty.

An example of a real-world policy decision. Let us consider a concrete policy example (this is a fictitious example, but closely resembles a policy decision that might be made in the real world). The government has built a dam and reservoir and every year must decide how much water to release from the reservoir to provide irrigation water for farmers. If the water is left in the reservoir, it will provide a healthy habitat for fish and birds, and be a place where people can enjoy outdoor recreation (hiking, canoeing, sport fishing, and birdwatching). Of course, the policymakers can release none, some, or all of the water from the reservoir.

How does an economist look at the issue? We use this example to illustrate a number of important points:

- *Every action has costs and benefits.* There are benefits (to farmers) from releasing the water—the farmers will use the water for irrigation that will increase crop yields and therefore increase the farmers' profits. But there are costs (to fish and wildlife and people who value these) associated with releasing the water.
- *Marginal benefits decline as the number of units consumed increases; marginal costs increase as the number of units produced decreases.* As described in Chapter 7, economists make frequent use of the concepts of "marginal costs" and "marginal benefits." Benefits to farmers increase as the quantity of

water released increases—but the benefits increase at a decreasing rate. The costs to the habitat increase as the quantity of water increases—and the costs increase at an increasing rate.

• *The "ideal" decision (as defined by an economist) is one that sets marginal cost equal to marginal benefit.* Just as the boy with the berries in Chapter 7 based his decision on equalizing marginal cost and marginal benefit, so too, say economists, should policymakers. To apply this rule to the hypothetical water-allocation example above, the ideal water allocation would be to continue to release water until the point is reached where the marginal benefits (the increased value of crops attributable to the last thousand gallons released) just equal the marginal costs (the additional value of environmental amenities lost by the last thousand-gallon release).

• *Under certain circumstances, an unfettered free market allocates resources in the optimal way.* One way to reach the optimum is to sell each increment of the water to the highest bidder. If a coalition of hikers, canoeists, conservationists, and environmentalists submits the winning bid, that increment of water stays in the reservoir; if a coalition of farmers submits the winning bid, the water gets released for irrigation. This will result in the optimal allocation of water: each increment of water is allocated to the use (farming or wildlife) for which it receives the highest social support. If the water were owned by a private owner whose selfish objective was to make as much money as possible, he would accomplish that objective by selling it to the highest bidder, and the water would be allocated in a way that is socially optimal. This is an expression of Adam Smith's "invisible hand," and explains why economists do not think of free competitive markets and private enterprise as inherently bad.

But economists recognize that there can be a number of problems with markets that cause the "market solution" to be different from the social optimum. Some of those problems are discussed in the next section.

Criticisms and Extensions of the Simple Policymaking Rule

Having laid out the general tenets of how economists define an "optimal policy," we now explore some finer points, including criticisms of the simple rule laid out in the preceding section.

Objectivity and Prejudices and the Role of Economics in Policymaking

In our simple example, there is a clear, apparently objective, answer to the question "What is the best policy?" If we were to present our example of marginal costs and benefits to twenty randomly chosen economists and ask them to identify the optimal policy, all twenty would likely come to the same conclusion.

But does this mean that economists never disagree with each other about policy? Clearly the answer to this question is a vehement "NO." Economists have prejudices or opinions about policy that influence their evaluations. An economist knows how to undertake an objective analysis, but if that analysis arrives at a conclusion that contradicts their prejudice, the economist may decide that their prejudice needs to be reexamined, *or* the economist may decide that there was some mistake in the analysis.

Of course, an easy way to cleave to the economist's dictum ("the optimal choice is where marginal costs equal marginal benefits") and come to a different conclusion about the optimal policy is to dispute the empirical basis of the decision. An economist with an environmentalist prejudice may say, "Release 2000 gallons?! That can't be right; it's much too high. How did my analysis lead me to this conclusion? Oh, I see. The numbers are clearly skewed in a pro-farmer way." We will examine some of the ways that numbers can be wrong in their basic development below. For now, suffice it to say that if we raise our estimate of the value of environmental amenities, then the "optimal" amount of water to release is decreased. In exactly the same way, an economist with a pro-farmer prejudice may be shocked that the analysis leads to a recommendation of such a small amount (in the economist's opinion) of water to be released, and may find "mistakes" in the underlying data—"Farmer benefits from water are grossly understated," might be this economist's claim.

On the one hand, this recognition of how economists actually analyze and debate policy undercuts their claim to scientific objectivity. However, this also illustrates a major strength of the economic approach: people on different sides of an issue (people with different prejudices) are forced to think in a careful and orderly way about what their opinions are based on; they are forced to define terms clearly; they are forced to produce and to defend empirical data supporting their position. In a policy debate among economists, sincere conviction and clever phrasing count for little.

Comparability of Costs and Benefits and the Monetary Valuation of Intangibles

You have read the preceding description of how economists think about policy. If you read it uncritically, you may not have noticed that we slipped something by you—we put the costs and benefits in monetary terms so that they could be compared to each other. This is the aspect of economic analysis that is most nettlesome to many thoughtful non-economists. Suppose in our example that, after some quantity of water has been released from the reservoir, a species of fish that lives only in that reservoir dies off and becomes extinct: how do we put a dollar value on that species? Or suppose that the food produced with the water used for irrigation saves ten people who otherwise would have died of undernutrition; how do we put a dollar value on those lives?

For many items, economists measure costs and benefits by prices determined in a competitive market. So, in our example, the value of increased production from irrigation can be measured using the market price of the crops grown. Even this has a controversial side, as we will discuss in more detail below. But for other goods that are not traded in markets, the problem of valuation is trickier. What, for example, is the value of a species of fish that might become extinct under certain policy choices? Economists have developed ways to assign monetary values to these nontraded commodities by conducting surveys that ask, for example, "How much would you be willing to pay to protect this fish from extinction?" or "How much would we have to pay you to compensate you for the loss of this species of fish?"

The underlying assumption that everything can be valued in monetary terms is troubling to many people. For example, there are those who would say to the fish extinction survey question: "It is wrong to take an action that would deliberately lead to the extinction of a species. In one sense, the value of protecting the species is very high to me; to compensate for the loss of the species you would have to pay me an infinite amount (or some arbitrarily high number). But I am opposed to the idea that I should be required to pay money to preserve the species; therefore I am not willing to pay anything to preserve the species."

Economist Tyler Cowen recognizes this weakness in the economic approach to policymaking:

> On the negative side, the economic approach considers only a limited range of values, namely those embodied in individual preferences and expressed in terms of willingness to pay. This postulate is self-evident to many economists, but it fails to command wider assent. It wishes to erect "satisfying a preference" as an independent ethical value, but is unwilling to consider any possible competing values, apart from preferences. It is hard to see why non-preference values should not be admitted to a broader decision calculus.
>
> Typically economists retreat to their intuition that satisfying preferences is somehow "real," and that pursuing non-preference values is religious, mystical, or paternalistic. The rest of the world, however, has not found this distinction persuasive. They do not see why satisfying preferences should be a value of special and sole importance, especially when those same preferences may be ill-informed, inconsistent, malicious, or spiteful. The decisions to count all preferences, to use money as the measuring rod, and to weight all market demands equally must themselves rely on external ethical judgments. For that reason, the economist has no a priori means of dismissing non-preference values from the overall policy evaluation. (2006:9)

The issue of putting a monetary value on things is especially troubling when it comes to human life. What is a human life worth? That may seem at first a horribly crass question to ask. But government policy has to deal with that question in many different contexts (the numbers in the examples below are entirely made up for purposes of illustration):

- Requiring every car to have a seatbelt and an air bag will increase car prices by $800 (or $800 million over the 1 million cars sold each year), but will save 20,000 lives in auto accidents. Do the benefits from the law exceed the costs?
- Requiring all cars to drive no faster than 10 miles per hour will reduce national output by $1 trillion, but will save 5,000 lives in auto accidents. Do the benefits from the law exceed the costs?
- Requiring all vegetables sold to be tested for pesticide residues will cost $50 billion per year, but will save 15 lives. Do the benefits exceed the costs?

Even more difficult are cases in which lives are saved by restricting people's liberty, or by forcing them to take actions they believe are wrong—for example, forcing parents to immunize their children when the parents have religious beliefs that prohibit immunization.

Market Prices, Market Allocations of Resources, and Distribution of Income

The market works, as described above, by allocating resources among their various alternative uses. Over the past hundred years, a lot of the labor force in the United States shifted from farm work in the early 1900s to factory work by the mid-1900s, to producing services and entertainment by the end of the century. This shift in resource allocation was driven largely by market forces. As profitable opportunities developed in manufacturing, the market directed more resources into factories. As profitable opportunities developed in the health services sector, the market directed more resources into hospitals.

If we take a step back, a troubling question arises: Why did the market direct resources toward the production of a television show or a sporting event, rather than toward the production of more food in a world where millions of people are undernourished? If we think of the market as an election in which goods are produced in amounts that depend on how many votes they get, the people with more money to spend have more votes than the people with less money to spend. Theoretical economists recognize that the "social optimum" achieved by perfectly competitive markets is an optimum that can be defined for a given distribution of income and that draws no conclusions about what distribution of income is appropriate.

Externalities and the Optimum

There is one set of circumstances that economists recognize as a common reason why competitive markets may not lead to a social optimum: when a decision made in the market by a buyer and a seller has benefits that accrue to or costs that are borne by others. Because these costs and benefits go to people outside the market transaction, they are referred to as "externalities," or "external costs" and "external benefits."

To return to our reservoir-water example, suppose that farmers bought a certain amount of water and it was released from the reservoir; homeowners along the river between the reservoir and the farmers would get external benefits from the release: they would be able to swim or boat or fish, and they would get these benefits without paying for them, but the benefits would only exist because someone else (the farmers) *did* pay for the water release. In measuring the costs and benefits of water release, the market has taken into account only the private benefits of the farmers (who participate in the market), and not the full social benefits (which would include the benefits to the river users). In a case where there are external benefits, the amount that is bid for the good is lower than its true social value, and the market price and the quantity provided are "too low" compared to what would be a social optimum.

An external cost might occur if someone who never visited the reservoir enjoyed watching birds that summered in the reservoir and then migrated many miles away to where the bird-watcher lived. Releasing water imposes a cost on this distant bird-watcher that would not be reflected in the bids for water by the reservoir users. In this case the social costs would exceed the private costs.

Economic Incentives and Human Behavior

A second insight of economics is that personal materialistic satisfaction is a strong motivation of human behavior. Of course, if you think about your own behavior, you will be able to identify a lot of other motivations: a sense of honor or a sense of duty, a desire to be liked or admired, and so forth. Some of these motivations are appealed to by advertisers to get us to buy more of their products. They try to convince us that if we buy their products, others will think we are "cool." But even marketing experts recognize that people respond to materialistic motivations, and so in order to get us to buy more of their products, they lower their prices.

In our study of the world hunger problem and policies to deal with that problem, this insight will enter in at least three ways:

1. We will see how economic incentives can be and have been used to achieve policy objectives. For example, policies that make it more expensive to have children have been successful in reducing population growth, and policies that make it more expensive to degrade the environment have been successful in reducing environmental degradation. In this context, we will see that assigning and enforcing property rights are often an integral part of creating economic incentives.

2. The production of goods also responds to economic incentives; the more that people are materially rewarded for producing a certain good, the more of that good will be produced. The implication of this is that the distribution of goods influences the quantity of goods available for distribution.

3. Production is a dynamic process, and methods of production change over time. As a commodity or resource becomes more scarce, or as increased demand for the commodity or resource results in increased price, people respond by finding ways to use the resource more efficiently, by finding ways to produce the commodity or make the resource available more cheaply, and by finding alternatives to the commodity or resource.

On the second point, a mistake that non-economists frequently make in discussing policy options is to conceive of the policy problem as one of how to distribute a fixed stock of goods. For example, we noted that there is sufficient food available for human consumption in the world such that every person could consume his or her caloric requirements. In responding to this assertion, many people think: "So the problem is just one of distribution. If people in developed countries just consumed less, the extra food could be used in the developing world, and the undernutrition problem would be solved." The first sentence is true: the problem can be thought of as a distribution problem. But the second sentence is false, or at least grossly misleading.

The amount of food that is available for human consumption depends on the amount of food that farmers worldwide produce. And the amount that farmers produce depends on the economic incentive—the price farmers receive for their output. If people in rich countries were to make a concerted effort to consume less food—if they were to spend less money on food and more on items other than food—then the price that farmers receive would drop and they would produce less food; in the world economy as a whole, resources would move from production of food into production of nonfood items. There would be a positive effect on the undernutrition problem in poor countries, but the effect would be much smaller than imagined by those who think, "If I consume 1,500 fewer calories each day, then that food can be given to people in poor countries, and three people there can each have an extra 500 calories per day." In order for this kind of redistribution to work, the rich person would have to continue to buy the food (or at least to pay for its production in some way) and then donate the 1,500 calories per day to the poor people.

The energy crisis of the 1970s provides excellent examples of the human responses to economic incentives. During that decade, oil prices shot up dramatically. Many people perceived this as an inevitable result of growing demand for a fixed resource, and therefore predicted that prices would continue to rise. But the high prices for petroleum products caused a number of reactions over time. People began to use the resource more efficiently: auto gas mileage increased, and people began to insulate their homes more effectively. Exploration companies discovered new sources of oil and developed ways to pump more of the oil out of the ground. Alternative energy sources—nuclear, solar, and wind—grew. As a result of these reactions, prices of oil and gas did not continue to rise.

Economics and the World Food Problem

How do these economic insights apply to the problem of worldwide undernutrition? What is the "optimal" nutrition policy? What kinds of government programs can be used to achieve the objective of reducing undernutrition?

Optimal Policy to Reduce Undernutrition

The benefits of reducing undernutrition are obvious: lives saved, health improved, productivity increased. For an individual case, the costs of achieving adequate nutrition are remarkably low. In countries with an average calorie deficit (see Table 6.5 for a partial list), an additional 250 calories per person per day would erase the deficit. (Though derived in a different way, this is consistent with the FAO's estimates of average calorie deficits among people who are undernourished, ranging from about 100 to 500 calories per person per day; see FAO 2000a, *State of Food Insecurity in the World.*) Two hundred fifty calories is about the equivalent of a peanut butter sandwich (two slices of bread and two tablespoons of peanut butter is 370 calories). The cost of a peanut butter sandwich is about 35 cents. If you put $5,000 into a bank account paying 2 percent annual interest, you could withdraw 35 cents a day from that account for seventy-five years. Thus, we can conclude that there are a substantial number of people whose lives could be saved at a cost of $5,000. Compare this to an estimated "value of human life" of $150,000 to $360,000 found in a study of Indian manufacturing workers (Simon et al. 1999).

Or compare this $5,000 figure to the estimated costs of saving a life implicit in policy choices made in the United States, shown in Table 16.1. Saving lives by means of improved nutrition is an incredible bargain. Economic analysis here serves only to raise the question: If the benefit-cost ratio is so favorable, why haven't policymakers leapt to make the investments necessary to substantially eliminate undernutrition?

There are a couple of possible answers. The first has to do with targeting. The peanut butter sandwich calculation assumes that the sandwich actually gets eaten by a person who is undernourished. But in reality, food donations are sometimes diverted to people who are not undernourished. A January 2004 report on the situation in North Korea states:

Table 16.1 Dollar Costs per Life Saved of Various Regulations in the United States

Government Action	Cost per Life Saved
Requiring seat belts and air bags in cars	\$100,000
Banning flammable sleepwear for children	\$1,200,000
Requiring seat belts in rear seats of cars	\$3,800,000
Restricting arsenic emissions from glass-manufacturing plants	\$40,200,000
Banning asbestos	\$329,000,000

Sources: Viscusi and Gayer 2002; Viscusi 1993.

> The current food crisis is a result of foreign donors refusing to contribute food for North Korea because the government has not allowed foreigners to observe where the donated food goes. Other witnesses have consistently reported that the donated food goes to the armed forces and is not sent to areas where there has been unrest, or where the government suspects there might be unrest (because a number of locals have fled to China or Russia). . . . New supplies will not arrive for several months. But after that, the food aid could dry up again if the North Korean government does not become more cooperative. (Strategy Page 2004)

It is natural to want to avoid being "conned"—tricked into making charitable donations to people who do not deserve our charity. From an economic standpoint, however, even if only one in ten of the donations hits its mark, the program would still be more cost-effective than any of the policy steps listed in Table 16.1. A more cynical answer is that the people who are dying from lack of seatbelts or from asbestos are "like us" and therefore it is worth the high cost to save those lives. We can empathize with the people who would die in the absence of the government policy, but we imagine that the people who are dying from undernutrition are "not like us"—we cannot imagine being that poor, therefore our empathy is low. Subramanian and Cropper (1995) cite other unfunded programs that would save lives at a low cost.

Finally, the \$5,000 figure is the cost of saving a single life, without taking into account any impacts on market prices that would occur if the policy were aimed at reducing undernutrition among many of the 800 million suffering from it. As subsequent chapters will explain, a large-scale program would increase food prices and the environmental costs associated with increased food production.

Policy Instruments to Reduce Undernutrition

Economic analysis has a lot more to contribute to the question: What kinds of policy actions can contribute to reduced undernutrition? Most of the rest of the book is devoted to some answers to this question. The supply-demand framework helps organize the discussion.

Chapter 7 emphasized the two elements of the food security equation—income and price. For the most part, our policy discussion can be broken down into policies that influence income and policies that influence price.

Policies to raise incomes of the poor. Chapter 17 will discuss policies that raise the incomes of the poor. There are two possibilities: redistributing income from rich to poor, or improving the rate of economic growth.

The main economic rationale for redistributing income from the rich to the poor is the belief in the declining marginal utility of income. We discussed previously the principle that as food consumption increases, unit by unit, declining marginal benefits accrue from adding an additional unit. Many economists accept the hypothesis that this principle can be extended to cover the consumption of all goods taken together.

A direct implication of this is that a dollar is worth more to a poor person than to a rich person. On the one hand, the idea is that a couple of more dollars in the hands of a poor person will be spent on "necessities"—items that are fundamental to life. On the other hand, taking a couple of dollars away from a rich person will cause that person to consume fewer frivolous things. The research relating income to happiness (see Box 9.2) is consistent with this hypothesis. A 10 percent increase in income has the same impact on happiness regardless of the income level, so taking $1,000 from a person making $100,000 (reducing their income by 1 percent) and giving that $1,000 to a person making $10,000 (increasing their income by 10 percent) will increase the poor person's happiness more than it decreases the rich person's happiness. Therefore, this transfer from rich to poor increases the "common good." This may explain why governments are motivated to adopt programs that have the effect of redistributing wealth from the rich to the poor.

The practical problem is that the "declining marginal utility of income" hypothesis implies that the appropriate policy is total and complete equality of income distribution. If one person in the country (or the world) earns slightly more than another, then money should be taken from the former and given to the latter. Most people reject this policy prescription. That rejection raises questions about whether this is the true explanation for policy concern about the poor and hungry.

There is much literature on the subject of what kinds of policies may promote general economic growth. We will provide a general overview in Chapter 17, which addresses the issue of "globalization" and whether integration into the global economy can be beneficial to growth rates in developing countries.

Policies to reduce the price of food. Chapters 18–22 will consider a variety of policies that reduce the price of food. Chapter 18 discusses population control. If population growth can be reduced, the demand for food will not increase as quickly, food supplies per capita will increase, and food prices will decrease.

Chapters 19–22 consider policies that target food prices more directly. These policies can be thought to operate in one of two ways they can distort the social equilibrium and cause a reduction in economic efficiency, or they can correct a distortion and increase economic efficiency.

If the aggregate supply-and-demand curves represent the true social costs and benefits, then policies that alter the equilibrium price are "distortionary"—they reduce economic efficiency. This is illustrated in Figure 16.1, which shows how a government policy can create a wedge between the price that consumers pay and the price that farmers receive. Perhaps the program is one that sells food to consumers at below cost, or perhaps the program pays subsidies to farmers, or perhaps the subsidy is paid to firms in the processing or marketing sector. In any case, the price received by farmers is higher than the price paid by consumers (the difference between the price at point A and the price at point D in the figure).

This type of policy achieves the direct objective we are looking for here: quantity increases (from the "market quantity" to the "quantity with the program" in the figure). But the policy reduces economic efficiency. As the quantity produced increases above the market quantity, the cost of producing an additional unit exceeds the value that consumers get from consuming the additional unit. The quantity of this efficiency cost is shown as the shaded triangle in Figure 16.1. Subsidies also have a direct cost paid by the government. This is the amount paid per unit (producer price minus consumer price), times the number of units. This cost is shown as rectangle ABCD in the figure.

Figure 16.1 Impact of a Policy That Subsidizes Production or Consumption

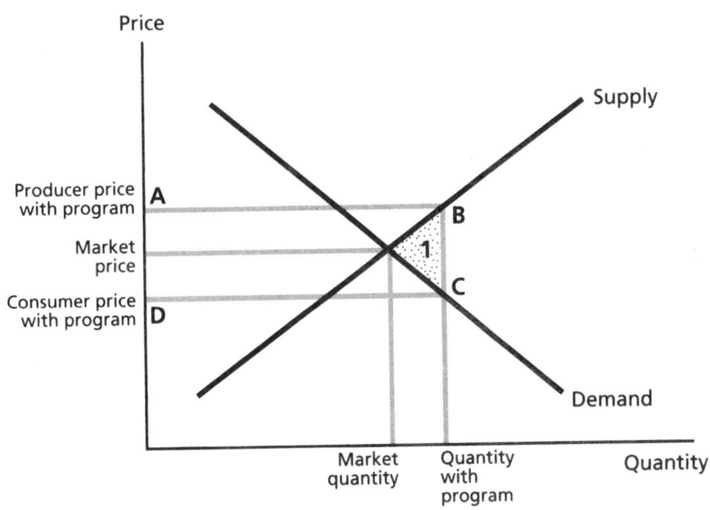

But policies that change prices can sometimes be seen as "corrective." For example, as discussed, there is the possibility that production or consumption of a good might create external benefits that are not reflected in the market supply-and-demand curves. In this case, the market equilibrium quantity will be lower than the optimum, and a government policy to subsidize consumption or production may correct the situation.

Economist Arnold Harberger (1983) suggested that we all (or at least many of us) suffer when any person in the world (or country or ethnic group or family) suffers from undernutrition or failure to meet "basic needs." When a hungry person is fed, a direct benefit goes to that person (which benefit is reflected in the market transaction), but also an indirect benefit goes to us, because we care about human suffering. This indirect benefit is external to the market. Because such a reduction affects our happiness, we would be willing to pay to see the incidence of hunger reduced or eliminated. The reduction of hunger is a good: the more it happens, the better we feel. But there is no market on which we can purchase this good. This is a case of missing markets, or *externalities.* Private action is unlikely to solve this problem. Because we know others also care, we may wait for *them* to take actions to reduce hunger, in which case *we* get the good (reduced world hunger) free of charge. Government action may be justified in creating an artificial market by collecting money from each of us who derives satisfaction from reductions in hunger, and using that money to reduce hunger.

Chapter 21 will examine another kind of corrective action: removing distortive policies that actually *reduce* food output and consumption. Chapter 22 will discuss policies that cause the aggregate supply to increase, thereby increasing output and reducing price. These policies include investments in research and development, to offset underinvestment in this sector by private markets. They also include government interventions to correct market failures through the provision of loans to farmers.

17

Policies That Raise the Incomes of the Poor

The world's hungry are hungry because they are poor. They cannot afford enough food to provide their basic needs. Policies to alleviate their poverty fall into two categories: policies that promote general economic growth and policies that redistribute income or wealth from rich to poor.

Promoting Economic Growth

There is no doubt that broad-based economic growth is one of the most effective antipoverty programs (Dollar and Kraay 2002). Economic growth creates jobs and raises the incomes of the poor. Figure 9.1 provided some evidence of the benefits of economic growth. To a limited degree, government projects can contribute directly to economic growth; but in general governments do not create jobs very efficiently. Developing countries must rely on growth in the private sector for their economic prosperity.

The history of the past four decades frames the debate over what kinds of policies best promote economic growth. A couple of things are obvious:

- *Economic growth is possible.* For example, per capita income in Taiwan grew from $1,256 in 1960 to $12,181 in 1998 (in constant, or "inflation-adjusted" dollars). South Korean per capita income grew from $904 to $9,454 over the same period. Of a comprehensive sample of fifty-eight countries with low incomes (per capita incomes of less than $1,500) in 1960, per capita incomes doubled in twenty of them.
- *Economic growth is not inevitable.* In these fifty-eight low-income countries, per capita incomes declined in twelve countries, and grew by less than 1 percent per year in nineteen others. Per capita income in the Democratic Republic of Congo fell from $489 in 1960 to $197 in 1998. Per capita income in Madagascar fell from $1,191 in 1960 (above the South Korean level) to $581 in 1998 (about 5 percent of the South Korean level).

A large economics literature deals with prerequisites and policies concerning economic growth. See the World Bank's website on economic growth (http://econ.worldbank.org/programs/macroeconomics/topic/22009) for a review of the issues and a guide to the literature. If there were a simple strategy that guaranteed economic growth, then we would see growth everywhere (and this would be a very short chapter). But there is no such simple strategy, and economists continue to debate which approaches are most likely to be successful.

Evolution of the Washington Consensus

In 1990, John Williamson published a paper that laid out the general consensus about what policies Latin American countries should adopt to promote growth. This set of recommendations became known as the "Washington Consensus" because it reflected the thinking of economists at the World Bank and the International Monetary Fund (IMF). The Washington Consensus is based on the observation that economic growth comes from three main sources:

- High savings leading to increased capital stock.
- High labor productivity.
- Adoption of new technology.

This translates into three broad sets of policy recommendations:

- Promotion of savings and investment through good macroeconomic policy.
- Promotion of labor productivity through education, health, and antipoverty programs.
- Market orientation to promote appropriate incentives to economic decisionmakers.

Rodrik summarizes the original Washington Consensus as prescribing "a simple and universal list of policy reforms: liberalize trade, privatize public enterprises, deregulate prices, bring down inflation through monetary and fiscal retrenchment," and notes that the consensus evolved to include "deeper institutional reforms: fight corruption, improve courts, enhance public administration, improve regulation" (2008:2). These recommendations were successfully followed by a number of countries, notably the high-growth Asian countries. However, they were not universally successful.

The recognition that "pro-growth policy" should be defined broadly enough to include initiatives to change underlying institutional arrangements led economists to expand their scope of analysis. Easterly and Levine (2003) did an extensive empirical analysis to see whether high rates of economic growth could be explained by (1) geography (climate, access to the sea), (2) institutions (property rights, democracy, individual liberty), or (3) policies (low inflation, openness to international trade, stable exchange rates). Their answer: institutions make the biggest difference. "Good" policies cannot overcome the disadvantage of "bad" institutions. (Easterly and Levine do find an indirect effect of geography on quality of institutions: European colonizers settled in larger numbers in colonies with moderate climates, and these colonizers brought their European institutions with them.) Institutions will be discussed in more detail below.

In 2008, the Commission on Growth and Development, under the leadership of Nobel Prize–winning economist Michael Spence, produced a report enunciating a "new consensus"—each country must discover its own path to economic growth:

> The Spence report represents a watershed for development policy—as much for what it says as for what it leaves out. Gone are confident assertions about the virtues of liberalization, deregulation, privatization, and free markets. Gone are the cookie cutter policy recommendations unaffected by contextual differences. . . . Yes, successful economies have many things in common: they all engage in the global economy, maintain macroeconomic stability, stimulate saving and investment, provide market-oriented incentives, and are reasonably well governed. [These] frame the conduct of appropriate economic policies. Saying that context matters does not mean that anything goes. But there is no universal rule-book; different countries achieve these ends differently. . . . The new policy mindset . . . is explicitly diagnostic and focuses on the most significant economic bottlenecks and constraints. Rather than comprehensive reform, it emphasizes . . . narrowly targeted initiatives in order to discover local solutions. (Rodrik 2008:1–2)

Institutions and Economic Growth

Economist Douglass North gives this definition of institutions: "Institutions are the way we structure human interaction—political, social and economic—and are the incentive framework of a society. They are made up of formal rules (constitutions, laws and rules), informal constraints (norms, conventions and codes of conduct), and their enforcement characteristics" (2005:A14). The importance of institutions to economic growth has at least three interrelated aspects: property rights, regulatory efficiency, and lack of corruption.

Property rights assign ownership of productive assets to individuals and give the owners the legally enforceable right to use those assets as they choose. Economist Hernando De Soto (2001) has been a leading voice about the importance of property rights to economic growth. He notes the following advantages that flow from a system of well-identified and clearly enforceable property rights: (1) the property or assets can be pledged as collateral, thus improving the availability of credit; (2) assets can be divided or consolidated in an efficient way among users, so that (for example) where small farms are more efficient, small farms can evolve by sale or rental of land that breaks up inefficient large farms; (3) assets can be owned and managed by the individual who is best suited; (4) information about asset value is more readily available, allowing investors from far away to evaluate investment opportunities; (5) tying an individual owner to an individual piece of property (especially real estate) makes it easier to identify, locate, and enforce agreements with that person.

De Soto cites numerous examples of problems that arise where property rights are not well developed, or not enforceable. Squatters in developing countries build homes on land they do not own; but the squatters cannot borrow money against the value of those homes, and the squatters have little incentive to improve the quality of the homes, since they could be thrown out at any instant. Tribal chiefs assign farmland to tribal members; but the farmers have little incentive to improve the land, since it may be reassigned next season.

Regulatory inefficiency can contribute to weak property rights. *The Economist* (2001) reports that in the Philippines, it can take 13–25 years for a squatter to complete all the legal steps required to obtain ownership of the land on which he lives; in Egypt, it can take 6–11 years for an owner of farmland to obtain legal permission to build a house on that land. Many other examples, and country-by-country indicators of regulatory efficiency, can be found in reports from the "Doing Business" website of the World Bank (http://www.doingbusiness.org/downloads).

Government bureaucracies can be inefficient, and regulations governing economic transactions can be cumbersome, even when the bureaucrats and regulators are impeccably honest. But the possibility of government corruption adds another layer to the problem of regulatory efficiency, as dishonest regulators may deliberately hold up the regulatory process in order to put pressure on people to "facilitate" the bureaucracy with a bribe or other emolument. The organization Transparency International (see http://www.transparency.org) publishes a "corruption perceptions index" and a "bribe payers index" that rank countries according to various measures of corruption. Mauro, in an empirical analysis of the relationship between corruption and economic growth, concludes that "corruption may have large, adverse effects on economic growth and investment" (1997:93).

One (imperfect) check on government corruption is a healthy democratic process and an adversarial free press. *The Economist* has developed a "democracy index" that ranks countries according to whether they have a competitive multiparty political system, universal suffrage, regularly contested elections, and media access (see Kekic 2007).

One measure of how important institutional arrangements are to economic prosperity comes from Hendricks (2002). He compares output per worker of immigrants to the United States with output per worker of people who remain in the migrants' home countries. Overall, he finds that migration to the United States makes workers about four times more productive. About one-third of that improved productivity is attributable to the fact that workers in the United States have better physical capital—better tools and equipment. The other two-thirds is attributable to the fact the institutional environment in the United States is more conducive to productivity.

Globalization and Economic Growth

Since the early 1990s, a vociferous debate has taken place about the desirability of "globalization." Though the term may mean different things to different people, it refers generally to a policy of increasing the integration of countries in the world economy. For developing countries, a policy that embraced globalization would entail the following:

- Opening borders to trade by reducing impediments to imports and exports, and subjecting these regulations to restrictions imposed by the World Trade Organization.
- Adopting macroeconomic policies required as conditions for loans from the International Monetary Fund.
- Adopting market-oriented industrial, agricultural, and sectoral policies, as a condition of obtaining IMF loans.
- Reducing restrictions or regulations that discourage foreign investment.
- Adopting labor and environmental policies that will attract foreign investment.

As this list makes clear, globalization promotes the same kinds of policies that make up the Washington Consensus. Critics of globalization raise the following objections:

- Policies that attract investment are policies that encourage or permit low wages, poor working conditions, and poor environmental quality.
- Fiscal policies imposed by the IMF require countries to reduce or eliminate health, education, and poverty alleviation programs.
- Policies imposed by the IMF and the WTO are antidemocratic, since these international organizations may countermand decisions made by democratically elected leaders and legislatures.
- There is also a suspicion on the part of globalization critics that the international organizations are controlled by multinational corporations, and that the entire globalization effort is intended to enrich these corporations without regard for harm done to ordinary people in the process.

Has increased globalization been good for economic growth of poor countries, and has it been beneficial to the poorest people in poor countries? Paul

Collier and David Dollar (2002) of the World Bank conducted a study in which they divided countries into three groups. In rich countries, income per capita grew at an annual rate of about 2 percent per year; in "more globalized" poor countries—poor countries with a relatively high proportion of international trade to national income—income per capita grew at an annual rate of 5 percent per year; and in "less globalized" poor countries, income per capita *declined* at a rate of 1 percent per year.

The comparative experiences of Asia (which experienced dramatic economic growth from the early 1980s to the late 1990s) and Africa (which had stagnant growth over the same period) suggest that integration into the global economy can be good for growth. Per capita incomes in Southeast Asia and sub-Saharan Africa were nearly identical from 1970 to the mid-1980s. Then incomes in Southeast Asia shot up, while incomes in Africa regressed. By 1997, incomes in Asia were more than three times higher than those in Africa. Court and Yanagihara (n.d.) provide evidence that this was related to the fact that Asian governments adopted policies that embraced world markets. Exports from Southeast Asian countries grew at over 12 percent per year from 1985 to 1995; exports from sub-Saharan African countries grew at only 3 percent over this period. In the mid-1970s, exports were about 30 percent of GNP in both regions; by 1997, exports were over 50 percent of GDP in Southeast Asia, but were still at 30 percent of GDP in sub-Saharan Africa. In the early 1970s, foreign direct investment was nearly zero in Malaysia, Indonesia, Ghana, and Kenya. By the late 1990s, it was still nearly zero in the African countries, but had grown to $4–6 million a year in the Asian countries.

Nor has economic growth only helped the rich elites. Dollar and Kraay (2002) find that incomes of the poor rise dollar-for-dollar with average incomes in a country, and that the rise is simultaneous, not lagging behind average income growth as one might expect from a "trickle-down" hypothesis.

The financial crisis in many Asian economies in the late 1990s reveals some of the weaknesses of the globalization strategy. Nobel Prize–winning economist Joseph Stiglitz (2002) points out that policies imposed by international bodies do not always take into account the special circumstances of each country. Private-sector solutions require the existence of an institutional and cultural infrastructure that may not exist in every case. The best macroeconomic policy is not the same for all countries at all times. The pace of globalization can influence its effectiveness. Stiglitz calls for the globalization process to be reformed so that it can help poor countries grow.

Agricultural Development

In addition to the preceding recommendations for growth, many economists would emphasize the importance of promoting growth in the agricultural sector (DFID 2002). The economies of almost all developing countries are dominated by the agricultural sector. One of the most important stimulants to economic growth and increased employment in these economies is increased agricultural production. This is important not only because increased agricultural production increases farm employment, but also because increasing the quantity of food supplied lowers its price.

Food is a wage good (Mellor and Johnston 1984). That is, the cost of food can substantially affect the wage rate. Consider two developing countries that are competing in the international marketplace to sell a labor-intensive product such as shoes. In country A the price of food is high, and in country B it is low. Even though a shoe manufacturer in country B pays lower wages than its competitor in country A, the workers in the shoe factory in country B can live as well as those in country A because they can buy food more cheaply. Low food prices stimulate employment.

As low food prices make low wage rates possible, employment is stimulated not only in the export sector but also in the domestic sector of the economy. Local manufacturers can compete more successfully with importers to manufacture goods. Low food prices reduce the proportion of the household budget that all people, middle- and upper-income people as well as low-income people, must allocate to food, thus releasing purchasing power for nonfood items. This raises the demand for nonfood goods and services and further increases employment.

Stimulating third world agricultural production will involve making policy shifts away from the large number of production disincentives now in place and toward production incentives. There are other avenues to stimulating increased agricultural production, such as government sponsorship of agricultural research and educational services or improved roads (see Box 17.1). Policy alternatives to spur agricultural production are discussed in more detail in Chapter 22.

Economic Growth and the Reduction of Undernutrition

Figure 9.1 showed that the percentage of population that is undernourished declines as a country's income per capita grows. Figure 17.1 provides a different perspective on that basic relationship. It shows that during the three decades of rapid growth in per capita incomes, Asia and the Middle East made enormous progress in reducing the incidence of undernutrition. In Asia, the percentage of people suffering from undernutrition dropped from 41 percent in 1970 to 12 percent in 2002. In Latin America, per capita income growth was a little slower, and progress on undernutrition was a little less (20 percent undernourished in 1970 and 10 percent in 2002). But in Africa, where per capita income growth was much lower, the reduction in undernutrition was also much lower (36 percent in 1970 and 32 percent in 2002). This reaffirms the conclusion that per capita income growth reaches the poorest of the poor and improves their nutritional status.

Box 17.1 Roads in Africa

Western travelers to sub-Saharan Africa are often amazed by the lack of adequately paved roads and highways. For example, a visitor to Zambia in the late 1990s would have found that one of the main highways leaving the capital of Lusaka was only one lane in each direction, and that those lanes were riddled with potholes, causing average vehicle speeds to fall to about 35 kilometers per hour. Experienced drivers would swerve into the wrong lane, or out onto an unpaved shoulder to avoid some of the worst of the potholes. And off of this main highway, "roads" were simply dirt tracks, sometimes made impassable by ditches carved by flooding rains.

The Economist ("The Road to Hell Is Not Paved," 2002) sent a reporter to ride along with a truck driver who was delivering beer to a town 500 kilometers from the brewery. The truck was stopped at police road blocks forty-seven times. Sometimes, the policeman only wanted to check the driver's papers; sometimes, the policeman wanted cash; sometimes, the policeman wanted a beer. Only two-thirds of the truckload of beer arrived at its destination. At one such stop, seven policemen and the three people in the truck spent three and a half hours negotiating the terms of passage (they settled on $12). One policeman, citing a hitherto-unknown law, was challenged by the driver and responded: "Do you have a gun? No. I have a gun, so I know the rules." Along one stretch of road, locked barriers had been erected to stop heavy trucks that tear up the dirt roads when they are wet. But the man with the key was nowhere to be found. When they found the man with the key and opened the barriers, they drove a few more miles and discovered that a bridge was out. All in all, the 313-mile trip took four days. (For comparison, consider that a truck driver on a modern highway without stops could expect to complete a 313-mile trip in about five hours.)

In 2006–2007, the West African Trade Hub project of USAID (2007) undertook a study of travel conditions between Ouagadougou in Burkina Faso and three other cities (Tema in Ghana, Bamako in Mali, and Lome in Togo). Truck drivers were outfitted with impeccable credentials and sent to drive the routes. On the worst stretch (417 kilometers in Mali), the average driver was stopped nineteen times, at a delay of over two and a half hours, and paid $105 in bribes.

An analysis of public investments in India (Fan, Hazell, and Thorat 2000) found that investments in road construction had a greater impact on poverty reduction than any other type of public investment.

While Figure 17.1 (or Figure 9.1) makes a compelling case at the national (or macro) level, the evidence from the household (or micro) level raises some interesting questions. Subramanian and Deaton review the state of debate:

> In the recent literature . . . , there is debate on the extent to which nutrition responds to income. For many years, conventional wisdom has held that hunger and malnutrition would be eliminated by economic growth. . . . [S]ome recent studies . . . have argued that the elasticity is close to zero, so that "increases in income will not result in substantial improvements in nutrient intakes" (Behrman and Deolalikar 1987:505). If this position is accepted, there are important implications for the way economists think about

Figure 17.1 Faster Growth in Income per Capita Leads to Larger Reductions in Prevalence of Undernutrition, 1970–2002

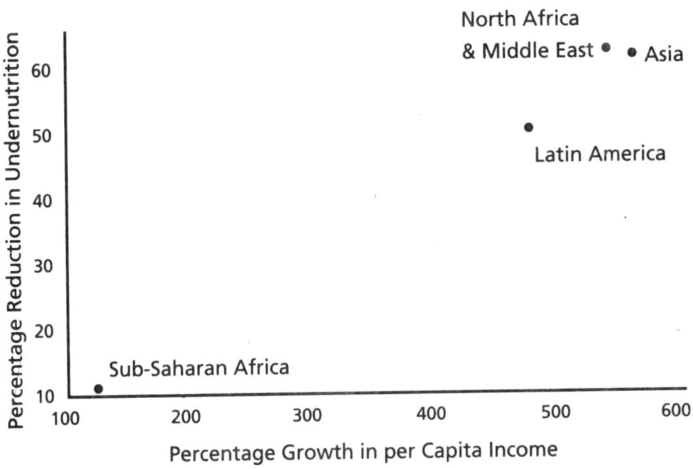

Sources: FAOSTAT 2008; World Bank.

> development. In accord with some popular beliefs, economic policies that are good for growth do not imply the elimination of hunger. Indeed, even policies that increase the incomes of the poorest may not improve their nutrition. [This] also creates a chasm between the way economists think about living standards . . . and the way living standards are often characterized by nutritionists and development practitioners, who see development largely in terms of guaranteeing that people have enough to eat. . . . To take but one example, economists think of substitution possibilities as welfare enhancing; if it is possible to substitute across a wide range of foods, consumers are well protected against changes in relative prices. To the nutritionist concerned only with adequate diet, welfare is decreased by voluntary substitution away from approved to disapproved food. (1996:134)

In their study of the Indian village of Maharastra, Subramanian and Deaton found that calories could be purchased quite inexpensively—a deficit of 600 calories per day could be eliminated by spending 4 percent of the daily wage—and that even the poorest individuals could afford adequate calories, but these individuals instead allocate their food expenditures to better tasting but less nutritious food. "If nutrition is a trap, it is one from which there is a ready escape" (p. 135).

Redistributing Wealth from Rich Countries to Poor Countries

Economic growth is one sure way to increase the incomes of the poor. But that may also be accomplished by redistribution. Foreign aid, or foreign development assistance, consists of money sent by governments of rich countries to governments of poor countries (or to nongovernmental projects in those countries). The purpose of this aid is not to put money directly into the hands of poor people in developing countries, but to help finance projects that will promote economic growth. Therefore, although foreign aid on the surface appears to be a redistribution from rich to poor, it is more closely related to the discussion in the preceding section about policies to promote growth.

Table 17.1 shows the largest donors of aid in 2007. Not one country donated more than 1 percent of its wealth to the poor, and the average was 0.28 percent. Table 17.2 shows the largest recipient countries, with Iraq and Afghanistan at the top of the list, and with the Palestinian administrative areas and Egypt appearing in the top 16.

The United States, in its Millennium Challenge Grant program, made a commitment to increase its foreign assistance to 0.7 percent of GDP, and to direct much of that new assistance to the poorest countries of the world. But recipient countries are required to meet certain standards ("challenges") in order to be eligible for this assistance.

There are some strong disagreements among economists about whether or not foreign aid can be effective. A good example of how economists disagree—and how those disagreements help to move a policy debate forward—is found in the question of whether foreign aid is effective.

Table 17.1 Foreign Development Assistance by Donor Country, 2007

Amount of Assistance Donated ($ millions)		Assistance as Percentage of Donor GNI	
United States	21,786.90	Norway	0.95
Germany	12,290.70	Sweden	0.93
France	9,883.59	Luxembourg	0.91
United Kingdom	9,848.54	Denmark	0.81
Japan	7,678.95	Netherlands	0.81
Netherlands	6,224.26	Ireland	0.55
Spain	5,139.80	Austria	0.50
Sweden	4,338.94	Belgium	0.43
Canada	4,079.69	Finland	0.39
Italy	3,970.62	France	0.38
Norway	3,728.02	Germany	0.37
Australia	2,668.52	Spain	0.37
Denmark	2,562.23	Switzerland	0.37
Belgium	1,952.83	United Kingdom	0.36
Austria	1,808.46	Australia	0.32
Switzerland	1,689.16	Canada	0.29
Ireland	1,192.15	New Zealand	0.27
Finland	981.34	Portugal	0.22
Greece	500.83	Italy	0.19
Portugal	470.54	Japan	0.17
Luxembourg	375.53	Greece	0.16
New Zealand	319.80	United States	0.16

Source: OECD, Stat Extracts, http://webnet.oecd.org/wbos/index.aspx.

Table 17.2 Largest Recipients of Foreign Development Assistance, All Sources, 2007

	Amount of Assistance Received ($ millions)
Iraq	8,965.25
Afghanistan	2,992.72
Tanzania	1,830.67
Cameroon	1,696.83
Sudan	1,666.14
Vietnam	1,488.37
Nigeria	1,385.22
China	1,331.23
Ethiopia	1,242.02
Mozambique	1,073.21
Uganda	1,002.46
Pakistan	976.41
India	903.19
Palestinian administrative areas	836.43
Kenya	824.09
Egypt	787.04
Congo, Dem. Rep.	771.66
Zambia	712.92
Ghana	708.46
Bangladesh	663.89

Source: OECD, Stat Extracts, http://webnet.oecd.org/wbos/index.aspx.

Jeffrey Sachs of Columbia University is confident that development assistance can be a practical way to eliminate worldwide poverty. Sachs's optimism is illustrated here:

> Malaria . . . is largely preventable and utterly treatable. There is no excuse for the millions of malaria deaths that will occur this year. . . . Just $2 to $3 per American and other citizens of the rich world would be needed each year to mount an effective fight against malaria. The rich world's actual spending to fight malaria is closer to 20 cents per person per year. . . .
>
> . . . [S]imilar steps would change the face of extreme poverty—indeed, put the world on a path to eliminate it in this generation. Yet these steps are not taken. . . . Americans . . . believe, erroneously, that corruption in poor countries blocks effective use of aid, even though dozens of impoverished countries are rather well governed yet still starved of help. (2005b:A17)

Sachs argues that a relatively modest increase in foreign aid could have a huge difference in world poverty.

William Easterly (2006) of New York University is pessimistic about the possibility that development assistance can be effective. He argues that grand plans such as those espoused by Sachs have been tried repeatedly and have never succeeded. "Economic development happens, not through aid, but through the homegrown efforts of entrepreneurs and social and political reformers." Easterly is not opposed to all foreign aid, but he favors small-scale projects with limited objectives. Aid agencies should stop trying to "achieve general economic and political development" and "start . . . fixing the system that fails to get 12-cent medicines to malaria victims." Aid fails, in Easterly's eyes, because of the lack of feedback and accountability. (For more debate between Sachs and Easterly, see Easterly 2005.)

Redistributing Wealth or Income Within a Country

Incomes of the poor and undernourished in developing countries can also be raised by direct policies to redistribute income within those countries.

Progressive Taxation

Taxes that take a greater percentage of income or wealth from the rich than they do from the poor are called *progressive.* Progressive taxation is one way of transferring income or wealth from the rich to the poor. In developed countries, the income tax is usually designed to be progressive, and the same features can be incorporated into the tax structures of the third world, as they often are.

Taxes that take a greater percentage of income or wealth from the poor than they do from the rich are called *regressive.* Sales taxes have a reputation for being regressive. In the developed world, the poor spend a greater proportion of their income compared to the rich, who save a greater proportion of theirs. However, in the third world, the poor are not generally as well integrated into the market economy as are the rich. The poor are much more likely to barter for and exchange goods and services and to grow some of their own food. In a situation like this, even a sales tax may be progressive.

The work of Tanzi and Zee (2000) on taxation in developing countries allows us to draw the following general lessons:

- Taxes in developing countries are low—about half the level of taxes in developed countries.
- Developing countries use taxes on exports and imports much more heavily than do developed countries. One explanation for this is that taxes on exports and imports are easier to collect (compared to income or consumption taxes) for a small group of customs agents monitoring movements into and out of the main port or ports of the country.
- Many developing countries impose consumption taxes (or value-added taxes).
- There is a wide variation about the extent of progressivity of income taxes in developing countries.

These characteristics imply that, on the whole, developed countries use their tax systems more aggressively to redistribute income than do developing countries.

The government's method of spending tax money can have distributive effects just as surely as does the method of collecting it. A government that taxes some rich people to provide services only to other rich people will do little to reduce undernutrition. Spending public money on programs to increase agricultural production (Chapters 21–22) or on programs that subsidize food consumption (Chapters 19–20) is what we think of as the expenditure side of redistribution through progressive taxation. A mere transfer of purchasing power from the rich to the poor may prove less effective as a way of improving the nutrition of the poor. For example:

> Take a simple case within a developing country, say India. If one rupee of purchasing power is taken away from a person in the top 5 percent of the income distribution, that will cause a reduction, in constant prices, of 0.03 rupee in food-grain consumption. That same rupee provided to a person in the bottom 20 percent of the income distribution will provide increased demand for 0.58 rupee of food grains. The one-to-one equality of financial transfers is matched by a nineteen-to-one inequality in the material transfers. Thus, a marginal redistribution of income is profoundly inflationary in driving up food prices. In this case, what the left hand of society gives to the poor, the right hand of the market takes away. (Mellor 1988:1003)

This is illustrated in Figure 17.2. Taking income from the rich shifts their demand for food back, but by a relatively small amount, since income elasticity of demand for food is low for rich people. Giving income to the poor shifts their demand for food out, by a relatively large amount, since their income elasticity of demand for food is large. Therefore, the shift in aggregate demand

Figure 17.2 Effect of an Income Transfer from the Rich to the Poor Is Partially Offset by an Increase in Aggregate Demand and a Resulting Increase in Food Price

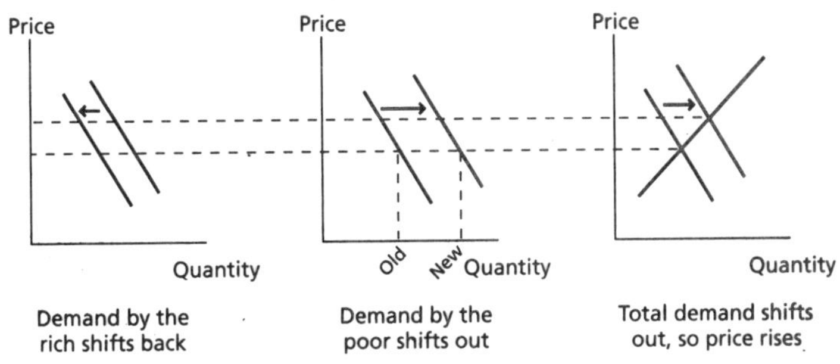

(combining the demands of rich and poor) is outward, driving up the equilibrium price. This partially offsets the impact of income on food consumption of the poor, though the new quantity is still above the old quantity.

Taxing land according to use-value is another way to implement a progressive tax. Through a properly executed land-use survey, farmland can be classified according to the value associated with its use potential. Often, third world taxes on good farming land are so low that large landholders can afford to keep their holdings while farming them inefficiently. By raising taxes and keeping them proportionate to the value of land use, farmers who are making poor or inefficient use of their land will be forced to sell it to those who would farm it better. The beauty of this system is that it readjusts resource use by weeding out the bad farmers without uprooting the good farmers, who may be doing a fine job for society.

Minimum-Wage Laws

It is frequently, and often emotionally, argued that minimum-wage laws are an effective way to improve the income of the poor. There is no question that, for those workers covered by minimum-wage legislation and whose wages are higher than they would otherwise be, minimum-wage laws yield a higher level of living. However, effective minimum-wage legislation, as it raises wages at the bottom end of the scale, motivates entrepreneurs to substitute capital for labor. This drives the labor covered by minimum wage out of the economy, and increases unemployment in the economy generally (Mincer 1976). If credit subsidies are available, the motivation to substitute capital for labor is even greater.

Minimum-wage laws are more easily enforced in urban than in rural areas. So one result of effective minimum-wage laws is an increase in the wage differential between the country and the city. In the third world, the difference in wage rates between farm and city has resulted in mass migrations of rural population to the city in search of jobs. Although the probability that an unskilled rural migrant will obtain an urban job may be small, the decision to migrate may be rational, because some migrants, in fact, do obtain good urban jobs (Todaro 1980). The waiting may take months, years, or even a lifetime, but if enough migrants obtain jobs, then generally the wait seems worth it for most people.

Effective minimum-wage legislation increases the ultimate reward from waiting for an urban job. It simultaneously increases the number of people in the queue, increases urban unemployment, wastes labor resources, and increases the number of family members accompanying unemployed migrants who may be subjected to undernutrition.

Land Reform

In Chapter 9 we discussed the possibility that income distribution can influence economic growth. Deininger and Squire (1998) showed that rates of growth are higher in countries where land is distributed more equally. Using a similar data set, Deininger and Olinto (2000) examined the experiences of sixty countries over the period 1966–1990. Of the thirty-five countries with Gini coefficients for land distribution of less than 72 (a lower Gini means a more equal distribution), twenty-one had growth rates that were higher than average. Of the twenty-five countries with land Gini coefficients higher than 72, only six had growth rates higher than average.

Since 1960, virtually every country in the world has passed land reform laws (De Janvry 1981:385). Land reform can mean many things, but typically it means at least one of the following:

1. *Redistributing the ownership of private or public land in order to change the pattern of land distribution and size of holding.* At one extreme this might mean creating small plots from large blocks of land and allocating these small plots to the poor. At the other extreme it might mean nationalizing all agricultural land and assigning it to large, state-owned farms.
2. *Changing the rights associated with land.* For instance, tenant farmers or sharecroppers can be made owners of the land they work. Lenders, too, can be prohibited from taking land from smallholders for lack of payment of debt.

Land reform can also consolidate individual fragmented holdings into contiguous blocks of land (World Bank 1975:2–21). This kind of reform is intended to improve productivity, not to redistribute wealth. In most cases of land reform, the hope is that the twin objectives of accelerated growth and increased equity can be accomplished.

The land reform program carried out by the US military government in Japan following World War II is widely credited with significantly helping the reconstruction of Japanese agriculture at the time. Similarly, a land reform program in Taiwan at about the same time is credited with stimulating greater productivity in Taiwanese agriculture.

There are two ways in which land reform can promote agricultural productivity. In Chapter 13 we pointed out that production per unit of land in the third world is typically higher on smaller farms, so dividing up large landholdings should result in increases in productivity. In addition, farmers with a permanent ownership interest in land have more incentive to make improvements on the land, and to work longer hours than would sharecropping farmers who share the fruits of their labor with absentee owners (Herring 1983).

Land redistribution can also create impediments for economic growth. For example, the Communist Revolution in China brought with it an agrarian reform that eliminated private ownership of land in the early 1950s (see Box 2.2). The move to communal ownership initially spurred agricultural production, but progress had stagnated by the late 1950s. Ultimately, during the 1980s, laws were changed to return agricultural land to private ownership, and landholdings became more concentrated (Ying 1996).

Other attempts at land reform have produced unforeseen consequences, some of which made the supposed beneficiaries worse-off than they might have been without the reform. One of the most pressing goals of the 1962 Algerian land reform was to provide employment for as many workers as possible. Yet it did not take long for the self-management committees on the newly nationalized large estates to realize that fewer workers on their farms meant more returns per worker, and an early study of the situation showed employment on the farms actually decreasing after the reform (Foster and Steiner 1964). Pfeifer concluded that the Algerian reform "promoted, rather than curtailed, the class differentiation of agricultural producers into successful commercial farmers and propertyless wage workers" (1985:81).

Likewise, an agrarian reform law passed in Peru in 1969 was intended to help correct the skewed landownership in that country, but the main beneficiaries turned out to be the relatively well-off permanent workers. Families living outside the sugar plantations received no benefits at all. Alberts concluded that "the agrarian reform did not accomplish a radical and lasting improvement in the degree of equity within the agricultural sector. The economic policies implemented by the military government were not conducive to agricultural growth nor did they accomplish anything toward reducing the urban-rural income gap" (1983:226).

In the early 1950s, Burma (Myanmar) passed a law requiring agricultural land to be worked by its owners. To keep from losing their lands, absentee owners began working it themselves, forcing their former tenant farmers off the land. The former tenants usually stayed on as laborers, but they no longer enjoyed some of the benefits that had been theirs as tenants. This same Burmese law made it illegal to foreclose on mortgages on agricultural land when the owners defaulted on their loans. Without land to pledge as collateral, farmers had trouble finding people willing to lend money to them, interest rates rose, and agricultural investment declined (Walinsky 1962).

In the late 1990s, a major land reform program was instituted in Zimbabwe. In 1997, President Robert Mugabe proposed a plan that would seize 10 million acres of farmland owned by about 1,500 large commercial farmers, and redistribute it. The initiative was a response to landownership patterns that had been established during colonialism, when blacks in Zimbabwe (then called Rhodesia) were legally prohibited from owning some of the best farmland. As a result, "whites made up 2 percent of Zimbabwe's population but own[ed] 70 percent of the nation's best land" (Duke 1998:A27). Implementation of the land reform has been harshly criticized for allowing extralegal bands of squatters to seize farms, and for contributing to the famine conditions in Zimbabwe in recent years.

De Janvry makes this observation: "With agriculture well advanced on the road to modernization . . . any drastic land redistribution is likely to nullify past technological achievements and imply shortfalls in production, at least in the short run. Where the population is increasingly landless and urbanized, the social cost of higher food prices [because of the inefficiencies resulting from land reform] may be more widespread than the welfare gains of land redistribution" (1981:389).

One final problem associated with land reform, or even the threat of land reform, is the chilling effect it may have on investment in agriculture relative to investment in other productive activities. Landowners who fear that land reform may be in the offing are understandably hesitant to invest heavily in productive improvements for their farms. In this view, land reform is part of a number of antiagricultural policies collectively referred to as *urban bias,* which we will discuss further in Chapter 21.

18

Policies That Address the Demographic Causes of Undernutrition

> One result of [population/resource] projections and their use in public discussion of population policy has been a shift in concern toward future generations. In China, as in most traditional societies, childbearing decisions were shaped by a desire by parents to be looked after in old age. By emphasizing future population/resource relationships in shaping family planning programs, government officials have shifted the focus of childbearing from the well-being of parents to the well-being of children.
>
> —Lester Brown (1983:38–39)

It may at first appear that population policy has little to do with food price. But Figure 18.1 illustrates how a policy that reduces population growth relaxes the upward pressure on price and increases food availability per person. A successful policy to reduce population growth will result in an aggregate demand curve for food that is closer to the origin. The resulting equilibrium is at a lower price and a lower quantity than the equilibrium that would exist if the population growth stayed high. But although the quantity of food drops, it drops by less than the decline in population; therefore, the food availability per person increases.

The usual way of thinking about population policy is that the cost of having children is not borne solely by the parents; the decision to have children imposes an external cost on society at large. A growing population puts pressure on the natural environment, makes the task of education more difficult, thins out the supply of capital per person, and tends to decrease equity. As described in Chapter 16, the existence of external costs provides an economic rationale for government to intervene to reduce the number of children. Following this logic, the present chapter examines policy alternatives for lowering the fertility rate.

We should note at the outset that, worldwide, fertility rates are already falling, and in parts of Europe have fallen far below the replacement rate.

Figure 18.1 If Population Grows More Slowly, Food Price Drops and Food Available per Person Increases

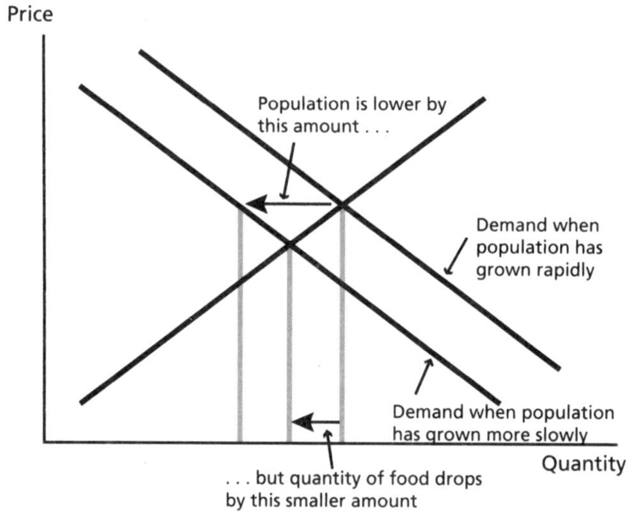

Columnist Mark Steyn (2006) argues that population growth may strengthen the geostrategic position of a country, and in this way the decision to have children may actually have external *benefits* to the society as a whole. In this case, the government's appropriate population policy would be to *increase* the fertility rate. Box 18.1 explores some European countries that have adopted or are considering pro-natalist population policies.

We begin with a review of the causes for the low fertility rates in Europe, since they may provide a road map to possible policies to reduce fertility in developing countries.

Ongoing Reasons for the Fall in Fertility Rates

Over the years, research has produced increasingly reliable and convenient methods of contraception, thus making it easier to limit fertility. But at the same time, many other things have been happening that have lowered human fertility rates (Caldwell 1983; Pullum 1983). It is a widely observed phenomenon, for example, that high income yields lower fertility, at least after the initial stages of extreme poverty are overcome. This relationship, discussed in Chapter 8, is illustrated in Table 18.1.

When very poor families experience an increase in income, their initial reaction is often to have more children. (The increased income, for instance, may

Box 18.1 Pro-Natalist Population Policies in Europe

In many European countries, fertility rates have fallen below the replacement rate, and governments have reacted in different ways.

The mayor of Laviano, Italy, started a program to pay women 10,000 euros (about $15,000) for bearing and raising a child within the city (Shorto 2008).

Norway has a program that pays mothers 80 percent of their salary during fifty-four weeks of maternity leave, and subsidizes childcare thereafter (Shorto 2008).

In Russia, one observer blames the low-fertility situation on the high cost of obstetric healthcare, and notes: "Politicians . . . speak of a demographic crisis in the country. They're saying that Russian women must be forced by any means to have not just one child but two or more" (Kakturskaya 2003).

France has succeeded in raising fertility rates: "While falling birthrates threaten to undermine economies and social stability across much of an aging Europe, French fertility rates are increasing. . . . France heavily subsidizes children and families from pregnancy to young adulthood with liberal maternity leaves and part-time work laws for women" (Moore 2006).

enable a marriage that otherwise might have been postponed for want of adequate dowry. Or the increased income might motivate the substitution of bottle-feeding for breast-feeding, with its consequent increase in the mother's fertility.) But as incomes rise, other factors mitigate to reduce fertility. Higher income is usually associated with better education, and better-educated parents tend to trade child quantity for child quality. They spend a greater proportion of their child-rearing resources on their children's health and education, with correspondingly smaller resources left over for raising more children. As income continues to rise, alternative uses of family resources open up: travel, more education, more leisure activities, for example. These, in turn, further compete with child-rearing resources in family decisionmaking, putting more downward pressure on fertility.

As education, income, and health improve, a decline in infant and child mortality rates normally occurs. After one or two generations of low infant and child mortality rates, people are less inclined to produce large numbers of children to ensure that at least some offspring will survive to maturity. Very high levels of income lead to a proliferation of private pension programs, or even government-sponsored social security programs, all of which lessen the pressure to have children because people are better able to care for themselves in their old age.

The education of women is particularly significant in reducing fertility. Educated women are more likely to postpone marriage in order to enter the work force, more likely to delay having children in order to remain in the work force, and more likely to know about and use contraception than are uneducated

Table 18.1 Population Growth, Fertility, and Mortality, Selected Country Groups, 2001

	GNI per Capita (PPP)	Population Growth, 1991–2001 (percentage)	Total Fertility Rate (births per woman)	Birth Rate (births per 1,000 population)	Death Rate (deaths per 1,000 population)	Rate of Natural Increase (percentage per year)	Life Expectancy at Birth (years)	Infant Mortality Rate (deaths per 1,000 live births)
World	7,376	14.8	2.70	20.9	8.8	0.1	63.8	53.1
Less developed countries	3,850	17.8	2.97	23.2	8.5	–0.4	62.3	58.3
More developed countries	26,989	4.0	1.58	11.2	10.1	1.8	75.9	8.8
Sub-Saharan Africa	1,831	28.6	5.42	39.1	16.1	–0.5	47.6	91.7
Asia	3,581	14.9	2.46	19.7	7.6	–0.1	66.0	50.4
Latin America and Caribbean	7,050	17.4	2.56	21.3	6.0	–1.6	71.2	30.2

Sources: Data for GNI per capita from UNDP 2003, tab. 1, grouped according to the following categories: "developing countries," "high-income countries," "Sub-Saharan Africa developing countries," "Latin America and Caribbean developing countries," and a population-weighted average of "East Asia and Pacific developing countries" and "South Asia developing countries." Remaining population data from US Bureau of the Census, International Database.

women (Anonymous 1988). The relationship between education and fertility for selected countries is shown in Figure 18.2.

Increased employment of women outside the home lessens their dependence on men (who sometimes are less motivated to limit family size than are women) and increases their tendency to favor use of contraception. At the same time, increased employment outside the home tends to change how women think of themselves. As women gain more and more equality with men in the work force and elsewhere in society (gaining the right to vote, to inherit property, to own land, to participate in the choice of a husband, etc.), they move away from thinking of themselves primarily as wives and mothers and toward thinking of themselves as playing multiple roles in life. And as they do this, they tend to have fewer children.

Public Measures to Accelerate the Demographic Transition

The phenomena we have been discussing will continue to put downward pressure on worldwide fertility rates. But in some regions of the world, the advantages of lower fertility rates are so great that governments want to accelerate the downward movement.

Most countries have a set of pro-natalist public policies left over from the days when wars, famines, and high mortality rates from uncontrolled infectious diseases, such as smallpox and the bubonic plague, regularly decimated their populations. Pro-natalist policies include tax deductions in proportion to the number of children in the family (common throughout much of the world), unlimited subsidized maternity leaves sponsored by government or private industry

Figure 18.2 Total Fertility Rate by Education of Wife, Selected Countries

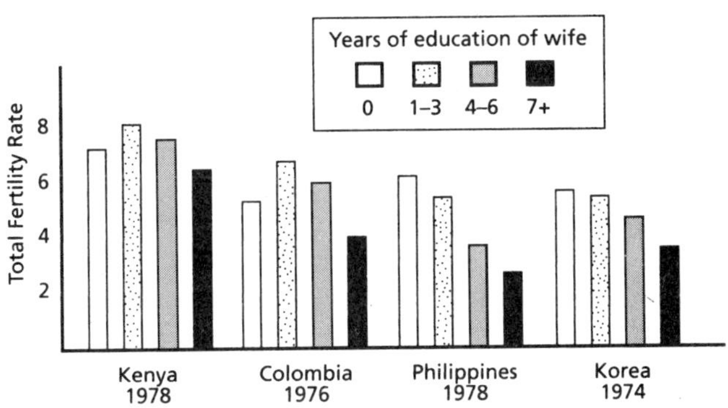

Source: Adapted from World Bank, *World Development Report*, 1984:110.

(again, common throughout the world), and childcare subsidies during the first several years of a child's life (found only in certain high-income countries, such as Canada, France, and Australia). For countries with such policies seeking to reduce the size of their populations, a first task in the direction of reducing fertility is to modify these policies so that they favor small families.

Economic Incentives and Disincentives

Economic incentives to reward low fertility, and economic disincentives to discourage high fertility, can both be used to motivate lower fertility.

One of the more imaginative incentive schemes involves family planning on tea estates in India. By law, the tea estates are required to provide substantial maternity and childcare benefits for their workers (tea-pickers are usually women). The benefits include hospitalization and medical care for the mother and infant, as well as long-term food, clothing, schooling, and medical care for the child. Each woman employee of childbearing age is offered a "savings account for family planning," the proceeds of which are available to her upon retirement, and into which the firm will pay the equivalent of one day's wages for each month that she is not pregnant. If the woman becomes pregnant, the company suspends payments for one year. For the third and successive pregnancies, the company not only suspends payments for a year, but also reclaims part of its past payments to help pay for its legally mandated maternal and childcare expenses. Women thus have a choice: maternity and childcare benefits for more children, or a better retirement program. Many women are opting for fewer children and more retirement benefits (Brown 1974:169; World Bank 1984:126, *World Development Report*).

Economic incentives for lower fertility are attractive, but they are expensive. In Bangladesh, a program was proposed that would provide a twelve-year bond with a maturity value of around $350 for women of childbearing age who had only two or three children and who underwent sterilization. Attached to the proposal was a scheme whereby couples who signed certificates to delay their first birth for three years after marriage, or who delayed their second and third births for at least five years, would be given $20 on presentation of their certificates after the agreed time, provided they had kept their pledge. It was estimated that to cover the entire population with both schemes would require about 10 percent of the annual government budget (World Bank 1984:126, *World Development Report*).

Not only are incentive payments for lower fertility expensive, but they can also waste public resources when people who would have had fewer children despite the incentive claim the financial benefits.

While economic incentives involve payments provided to delay or limit childbearing, economic disincentives usually involve the withholding of social benefits from those couples who produce more than a targeted number of children.

In the early 1980s, a series of economic disincentives to large families was in use in Singapore. The system (which was discontinued subsequent to a decline in birth rates) included incentives for families to have two children but disincentives for more than two. The system was summarized (Salaff and Wong 1983:16) as follows:

- Paid maternity leave for the first two children, but not for third and subsequent children.
- Preference in the choice of primary school given only to the first two children, with highest preference to the two children of a parent who has undergone sterilization before age 40.
- Removal of the large-family priority in the allocation of subsidized housing; only families with three or fewer children were allowed to rent rooms in public housing units.
- Escalation of delivery fees in public hospitals for the third and successive births, as well as fees for prenatal care (fees were remitted if sterilization followed delivery).
- Full tax relief only for the first two children, partial for the third, and none for fourth or subsequent children.

During this same period, a somewhat more draconian set of rewards and punishments, using "Glory Certificates" to promote the one-child family, was in place in China (see Box 18.2).

The government can also influence infertility indirectly by adopting policies that reduce benefits to parents who have more children, and raise the costs of child-rearing. For example, policies described by the catch-phrase "empowering women" work to reduce fertility in a variety of ways. First, as women feel that society and culture give them greater permission to participate in childbearing decisions, the costs to women of childbearing and child-rearing are more fully taken into account. Second, as women become better-educated, they become more aware of birth control techniques. Third, as women become better-educated, their value as workers increases; thus they see higher costs of foregone earnings or production as they devote time and attention to childbearing and child-rearing. Fourth, as women become more socially accepted in the labor market, their value as workers increases. Fifth, better education of women is likely to lead to reduced infant and child mortality rates, so that fewer births are necessary to achieve the desired number of surviving children. Sixth, better-educated women are more likely to want a good education for their children; this reduces children's availability for the labor force and therefore reduces the economic benefits and raises the costs of having children.

Box 18.2 The Chinese "Glory Certificate" System

Lynn Landman

To stimulate acceptance of the one-child family, the Chinese have instituted a system of incentives. Those who contract to limit their families to one child receive "Glory Certificates." Such couples are widely publicized and held up as models for their countrymen, as are the rewards they earn. In general, these may include free and priority medical care for the child; priority admission to nurseries, kindergartens, and primary schools; allotment of larger housing accommodations; and bonuses for city workers and increased work points, as well as large private plots and larger housing, for peasants. All these benefits remain in effect until the child reaches fourteen.

If holders of Glory Certificates renege on their commitment, they must return all the benefits and, in addition, their annual income may be reduced by 5 to 15 percent for varying lengths of time. Those who already have two children and go on to have a third must pay all the expenses associated with childbearing, and will not be entitled to paid maternity leave. Salaries are reduced and the usual subsidized grain allotment is not provided for the third child, obliging parents to purchase it on the open market at higher prices. Job promotions may be withheld for a time and mothers sometimes are fired from their jobs.

In a country where per capita annual income is only about $235, where housing is in short supply, and schooling and jobs are not guaranteed, these incentives and disincentives may be presumed to carry considerable clout.

Source: Landman 1983:9.

Note: In 2008, the fertility rate in China was estimated to be 1.8 children per woman (US Bureau of the Census 2008).

Moral Suasion and Regulation

The average marriage age for women in Bangladesh is 16. Half the women in South Asia and sub-Saharan Africa are or have been married by the time they are 19. The younger a woman is when she marries, the longer she is exposed to the risk of conception. So, to reduce fertility, it seems reasonable to try to persuade people to postpone marriage, such as through legislation that establishes minimum marriageable age. Of those countries that have tried such legislation, China seems to have been most successful. In 1980 the Chinese government raised the legal minimum marriageable age to 20 for women and 22 for men.

India has long promoted a vigorous advertising campaign to encourage the small family, with government-sponsored advertisements appearing on billboards and buses, at movie theaters, in magazines and newspapers, and on radio and television. (India's total fertility rate in 1986 was 4.4 children per woman. By 2000, it had fallen to 3.1.)

Some countries have tried intense community pressure on couples of childbearing age to limit their family size. Examples of such efforts in Indonesia and China are described in Box 18.3.

Box 18.3 Community Pressures to Lower Fertility

Rodolfo A. Bulatao

Pressures can be exerted by the community, or by major sections of it, to promote lowered fertility. Two cases will illustrate group pressures: *banjars* in Bali and production teams in China.

Banjars—traditional units of local self-government, which serve as centers for mutual aid and cooperative work—consist of all the male household heads in a hamlet or subvillage. The form is centuries old. The traditional head of a banjar is democratically elected but has no official standing. Instead, the banjar also has a second, official head, who may be appointed and may have charge of more than one banjar (Hull 1978).

Banjar meetings may be held every month (or thirty-five days), usually with perfect attendance (there is a system of fines for absence or lateness), and typically discuss development of the community and religious affairs (Astawa 1979).

Since 1974 these meetings have also included discussion of the family-planning status of each family. Each member is asked what he and his wife are doing about family planning. A register is kept, and a color-coded map of the community indicating eligible couples and their contraceptive status is prominently displayed in the banjar hall (Meier 1979).

The decline in marital fertility in Bali of about 30 percent in less than a decade has been dramatic enough to be labeled a "demographic miracle" (Hull et al. 1977). How much of the change has been due to the community pressures exerted through the 3,700 banjars is a difficult and probably unanswerable question. Other elements of the Balinese situation, such as acute pressures on the land, the penetration of modern influences (through such means as consumer goods, communication, and transportation systems, and Western-style schooling and tourism), and cultural factors such as the relative independence of young couples—which may facilitate contraceptive decisions—may encourage the decline in fertility. Furthermore, the effective logistical system of the family-planning program and creative uses of native art forms to communicate family-planning messages, and a stable, supportive government, may be influential.

Production teams in China, which are usually the effective unit in rural areas for production and income sharing, consist of thirty to forty households in a small village, within which kinship ties may be strong. Production teams assume important responsibility for the fertility of their members. As part of the national *wan xi shao* campaign (named for the reproductive norms of later marriage, longer birth spacing, and fewer births), the production teams were responsible for deciding which couples could have births, in line with the reproductive norms and with team quotas set from above (Chen and Kols 1982). The team birth-planning leadership group (the leaders all being local residents) might call all eligible couples to a meeting, at which their individual birth plans could be scrutinized and allocations made. Under the one-child campaign, which replaced the wan xi shao campaign in 1979, community birth planning still takes place, although allocation of birth quotas follows different norms. As couples become familiar with the system, the time-consuming meeting to adjust birth plans may be dispensed with, and the leaders may simply notify couples of their decisions.

(continues)

Box 18.3 continued

Adherence is in theory voluntary, resting on persuasion and education. Such elements as adult study groups and visits from birth-planning delegations maintain the peer pressure (Chen 1981).

As with the Balinese banjars, it is not possible to determine the specific impact of the social pressures exerted through production teams, which are only one element in the Chinese population program.

Source: Bulatao 1984a.

Subsidizing Family-Planning Services

Use of contraceptives among married women of childbearing age varies widely in the third world, from less than 10 percent in sub-Saharan Africa to around 40 percent in Latin America to as high as 70 percent or more in China and Singapore (World Bank 1984:128, *World Development Report*). Controlled experiments conducted in Mexico, India, Bangladesh, Korea, and the Philippines have all demonstrated that the provision of family-planning advice, technology, and materials significantly reduces fertility.

In a study of thirty-one countries, Bongaarts (1982) looked at determinants of fertility decline. He found that, in the countries studied, higher age at marriage reduced total fertility by 1.4 children. Increased use of contraception reduced fertility by 4.5 children. Greater use of induced abortion accounted for a reduction of 0.5 children, for a total reduction of 6.4 children. Reduced breastfeeding, of course, works the other way around, and accounted for an increase in fertility of about 1.5 children. In the bottom row of Table 18.2, these data are expressed as percentage contributions to reduction in fertility decline.

McDevitt (1996) offers a comprehensive review of information about contraceptive use in developing countries. In most third world countries, a substantial gap exists between women who would like to limit their fertility, and their access to modern contraceptive methods (see Table 18.3). However, the information from McDevitt for the late 1990s shows considerable improvement over similar information from the early 1980s published by Galway and colleagues (1987). McDevitt shows that fertility levels and contraceptive use are strongly (negatively) correlated across countries. Countries with fertility rates of 6 births per woman have contraceptive use rates of 10 percent or less for the most part, while countries with fertility rates of 3 have contraceptive use rates in the 45–60 percent range. Simmons and Lapham (1987) note that the impact of contraception use on fertility varies with programmatic and environmental factors. For instance, the availability of multiple public and private

Table 18.2 Accounting for Fertility Decline in Selected Third World Countries

	Total Fertility Rate (births per woman)			Percentage of Fertility Reduction by Contributing Factor				
	Initial	Final	Decrease	Older Age at Marriage	Reduced Breast-Feeding	Increased Use of Contraception	Increased Use of Induced Abortion	All Other Factors
India (1972–1978)	5.6	5.2	0.4	41	–58	114	n/a	3
Indonesia (1970–1980)	5.5	4.6	0.9	41	–77	134	n/a	2
Korea (1960–1970)	6.1	4.0	2.1	50	–38	53	30	4
Thailand (1968–1978)	6.1	3.4	2.7	11	–17	86	16	4
Composite of 31 countries (long-term)	> 6.0	< 3.0	5.0	28	–29	90	10	1

Sources: Data for composite of 31 countries from Bongaarts 1982; all other data from Bulatao 1984b:38.

Notes: Data for the composite of 31 countries account for the decline in total fertility that is typical of countries that start with rates above 6.0 (the predecline phase) and end with rates below 3.0 (the postdecline phase). The difference between the pre- to postdecline phases among these countries amounts to almost five children.

n/a = not available.

channels for the delivery of services increases the effectiveness of national programs. A recent contraceptive effort in Rwanda is described in Box 18.4.

The Complementarity of Fertility Reduction Policies

Fertility reduction policies often complement each other. For instance, providing subsidized family-planning services not only makes the technology available for reducing fertility, but also sends a message to the community that government supports the idea of fertility regulation. Joel Cohen (1996b) recommends, as part of any population control policy, "doing everything at once."

Table 18.3 Unmet Need for Contraceptives in Selected Countries

	Percentage of Women Without Access to Contraception Who Want		
	No More Children	Fewer Children	Total
Haiti	23	17	40
Yemen	21	18	39
Ethiopia	14	21	35
Uganda	14	20	34
Pakistan	17	11	28
Philippines	12	8	20
Nigeria	5	13	18
India	8	8	16
Indonesia	5	4	9
Brazil	5	2	7

Source: McDevitt 1996: fig. 54.

Box 18.4 Norplant Distribution in Rwanda

Rwanda's fertility rate (6.1 children per woman) is among the world's highest. Its population doubled between the mid-1980s and 2008 and is projected by the US Census Bureau to double again by 2038. In early 2007, Rwandan president Paul Kagama sat down with a *New York Times* reporter (Kinzer 2007) to discuss his plan for a national population control program.

Family-planning counseling will be required for every patient at a hospital or health center, regardless of their ailment. Free Norplant devices will be provided to all women of child-bearing age. Norplant is a small device that is implanted under the skin and that releases contraceptive hormones. It is an effective contraceptive for up to five years after implantation. The cost of the project is being underwritten by donations from the US government.

However, fertility reduction policies are often also complementary with programs that help to reduce undernutrition. For instance, successful promotion of prolonged breast-feeding not only reduces fertility but improves childhood nutrition and health. Persuading couples to marry later in life not only reduces fertility but, because women remain in the work force longer as a result, raises per capita income, thereby improving nutrition. Increasing the educational level of women not only decreases fertility but also increases their future productivity and undoubtedly improves the quality of the childcare they deliver. Cohen (1996b) quotes economist Robert Cassen as saying, "Virtually everything that needs doing from a population point of view needs doing anyway."

Demographer John Bongaarts, writing in 1994 on the eve of a worldwide conference on population, recommended the following steps to reduce population growth:

1. Reduce unwanted pregnancies by improving contraceptive education and availability.
2. Reduce the demand for large families by investing in education and reducing infant and child mortality.
3. Address population momentum by increasing the age of marriage and lengthening intervals between births.

19

Policies That Reduce the Price of Food Through Subsidized Consumption

> There is no greater scam in India at this time than the so-called food subsidy. Under the cover of "food security," the Government is keeping millions of tonnes of food out of reach of poor people.
>
> —Jean Dreze (2001)

In this chapter we survey policies aimed explicitly at lowering the food prices paid by consumers. To many, this is the most obvious and straightforward approach to the problem of undernutrition: give food (or sell it at subsidized prices) to the hungry. Here we will describe the mechanisms of subsidy that have been tried, and some of the problems that have arisen.

Rationale for Explicit Food Subsidies

There are a number of reasons why food consumption subsidies have been so popular. Developed countries have generally followed farm production policies that have led to burdensome agricultural surpluses. Furthermore, large numbers of people are hungry now and it is tempting for policymakers to feed people today rather than to sponsor programs, such as enhanced agricultural production research, that may take months or years to produce obvious benefits. And donors of famine relief are more charitable when their donations are directed to the most needy.

There are other reasons for the popularity of food consumption subsidies. Generally, rich people prefer to give hungry people food rather than cash (see the discussion of "basic needs" in Chapter 16). Further, particular groups of rich people derive benefits from food distribution programs: these groups include food processors and input suppliers in food-exporting countries; farmers, who see the demand increased for their products; grain elevator operators, who hope to store the food before shipment; the sea-freight shippers; and finally, the private voluntary agencies, such as CARE and Catholic Relief Services, that assist in distributing surplus food.

In developing countries, political leaders are interested in creating or continuing such programs. The groups most likely to influence political power are the military, civil servants, urban labor, and industrial interests. All of these groups are happy to be the recipients of cheap food, and political leaders are generally happy to curry favor among them, even at the expense of the country's rural sector (Hopkins 1988). Of course, the broader the class of subsidy recipients, the higher the cost of the subsidy program. And these costs can escalate rapidly, as explained in Box 19.1.

Proof of the political popularity of food price subsidies can be found in the morning newspaper. In 1996, after Jordan cut food price subsidies due to pressure from the International Monetary Fund, bread prices doubled and angry

Box 19.1 The Increasing Cost of Increasing Food Consumption Through a Subsidy

Shlomo Reutlinger

If households allocate only an increasingly smaller share of additional income to the augmentation of their energy intake, then the marginal cost of inducing energy augmentation through public intervention rises sharply as higher levels of intake are sought.

As an illustration, consider a nation in which 5 million people have average daily energy intakes of 1,500 calories, 15 million of 1,600 calories, 10 million of 1,700 calories, and the remainder of 1,800 calories and more. Let us further assume that, with declining income elasticity of demand as income rises, the additional (annual) income required to increase daily energy intake by 100 calories is \$10, \$15, and \$25, respectively, at the level of intake of 1,500, 1,600, and 1,700 calories. If the goal of the public intervention is to assure the entire population a minimum energy intake of 1,600 calories, 5 million people at very low levels of intake would have to get a total cash transfer of \$50 million. If the goal were to assure a minimum of 1,700 calories in the population, an additional \$15 per capita would have to be provided to the 5 million people with the lowest energy intake as well as to 15 million more people. The additional cost would be \$300 million more. If a minimum energy intake of 1,800 calories were to be assured, the additional cost would be \$1 billion. The marginal cost of raising minimum energy intakes from 1,700 to 1,800 calories is twenty times the marginal cost of raising minimum intakes from 1,500 to 1,600 calories.

The above calculations are illustrative, but not unrealistic, given what we know about the declining marginal propensities of households, at different levels of energy intake and income, to allocate additional income to energy intake. The marginal cost of public interventions to increase energy intake rises sharply as higher levels of intake are sought.

Source: Extracted from Reutlinger 1985:10–11.

demonstrators demanded that the prime minister be removed from office (Reuters 1996). In Zimbabwe, an economic crisis led the government to increase prices for food staples by 30 percent in 2000. Riots began in the capital city of Harare, and 160 people were arrested (Shaw 2000). In 2007, high corn prices—and thus high tortilla prices—caused riots in Mexico (Watts 2007).

Food Subsidies in Asia

Sri Lanka

When a food subsidy lowers prices to the general population, the costs of the program become so high that access must be limited through a system of rationing. This is illustrated by the experience of Sri Lanka. During World War II, when rice supplies were limited, the government instituted a program under which rice was sold at a subsidized price, but the quantity to each consumer was rationed (Edirisinghe and Poleman 1983).

In 1953, the costs of the rice program became too high, and price increases of nearly 300 percent were announced. A massive protest stopped the price increases and forced the resignation of the prime minister. From 1954 to 1966, Sri Lankans could buy rice at prices substantially below the world market price, but, through rationing, access was restricted to 4 pounds per week (equivalent to about 1,000 calories per person per day) (Edirisinghe 1987:12–13). In 1966, the basic weekly ration was cut in half, but issued at no charge. Two things were significant here: (1) because government did not need to purchase as much rice overseas, substantial foreign exchange savings accrued, and (2) there were no food riots.

In 1978, the ration system was targeted to the lower end of the income range through a means test. A year and a half later, food stamps were substituted for the ration cards. The food stamps carried a fixed rupee value; therefore their purchasing power declined with inflation, resulting in an automatic reduction of the inflation-adjusted costs of the food subsidy with no further government action. In 1985, targeting was restricted further, so that only the poorest quarter of the population was eligible for food stamps (Sahn and Edirisinghe 1993). The policy reforms succeeded in reducing government costs: food subsidies amounted to 23 percent of government expenditures in 1970, 19 percent in 1978, and 4 percent in 1984.

Bangladesh

Bangaldeshi food subsidies began during World War II. By the 1980s, the program had evolved into a complex series of measures including rural food rationing, food-for-work, public purchases and sales to stabilize prices, and government control of imports. The International Food Policy Research Institute began an evaluation of the rural food-rationing program. This program provided limited amounts of low-cost food to all rural households. The IFPRI concluded that the program was poorly targeted—70 percent of the subsidized food went to households who did not need government assistance. The IFPRI recommended that the money would be better spent on a "food-for-education" program that provided subsidized food to poor households who agreed to send their children to school. (For more information on food subsidies in Bangladesh, see Ahmed and Goletti 1997.)

India

The Indian food subsidy program (the subject of this chapter's opening epigraph) was intended to operate by using government funds to purchase food grains and then selling the grains (at a loss) through government shops. In practice however, the program bought more food than it sold; as government-owned food stocks rose, stored food began to rot, which resulted in huge administrative costs and price increases. According to Dreze: "Ordinary households . . . benefit very little from this 'subsidy.' . . . What they gain on one side from subsidized food . . . pales in comparison to what they lose as a result of having to pay higher food prices on the market. . . . [E]ven [poor] households see little advantage in purchasing food from ration shops . . . because the price differential is too small to compensate for the quality differential" (2001). So who gains from the program? Lakshmi reports the details of how corrupt dealers take advantage of the system:

> [T]he poorest people of India receive ration cards that they can use to buy wheat from government run "ration shops" at very low prices. But in the village of Kelwara, ration shopkeepers turned card holders away, saying the shops had received no wheat supplies from the government. In fact, the ration shops had received government wheat, but had sold it at higher market prices to people who did not qualify for ration cards. The shopkeepers covered their tracks by keeping a fraudulent set of books, that showed sales to ration card holders, when in fact those sales had not occurred. (2004:A17)

Food Subsidies in Africa

Egypt

The Egyptian government has a history of intervening in the food-marketing system that dates back to biblical times when Joseph, interpreting the pharaoh's dream, recommended storing grain during seven fat years to prepare for the seven lean years that he prophesied were to come (Genesis 41). Since the mid-1970s, the Egyptian government has taken on a substantial burden of public expenditures for food subsidies, with the share of the government expenditures for this purpose running as high as 17 percent (Alderman and von Braun 1984:12).

Additional costs of the Egyptian food subsidy have been borne by North American and European governments, which have provided substantial quantities of food at below-market prices. Indeed, the availability of such programs may be one of the reasons that Egyptians embarked on such an ambitious marketwide food subsidy.

As of the late 1980s, the Egyptian government was handling the major share of the sales of bread, flour, pulses, sugar, tea, and cooking oil in the country, making these commodities available to householders at prices significantly below world prices. Farm-gate prices (prices the farmer receives at the gate, before paying transportation costs to market) deviated less from world prices than did retail prices (see Table 19.1), but both sets of prices demonstrated a priority goal of Egyptian policy: cheap food for all.

The policy is widely credited with keeping the Egyptian rate of undernutrition low. Average calorie consumption exceeded requirements even among the poorest 12 percent of the population as a whole (USDA 1984:9), although significant numbers of urban households in the lowest-income quartile were found to be calorie-deficient (Alderman and von Braun 1984).

Despite the apparent success of the Egyptian food subsidy, it has been criticized as inefficient. Sources of inefficiency include:

- *Waste.* With bread as cheap as it is in Egypt, farmers purchase significant quantities of it for livestock feed. The resources spent processing the wheat into bread are a deadweight loss to society when the bread is fed to livestock.
- *Underinvestment in industry.* The more foreign exchange that is spent on a food subsidy, the less that is available for industrial investment. One study estimated that a 10 percent increase in available foreign exchange would increase industrial investment by 6 percent and industrial output by 4 percent (Scobie 1983). High rates of government spending on imported food could adversely affect industrial employment among the poor.

Table 19.1 Farm-Gate and Retail Price of Selected Agricultural Commodities, Egypt, 1982

	Price as Percentage of World Price	
	Farm-Gate	Retail
Wheat	64.5	36.8
Rice	26.6	17.7
Sugar	46.0	27.3
Beans	75.4	49.0
Cotton	27.2	41.3

Source: Computations by John Rountree, 1985, at the University of Maryland, based on data provided by Egypt's Ministry of Agriculture, Central Agency for Public Mobilization and Statistics, and Ministry of Supply.

• *Consumption inefficiencies.* Because of the depressed price of wheat, Egyptians eat more wheat than they would if they were paying the world price. A loss to Egyptian society associated with this overconsumption results because government pays more for wheat bought at world market prices than Egyptian citizens are willing to pay for that wheat. The amount of this cost above worth is represented by triangle 1 of Figure 19.1, which illustrates the case where the government uses subsidies to keep both the consumer price and the producer price below equilibrium, and makes up the difference in quantity demanded minus quantity supplied with (donated) imports from abroad.

Maize Subsidies in Southern Africa

Maize is the staple food crop in many countries of southern Africa. For example, in Zambia, maize makes up about two-thirds of calories consumed. During the 1980s, countries in the region experimented with government monopolies in the marketing of maize. Farmers could only sell to the government, and consumers could only buy from the government. By operating these monopolies at a loss, the government effectively subsidized both consumption and production, as illustrated in Figure 19.2.

(Notice the difference between Figure 19.1 and Figure 19.2. Figure 19.1 describes a policy that sets domestic price lower than the world price, but domestic consumers and producers face the same low price. The quantity consumed at this price is higher than the quantity produced; this difference must be made up with imports. Figure 19.2 describes a policy that allows the price farmers receive to be higher than the price consumers pay, but the quantity consumed is equal to the quantity produced.)

However, the costs of the subsidy program were enormous (Mwanaumo, Preckel, and Farris 1994). For example, in 1990 in Zambia, the cost of the maize subsidy program accounted for over 10 percent of the total government budget. The food subsidy programs in southern Africa contributed to overall government budget deficits (the Zambian government ran a deficit equal to 20 percent of its expenditures in 1986). As the countries sought help to finance these budget deficits, the IMF made eliminating the subsidies and privatizing the parastatals a condition of loans. Between 1990 and 1996—before and after the elimination of the food subsidy programs—the percentage of population suffering from undernutrition *declined* in three of the four countries involved, and increased slightly in Zambia. Jayne and colleagues (1995) conclude:

> Consumer subsidies on refined maize meal in [the four countries] have not necessarily promoted food security, because they have entrenched a relatively high-cost marketing system and impeded the development of lower-cost channels from developing. The negative effects of eliminating subsidies . . . have been partially or wholly compensated by relaxing controls on private

Figure 19.1 Cost Above Worth and Producer's Surplus Lost Due to a Marketwide Explicit Subsidy

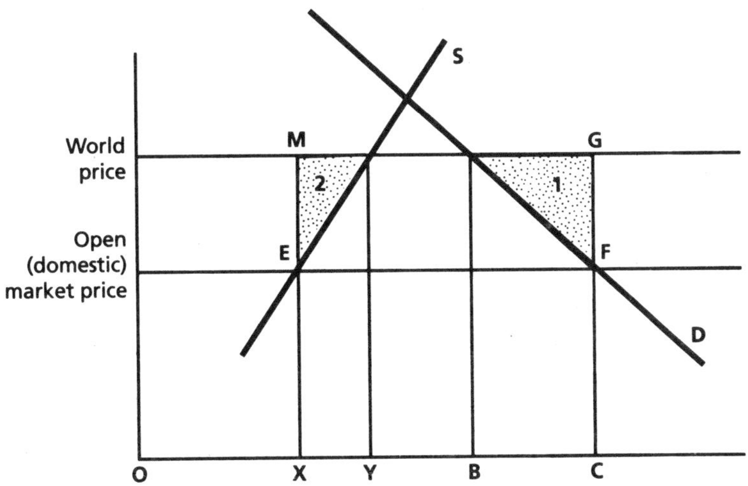

Note: Cost above worth – Suppose that D represents the demand curve for wheat and that OC represents the amount of wheat consumed, given the domestic and market price. If quantity BC is imported, then the area of triangle 1 is the loss to society, since government paid more for this wheat than it was worth to consumers.

Producer's surplus lost – Now suppose that S represents the supply curve for wheat, assuming no concessionary sales were available. Quantity OX represents wheat produced in Egypt given a depressed, domestic market price. Quantity XY represents wheat imported that could have been produced locally had the local price of wheat been equal to the world price. The area of triangle 2 is the loss to the Egyptian farmer, because it is a producer's surplus he could capture were he getting the world price, but which he now misses out on. Notice that the consumer would not care whether he paid the world price to the farmer or to a foreigner. But the Egyptian farmer cares, because he can produce that quantity of wheat with fewer resources than can the foreigner. And the economy cares, too, because triangle 2 is a loss to the Egyptian economy.

grain trade, which has raised consumers' access to less expensive whole maize meal distributed through the emerging informal markets. A 53 percent rise in the price of refined meal in Kenya (due to subsidy removal) has been estimated to raise household expenditures by less than 1 percent of total income for low income groups, due to the widespread availability of cheaper whole meal.

Zimbabwe

In 1996, Zimbabwe's food distribution program was being cited as a good example of how to help the poor improve their nutritional status (Jayne et al. 1996). The changes admired at that time included:

Figure 19.2 Impact of a Government Subsidy on Consumption: Lower Consumer Price, Higher Producer Price, Higher Quantity Consumed

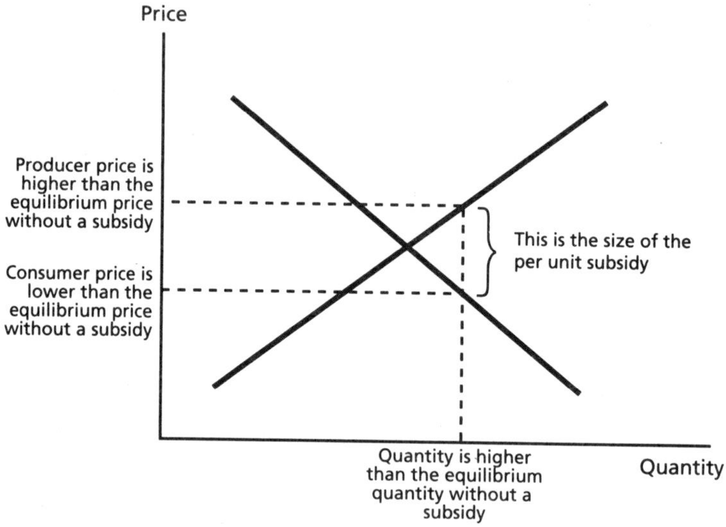

1. The elimination of a policy under which the government had a central role in maize marketing. That policy
   a. required maize farmers to sell their output to the government, and
   b. required the maize to be sold at subsidized prices to privately owned, large-scale, relatively high-tech "roller mills" that produced highly refined maize meal.
2. This elimination allowed the emergence of small-scale, relatively low-tech "hammer mills" that produced more coarsely milled maize, but at much lower cost.
3. Therefore, the elimination of the government subsidy did not hurt poor consumers, because they now had the option of low-cost hammer-milled maize meal to replace the subsidized, refined maize meal.
4. The subsidy elimination saved the government significant costs of a program that had provided a lot of benefits to the relatively high-income consumers of refined maize meal. Therefore, the change reflected a move from a "nontargeted" subsidy to a system that provided the new option of coarsely milled maize meal—a product consumed mostly by low-income consumers.

But in 2001—in part because of a regional drought, and in part because of government land redistribution policies that discouraged domestic production—food shortages became widespread and food price increases led to riots in the capital city (see Shaw 2000). The government reinstituted central control over the maize market. The new program did not solve the food shortage problem (Mudimu 2003).

By 2005, the government-run maize distribution program had become a corrupting force in the country's politics. As reported by Timberg (2005):

> Hundreds of bags of cornmeal were stacked in front of a bar. . . . The officials first held a rally. . . . The next day, as hundreds of people from surrounding villages gathered to collect the 110-pound bags they had ordered and paid for months before, ruling party officials announced that only their supporters were eligible. When the names of opposition voters were called, they were simply handed back their money, according to several people who were turned away. The leftover bags went on sale hours later for twice the price.

Next, the government tried to restore economic growth by pumping vast sums of currency into the economy; the subsequent massive inflation then inspired Zimbabwean president Robert Mugabe to impose economy-wide price caps, with predictable results:

> Not even an unchallenged autocrat can repeal the laws of supply and demand. One month after Mr. Mugabe decreed just that, commanding merchants nationwide to counter 10,000-percent-a-year hyperinflation by slashing prices in half and more, Zimbabwe's economy is at a halt. Bread, sugar and cornmeal, staples of every Zimbabwean's diet, have vanished, seized by mobs who denuded stores like locusts in wheat fields. Meat is virtually nonexistent, even for members of the middle class who have money to buy it on the black market. . . . Zimbabwe's vast underclass, the majority of its 10 or 11 million people, has long been unable to afford most food, so the rural poor survive on whatever they can grow. Urban and rural poor alike stay afloat with food and money sent by the two million or more Zimbabweans who have fled abroad. . . . Mr. Mugabe has cast the price cuts as a strike not against hyperinflation, but against profiteering businesses that he says are part of a Western conspiracy to reimpose colonial rule. . . . The government took over the nation's slaughterhouses in early July after meat disappeared, [but] the takeover . . . seems ineffectual: this week, butchers killed and dressed 32 cows for the entire city. Farmers are unwilling to sell their cows at a loss. (Wines 2007)

The stark failure of government food policies is illustrated with this story:

> Less than ten miles from Zimbabwean President Robert Mugabe's mansion in Harare—the largest private residence on the African continent—Cleophus Masxigora digs for mice. On a good day, he told me, he can find 100 to 200. To capture the vermin, he burns brush to immobilize them, then kills them with several thumps of a shovel. This practice has become so widespread in Zimbabwe that . . . state-run television has broadcast warnings against citizens setting brush fires. Masxigora began hunting mice to support (and feed) his wife and three children soon after Mugabe began confiscating thousands of productive, white-owned farms in 2000, a policy that has since led to mass starvation. (Kirchick 2007)

Food Subsidies in Latin America

Brazil

During the period 1966–1982, the government of Brazil attempted to achieve self-sufficiency in wheat production and at the same time provide cheap wheat to its consumers. As part of its attempt to achieve these goals, the government became the sole seller and buyer of both domestically produced and imported wheat. The prices of wheat and wheat products were rigidly controlled throughout the economy. Farmers were encouraged to increase wheat production through a price support subsidy, and millers were provided with wheat at a price substantially below that paid to the producer, with the government making up the difference out of the general tax till.

In their study of the Brazilian wheat policy, Calegar and Schuh (1988:9–10, 43–45) determined that 86 percent of the subsidy went to consumers. This means that only 14 percent of the subsidy costs went to administration or were lost through slippages such as manipulations by the millers. Even so, only 19 percent of the total subsidy went to the true target group, the low-income consumers. Furthermore, gains in consumer welfare were slightly biased toward the high-income population groups (they bought more bread per capita than did the low-income groups). Calegar and Schuh conclude that the marketwide wheat consumption subsidy was not an effective policy for redistributing income and suggest that a preferred policy would be to target the food subsidy specifically at low-income groups.

Venezuela

Like President Mugabe in Zimbabwe, President Hugo Chávez of Venezuela adopted a policy of price controls, while continuing to rely on the private sector to distribute food. The results were exactly as predicted by economic theory: by setting the controlled price below the market equilibrium price, the government created a situation in which quantity demanded was larger than quantity supplied, and shortages developed. Rather than using market price to "ration"—to allocate food among the citizens of Venezuela—this system ended up using long lines to allocate the food to those first in line. In addition, people in the private sector began to devote considerable effort to figuring out how to "game" the system—how to use the price control program to make more money. Some people decided to break the law and buy and sell on a black market at prices higher than the price control program allowed. Others, betting on the possibility that the perception of food shortages would force the government to raise the prices, decided to hold on to food stocks ("hoarding" food, in the eyes of government regulators) until after the expected price increases went into effect. This behavior further exacerbated the food shortages.

The situation in 2007 is described by Pearson (2007):

> The state runs a nationwide network of subsidized food stores, but in recent months some items have become increasingly hard to find. . . . "They say there are no shortages, but I'm not finding anything in the stores," grumbled Ana Diaz, a 70-year-old housewife. . . . "There's a problem somewhere, and it needs to be fixed."
>
> Gonzalo Asuaje, president of the meat processors association . . . , said that costs and demand have surged but in four years the government has barely raised the price of beef, which now stands at $1.82 per pound. Simply getting beef to retailers now costs $2.41 per pound without including any markup, he said. "They want to sell it at the same price the cattle breeder gets for his cow," he said. "It's impossible."

Targeted Subsidies

Two common problems appear in the preceding descriptions: subsidies are expensive because they are not well targeted toward those in need, and subsidies are inefficient because of corruption or mismanagement. There is not much to say beyond the obvious about how to eliminate corruption and mismanagement. But there are a number of ways to target subsidies toward those who are food-insecure.

The Sri Lankan program (as it had evolved by the 1980s) illustrates one approach: target subsidies by requiring recipients to prove that they are undernourished or poor. A study of the Philippines advised a two-step procedure for targeting: (1) identify target villages having high concentrations of underweight preschoolers; (2) within the selected villages, identify households having preschoolers whose anthropometric measurements indicate high risk for undernutrition (Garcia and Pinstrup-Andersen 1987:78). In other cases, a maximum income or wealth level is established, and individuals must fall below that level to participate in the subsidy. It is difficult to enforce these targets. For example, the Sri Lankan program in 1978 restricted participation to households with annual incomes below 3,600 rupees (about $240). A survey of household income indicated that only 7.1 percent of households had incomes below this eligibility level. Yet almost half the population managed to qualify for the program (World Bank 1986:93, *World Development Report*).

Self-Targeting

The easiest way to target a food subsidy is to subsidize foods that have negative income elasticities of demand: the "inferior goods," to use the economists' jargon introduced in Chapter 7. Inferior foods vary from culture to culture but are typically starchy staples such as cassava, yams, maize, sorghum, or millet. As income increases, people usually eat less of these foods.

The government of Bangladesh experimented with this idea by subsidizing sorghum consumption, but the experiment, although supposedly successful, was not implemented countrywide (Karim, Majid, and Levinson 1984; Ahmed 1988:226).

Jayne and colleagues (1995), in their previously cited review of maize meal subsidies in four countries in southern Africa, concluded that a distinction should be made between programs that subsidize refined maize meal and programs that subsidize whole maize meal. As Table 19.2 shows, a majority of households in the poorest 20 percent of the income distribution in Kenya consume whole maize meal, a less expensive type of meal produced by small hammer mills. The richest households in the income distribution predominantly consume refined maize meal produced by large-scale roller mills.

Direct Distribution

Affluent countries are familiar with direct food distribution programs carried out through school lunch programs or by soup kitchens set up in low-income urban areas. In the third world, direct distribution of food is more likely to take the form of supplemental feeding programs targeted at the groups most vulnerable to undernutrition: pregnant and lactating women, infants, and preschoolers. Despite the popularity of such programs, the results have been disappointing (Kennedy and Knudsen 1985).

Beaton and Ghassemi (1982) found that in the eight supervised feeding programs and thirteen take-home food programs for which they had data, the net increase in food intake by the target recipients ranged from 45 to 70 percent of the food distributed, with one program showing a net effect of only 10 to 15 percent. Some of the reasons for these disappointing results are discussed in Box 19.2.

Table 19.2 Targeting Food Subsidies for Maize Meal in Kenya

	Percentage of Households Who Consume	
Income Group	Refined Maize Meal	Whole Maize Meal
Poorest 20 percent	38	59
Second poorest 20 percent	53	44
Middle 20 percent	74	25
Second richest 20 percent	76	22
Richest 20 percent	80	18

Source: Jayne et al. 1995.

Box 19.2 Supplementary Feeding

Eileen T. Kennedy and Per Pinstrup-Andersen

Supplementary feeding programs distribute foods through noncommercial channels to pregnant and lactating women, infants, and preschoolers. These programs are the most common form of nutrition intervention in developing countries.

There are three common forms of delivery: (1) on-site feeding, (2) take-home feeding, and (3) nutrition rehabilitation centers (NRCs). NRCs include both residential facilities and programs in which children are cared for during the day but return home at night.

Data from more than 200 supplementary feeding projects indicate that many supplementary feeding programs have had a significant and positive effect on prenatal and child participants (Anderson et al. 1981; Beaton and Ghassemi 1982). Despite the significant, positive effect, however, the benefits are usually small. Increments in birth weights attributed to the supplementary feeding programs are typically in the range of 40–60 grams. Similarly, the increases in growth seen in preschoolers, although significant, are small.

Several reasons are given for these small but significant effects. First, it appears that only a part of the food given is actually consumed by the target population. "Leakages" occur when the food is shared by nontarget family members or when the food is substituted for other food that normally would be consumed. Other factors, such as the timing of supplementation, duration of participation, nutritional status of recipients, and related services available, all influence the effectiveness of supplemental feeding.

Timing of supplement. Pregnancy and the period from six months to three years of age are the most nutritionally vulnerable times. Studies indicate that it is the last trimester of pregnancy that is the most critical for supplementation. Preschoolers below the age of three are also at special risk. Inappropriate weaning practices, delayed introduction of solid foods, food taboos, and infection all contribute to a higher prevalence of second- and third-degree malnutrition in this group.

Duration. For prenatal women, there appears to be a minimum participation of 13–15 weeks needed to produce significant changes in birth weight. For infants and children, the minimum level of participation needed to affect growth depends heavily on the type of delivery system used.

Nutritional status of participants. Children with second- or third-degree malnutrition exhibit greater benefits from supplemental feeding than do marginally undernourished children. The same is true for pregnant women.

Other services. Inadequate intake of food is only one of several factors that contribute to undernutrition. Undernutrition and infection often occur simultaneously. It is not surprising, therefore, that the most successful supplementation activities have been those with strong ties to primary healthcare programs.

Source: Extracted from Kennedy et al. 1983:35–40.

Rationing

A subsidized food-rationing system allows a consumer who holds a ration card to purchase a specific amount of some food or foods in a given time at a price lower than the market value.

A subsidized food-rationing system requires that the government set up either a marketing system of its own (India's food ration shops), or a system for reimbursing commercial retail outlets for the discounts that they give for the rationed food (food stamps, described later). In either case, the government must employ auditors to monitor the system to minimize cheating. For example, as described previously, operators of food ration shops in India diverted food supplies from their stores and sold them at higher prices in public markets.

A 1983–1984 experiment in the Philippines provides a case study of costs and benefits from a real-world, subsidized food ration scheme. The experiment was set up so that all households in seven villages, known for a high incidence of undernutrition and poverty, were provided subsidized food. These villages were matched with seven control villages. The program did increase food consumption among the target villages. Although distribution of the extra food within the household favored adults, preschool children also consumed more and showed improvements in their nutritional status. If only weight gains among the undernourished were counted as benefits, the cost of adding 1 kilogram to the weight of an undernourished preschooler was estimated at $101 per year. (Edirisinghe [1987:70], in his study of food subsidies in Sri Lanka, found that discrimination against younger family members diminished when the more productive members of the household had at least 80 percent of their energy requirements met.)

The researchers in the Philippine experiment estimated that the cost-effectiveness of the program compared favorably with that of other programs. Costs were kept low through careful targeting, through the cooperation of the local bureaucratic structure in administering the program, and by using existing retail outlets instead of a parallel, government-operated marketing system (Garcia and Pinstrup-Andersen 1987:9, 78–79).

Although the Philippine effort was targeted at rural villages, it has been found that nationwide subsidized ration schemes generally show an urban bias. For instance, the subsidized wheat ration system in use in Pakistan was found to contribute about 11 percent of household income for urban households with incomes below the median. Rural households gained less than 1 percent of their income from the system. The reasons for the difference are that rural households are less likely to participate in the program, smaller quantities of rationed food are available there, and wheat is not sold in many rural areas (Rogers 1988c:247).

Food Stamps

Food stamps are somewhat different from ration coupons for purchasing subsidized food. Food stamps have a face value that can be used in any food store to purchase food at the market value. In addition, people are often required to purchase their food stamps. Since a food stamp plan does not require government to set up a parallel marketing system for the subsidized food, the system may be cheaper than rationing.

The first food stamp plan ever was introduced in the United States just before World War II, but it is the 1961 revision of the plan that economists like to talk about. In that version, eligible families could purchase stamps with a cash value depending on household needs for food. They paid varying amounts for the stamps depending on their income level. This arrangement made it possible to vary the food-linked income transfer according to need and therefore extend the limited government food welfare expenditures to a broader segment of the population.

In his study of the food stamp program in Sri Lanka, Edirisinghe (1987:55) found that the caloric intake response to an additional rupee from food stamps was exactly the same as from an additional rupee of income. Because of decreasing income elasticity of demand as income rises, the cost of providing 100 additional calories through food stamps increases as income increases. Despite this finding, food stamp programs will probably continue, simply because they are more acceptable politically than straight cash transfers.

Food-for-Work

Adding the requirement that recipients of food aid work in exchange for the food-linked income transfer is an interesting twist. Food-for-work has the potential to increase the productivity of the region in which it is applied and, at the same time, provide productive activities for recipients who would otherwise be unemployed or underemployed (Mellor 1988:1004). Food-for-work projects typically improve rural infrastructure through building farm-to-market roads, constructing irrigation canals, and so forth. They have also been used in improving squatter settlements or in erecting community buildings (Jackson and Eade 1982:24).

During the early 1980s a food-for-work project in the Rift Valley of Kenya employed low-income farmers on local public works projects, particularly for erosion control and water-harvesting devices. The project had two positive economic outcomes: first, a good deal of farmland was improved and its access to irrigation water was enhanced, and second, the participating farmers used some of their food-linked income transfers for capital investments on their farms and thus increased their own productivity. In fact, during the second year of the program, the farmers devoted fewer hours to food-for-work activities, apparently in part because of a greater need to tend their own farms (Bezuneh, Deaton, and Norton 1988).

This success story is heartwarming, yet at the same time introduces one of the problems with food-for-work: the benefits often go mainly to those who possess land. Typically, the recipients of the food are not landowners but the landless unemployed and underemployed. If their projects improve the productivity of land owned by others, the inequality of asset distribution in the area could increase. In one food-for-work tree-planting project in Ethiopia, the workers became so resentful that their work was enhancing the private property of already powerful landed people that they planted all the trees upside down (Maxwell 1978a:40).

Another problem stems from the growing number of female food-for-work laborers. The extra time they put into food-for-work programs may detract from the quantity and quality of care that they give their children. Typically, they leave their infants and preschoolers to be cared for by older siblings (Kennedy et al. 1983:28).

Food Aid and the Costs of Explicit Food Subsidies

Explicit food subsidization is an expensive way of improving nutritional status. This is especially true of food aid—subsidized food that is sold (or given away) by food-exporting countries (e.g., the United States, Canada, and Europe). The quantities of food aid shipments from 1981 to 2006 are shown in Table 19.3.

Oxfam, one of the leading voluntary agencies involved in distributing surplus food to third world countries, commissioned a report on food aid for such purposes as disaster relief, food-for-work, mother and child health, and school feeding programs. The report's authors, Jackson and Eade (1982:65), found that the cost of the sea-freight to the US food aid program amounted to 53 percent of the value of the food. When the food arrives at a third world port, there are other costs—warehousing, transportation, and administration as the food is distributed to the needy. Jackson and Eade found that the sea-freight plus within-country costs of the US food aid program in one country, Guatemala, amounted to 89 percent of the original cost of the food.

Jackson and Eade cite a number of disturbing studies suggesting that there are sometimes ways to improve third world nutrition more cheaply than with

Table 19.3 Quantity of Food Aid from All Donors, 1981–2006 (metric tons)

	Cereals	Noncereals	Total
1981	9,329,869	747,479	10,077,348
1986	13,127,498	1,019,363	14,146,861
1991	13,960,774	1,307,569	15,268,343
1996	5,057,746	921,819	5,979,565
2001	7,413,712	1,434,762	8,848,474
2006	3,737,011	834,952	4,571,963

Source: FAOSTAT 2008a.

explicit food subsidies. For example, a study in India found that it cost 1.5 times as much to prevent a child's death through supplementary feeding as it would to provide basic medical services, and that "for children aged 1–3 years, nutrition supplementation was up to 11 times more expensive in terms of lives saved than medical services." The study concluded that "even where it has been nutritionally effective, supplementary feeding has not proved to be cost-effective" (Maxwell 1978b:295 n. 36, 297).

Despite their expense, these programs can provide a solution to the most immediate of third world nutritional needs, such as famine relief. Note, however, that to be effective, they must be well administered, and this in itself is expensive. Note also that these programs do not become self-sustaining. Long-term, self-sustaining solutions to the hunger problem will need to address population growth rates, purchasing power of the poor, income distribution, and health.

An IFPRI evaluation (Hoddinott, Cohen, and Bos 2003) of food aid draws the following conclusions:

- In recent years, the magnitude of food aid to developing countries has declined to about one-third its level three decades ago.
- A higher percentage of food aid is now going to the poorest countries (rather than recipient countries chosen for their geopolitical significance), but food aid per capita has declined in all country groups.
- Food aid is higher in countries that have low per capita incomes, civil conflicts, and natural disasters.

The IFPRI report recommends that food aid can be a useful tool in cushioning the effects of natural disasters, or human conflicts, but should not be relied upon as a permanent source of food in poor countries.

Since 2002, the United States has donated food to schools and education projects in developing countries under the McGovern-Dole Food for Education program (for details, see USDA, Foreign Agricultural Service 2006).

Effects of Explicit Food Subsidies

Explicit food subsidies succeed in transferring income, but they are expensive. Third world governments seldom have the resources to sponsor explicit food subsidies on their own. Therefore, the direction of income transfer through these programs has been mainly from the developed to the underdeveloped world, chiefly through the US food aid program PL480 and the World Food Programme.

These subsidies increase food consumption. Because of the fungibility of the food transferred (commonly grains or grain products) in the third world setting, the food received is usually treated as the equivalent of cash. Therefore, some of the resulting increased purchasing power is spent on nonfood items.

Marketwide subsidies usually benefit urban consumers far more than rural consumers. This is due in part to the difficulty of operating a subsidy program in rural areas and in part to the greater political clout of urban special interest groups. In these untargeted subsidies, the rich enjoy a greater income transfer than the poor because the rich purchase more food. Even so, the poor may well benefit from a greater percentage increase in income from the subsidy.

In very low-income households, the lion's share of the increased food consumed may go to the productive adults unless the subsidy is sufficient to approach food adequacy among those adults.

Even with foreign assistance, explicit subsidies can be expensive to third world governments, often claiming more than 10 percent of their annual budget expenditures. The question must be raised whether the same amount spent on other programs would accomplish more for the poor. Careful targeting of the subsidy can save considerably on costs.

Food subsidies put downward pressure on wages, which partially offsets the real-income transfer. Still, lower wages may increase employment among the poor.

If the food for an explicit food subsidy is purchased in the same country where it is dispensed, the demand for food is increased, because the poor are now eating more than they otherwise would. This results in higher food prices, which work as an incentive to agricultural production. The US food stamp and school lunch programs thus provide an incentive to US agriculture.

Conversely, if the subsidized food is purchased in a developed country and dispensed in a third world country, the effect is to raise farm prices in the developed country and lower them in the third world country. The program thus acts as an incentive to agriculture in the developed country but as a disincentive to agriculture in the third world country. In the developed world, special interest groups who benefit from food surplus disposal programs are likely to insist that their donated food does not depress third world farm prices. We find it hard to understand how they can claim the incentives to the developed world's agriculture without recognizing the corresponding disincentives to agriculture in the recipient countries.

If third world farmers bear some of the costs of explicit food subsidies through the price disincentives described above, those costs are small compared to the costs that may be incurred from implicit food-linked income transfer programs.

20

Policies That Improve Access to Food: It's All About Distribution (Isn't It?)

In Chapter 8, we saw that the world food supply is sufficient to allow every person in the world to consume an adequate number of calories. It seems logical to ask, therefore, "So, it's really all about distribution, right?" Serious students of the world food problem draw similar conclusions: "Enough food is available to provide at least 4.3 pounds of food per person per day worldwide. The problem, therefore, is not of production but clearly of access and distribution" (Mittal 2002:304). The purpose of this chapter is to take seriously this line of argument, and to demonstrate that this conclusion is grossly misleading.

There Is Sufficient Food, But . . .

Let us begin by reviewing the numbers. The worldwide requirement for food is about 2,350 calories per person per day. This reflects an average of requirements that are smaller for children (and the elderly) than for adults, and smaller for women than for men, as shown in Table 3.1. Currently, food supply is about 2,800 calories per person per day. The "average food surplus"—the difference between average supply and average requirement—is about 450 calories per person per day.

A more detailed exploration of the world's "food balance sheet" (FAOSTAT 2008b) suggests several ways that the average food surplus might be considered to be even larger. More than 900 calories per capita per day are lost in processing or waste. Of course, it is impossible to eliminate all such losses, but even reducing these losses by half would double the average food surplus. Another 1,000 calories per (human) capita per day are fed to animals, but animal products provide the average human with only about 450 calories per day. That reflects an additional 550 calories per person that could be obtained without increasing food production.

Diet for a Small Planet: Redistribution Through Voluntary Restraint

This last fact has led some to conclude that voluntary changes in diet in developed countries might succeed in reducing or eliminating hunger in the developing world. The Hearts and Minds website (2003) on "socially responsible food" states, for example, that "if the USA reduced meat consumption by 10 percent, we would free more than 12 million tons of grain a year—enough to feed 60 million starving people."

We can analyze the assertion as follows. Table 20.1 shows calories per day, per capita and in total, for the United States and for the world in 2001. This provides a basis for understanding the Hearts and Minds claim quoted above. First, assume that the United States reduces all calories from animal products (not just meat, but also dairy, eggs, and fish) by 10 percent. The reduction is 102.9 calories per day per person, or 29.42 billion calories per day for the country as a whole, or 10.74 trillion calories per year. A metric ton of grain has about 3.15 million calories, so 12 million metric tons of grain would provide 37.8 trillion calories per year, enough to provide 60 million people with 1,726 calories per day.

The ratio of feed to meat used here (37.8/10.769 = 3.5) is larger than the ratio of feed to animal products in the preceding section (1,094/460 = 2.4), but that difference is not the primary reason why the conclusion in wrong. There are two fundamental flaws in the reasoning. First, the world food supply is not a fixed quantity, so that if less is taken by one group, that difference is available to another group. Farmers produce food because consumers buy it. If one group of consumers were to reduce their food demand, prices would drop, other consumers would increase their consumption, and producers would react by producing less food. Second, the increase in grain available is not automatically allocated to the world's hungry.

Figure 20.1 illustrates the impact of a small backward shift in aggregate demand for calories for animal products in the United States. What is the relative

Table 20.1 Food Consumption Patterns in the United States and the World, 2001

	United States (population 285,926,000)		World (population 6,110,437,000)	
	Per Capita	Total (billion calories)	Per Capita	Total (billion calories)
Calories per day	3,764	1,076.225	2,807	17,151.997
From animal products	1,029	294.218	460	2,810.381
From vegetable products	2,735	782.007	2,347	14,341.616

Source: FAOSTAT 2008b.

Figure 20.1 Impact of a Reduction in per Capita Demand for Food in the United States on Consumption in the Rest of the World

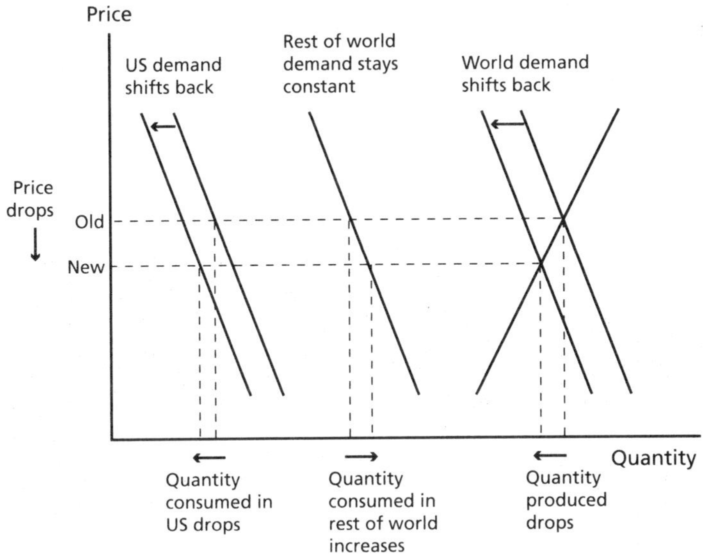

size of these effects? Imagine that US consumers reduce their calories from animal products by 10 percent (by 29.42 billion calories) at current price levels. This is 1.047 percent of world demand for animal calories. Assuming an elasticity of demand of –0.5, and an elasticity of supply of 0.6, to close the gap of 1.047 percent of quantity supplied, price will drop by 0.952 percent. Quantity of animal products demanded in the rest of the world will increase by about 11 billion calories per day (from 2,516 to 2,527 per capita), quantity demanded in the United States will drop by about 27 billion calories per day (the 10 percent decline is eroded slightly by the price decline), and quantity supplied will drop by about 16 billion calories per day.

This net drop of 16 billion calories per day in production of animal products will shift the supply of grain to the right by 38–56 billion calories per day, or 14–20 trillion calories per year (about half of the 37.8 trillion calories per year projected by Hearts and Minds). (Recall the difference between the FAO food balance sheet and the Hearts and Minds assertion about whether the proper conversion rate is 2.4 or 3.5 calories of grain for each calorie of animal products.)

If there were some way to direct this "freed-up grain" into the hands of the most hungry, it would be sufficient to feed 20–35 million people a diet of 1,700 calories per day. But the market mechanism for allocating goods relies on price. As a gap appears between aggregate quantity available and aggregate quantity demanded, the price will fall; producers will cut their production and consumers will increase their consumption, as shown in Figure 20.2.

The 38–56 billion calories per day of grain added to the market is 0.26–0.39 percent (less than 1 percent) of the total calories from plant (nonanimal) products. In the calculations below, we assume that the elasticity of supply for plant calories is 0.40, and that elasticity of demand for plant calories is –0.15 in developed countries and –0.35 in developing countries. As an additional 0.26–0.39 percent of plant calories are made available for human consumption, the price will drop by 0.38–0.55 percent. Consumption in developed countries will increase by 15–22 billion calories per day. This would provide each resident of the developing world with 3–4 additional calories per day. See Table 20.2.

Let us review the logic by which the 37.8 trillion calories per year (or 1,700 calories per day for 60 million starving people) projected by Hearts and Minds has shrunk to 3–4 calories per day:

- The direct impact of a reduced meat consumption by some people is partially offset by increased meat consumption by others, responding to a lower meat price. If changes in diet create a direct reduction of 10.74

Figure 20.2 Impact of an Increase in Proportion of Grain Produced for Human Rather Than Animal Food

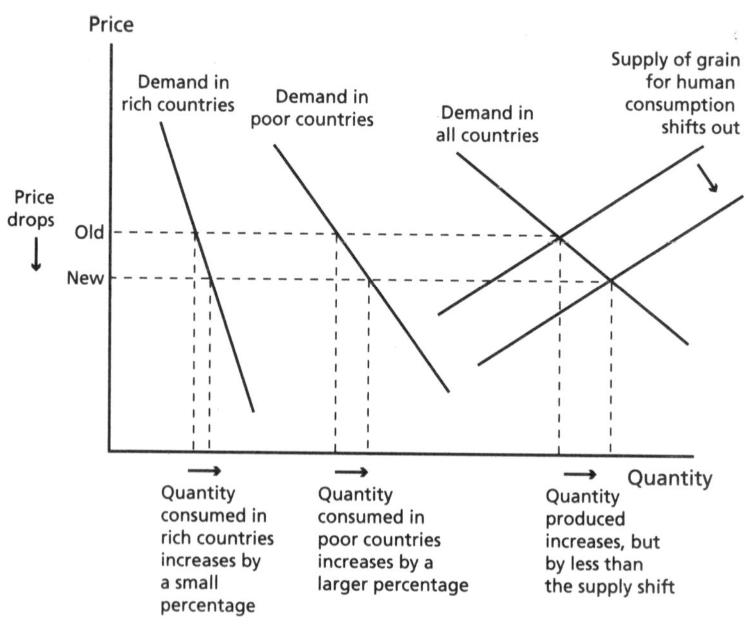

Table 20.2 Food Consumption Patterns in Developed and Developing Countries, 2001

	Developed Countries (population 1,317,872,000)		Developing Countries (population 4,792,565,000)	
	Per Capita	Total (billion calories)	Per Capita	Total (billion calories)
Calories per day	3,283	4,326.574	2,675	12,820.112
From animal products	855	1,126.781	350	1,677.398
From vegetable products	2,428	3,199.793	2,325	11,142.714

Source: FAOSTAT 2008b.

trillion calories per year in animal products, about half of this is offset, so the net impact is to reduce animal calories consumed by 5.84 trillion calories.

- The number of calories of feed "freed up" for each calorie of animal products may be 2.4 rather than 3.5, so 5.84 trillion of feed converts to 14.02–20.44 trillion.
- The direct impact of an increase in available plant calories drives down the price and is partially offset by reduction in supply of plant calories that would otherwise have been available. Of the 14–20 trillion calories of animal feed newly available for human consumption, about 60 percent of this is offset, so the net impact is to increase plant calories consumed by humans by 6.0–8.7 trillion calories.
- Some of this increase in plant calories takes place in developed countries: of the 6.0–8.7 trillion increase in calories, about 5.4–7.9 trillion occurs in the developing world.
- Within the developing world, consumption increases among the adequately nourished as well as the undernourished. Dividing the 5.4–7.9 trillion calories among the 4.7 billion people in the developing world yields 1,150–1,680 additional calories per person *per year,* or 3.2–4.6 calories per day.

This demonstrates that voluntary restraint in the developed world will do little to improve the world undernutrition problem. Complete elimination of animal products from the diets of all people in the developing world would increase calories per day in the developing world by 120–180. This would make substantial inroads into the undernutrition problem, but would not eliminate undernutrition. The FAO (FAOSTAT 2008b), in estimating the "depth" of undernutrition, lists many countries in which the average calorie deficiency among the undernourished exceeds 200 calories per day.

Policy Approaches

Of course, those who argue that we can solve the world food problem through redistribution are not restricting themselves to purely voluntary measures. What kinds of programs would be a part of a redistribution solution?

- "Overconsumption" in developed countries must be reduced.
- Overall production must be maintained at current (or close to current) levels.
- "Underconsumption" in developing countries must be reversed, and the increased consumption must be targeted to those who are undernourished.

Reducing overconsumption is conceptually quite simple, but impossible in practical, political terms. Food consumption will drop in response to a tax on income, or a sales tax on food. But huge taxes would be needed to achieve a substantial drop in consumption. In the United States, a reduction in average daily calories consumed from the current level of 3,750 to 3,000 (still a little above calories produced per person worldwide) is a 20 percent drop, too large to apply standard elasticity calculations. However, we can make some inferences about the size of the taxes needed by looking at the historical experience in the United States. The US populace last had an average consumption of 3,000 calories per capita per day in 1968. In 1968, real (inflation-adjusted) disposable income per person was $17,266, about half of the current income of $32,350. Food prices have risen more slowly than prices in general since 1968. Therefore a sales tax on food of about 4 percent, combined with an income surtax of about 45 percent, would be needed to return the United States to the consumption patterns of the late 1960s.

The second policy objective of the "redistribution solution" requires that while consumption in developed countries is reduced, overall production remains unchanged. Of course, the tax revenues collected under the programs to reduce overconsumption may well provide government revenues to operate such a program. (Undoubtedly, an income tax surcharge of the size projected here [45 percent] would have substantial incentive effects. Those are ignored here.) A program of government purchases of food could be used to bridge the gap between current levels of production and the reduced levels of consumption attained by the taxes. Assuming that the government purchased 20 percent of farm marketings (current value about $240 billion), the program would cost the US government about $50 billion (about 2.5 times the cost of current commodity programs).

The third part of the "redistribution solution" would be donations of the food purchased by the developed-country governments to countries with undernourished people. The US foreign aid budget is currently about $11 billion and less than half of this goes to low-income countries. Thus, the kind of food distribution envisioned here would mean not only a huge change in the scope of foreign assistance (increasing it by a factor of five), but also a radical change in the targeting of foreign assistance.

A fourth part of the "redistribution solution" would be a set of programs to ensure that the food donated by the developed countries actually reaches the undernourished people in the recipient countries. Here again, it is not difficult to conceive of programs to target the undernourished, but it may be problematic to get these programs adopted and implemented administratively.

Of course, the kinds of policies considered here are policies that would be adopted within a market-oriented system. Two alternatives to a market-oriented economic system might be considered: centrally planned production and consumption, and household food self-sufficiency or subsistence agriculture. Both of these alternatives have been tried (or are being tried), neither with notable success in eliminating undernutrition.

Why Does This Matter?

The main conclusion that we draw from the previous section is that a policy to solve the world's food problem solely through redistribution is politically infeasible. Huge taxes to reduce food consumption and huge increases in government expenditures for foreign assistance are not realistic policy proposals. But does it matter?

A sage once said: "In policy debates, never let the obvious go unstated." The reason that it is worthwhile to examine the "redistribution solution" argument seriously is that the conclusions drawn from the argument may actually impede progress toward solving the world food problem. The "dangerous" conclusions are:

- Protection of natural resources can be achieved by cutting back on (or at least halting the growth of) food production.
- It is unnecessary to develop and adopt new technology to increase food production.

The first of these conclusions is illustrated by Rosset, Collins, and Lappe (2000): "Where dominant technology destroys the very basis for future production, by degrading the soil and generating pest and weed problems, it becomes increasingly difficult and costly to sustain yields. Under these . . . conditions, mountains of additional food could not eliminate hunger. The alternative is to create a viable and productive small farm agriculture using the principles of agroecology."

The second of these conclusions is also illustrated by Rosset, Collins, and Lappe: "We must be skeptical when Monsanto, DuPont, Novartis and other . . .

companies tell us that genetic engineering will boost crop yields and feed the hungry. . . . [A] second Green Revolution they promise is no more likely to end hunger than the first." Or consider Andrew Kimbrell's seventh "deadly myth": "biotechnology will solve the problems of industrial agriculture." Kimbrell concludes: "If biotech corporations really wanted to feed the hungry, they would encourage land reform, which puts farmers back on the land, and push for wealth redistribution, which would allow the poor to buy food" (2002:62).

Underlying these conclusions is a deep-seated suspicion that technology causes problems, it does not solve problems. And if agricultural production cannot be increased substantially without new technology, it is reassuring to believe that increased production is not necessary to solve the world food problem.

A clearer view of what kinds of policy changes would make up a "redistribution solution" suggests that improved technology must be a part of any solution to the world food problem. The enormous improvement in the world food situation in the past four decades (the percentage of people suffering from undernutrition in the developing world has dropped from 35 percent in the early 1960s to less than 20 percent today) is undoubtedly attributable in large part to new technology. In addition to boosting yields, new technology (low-impact tillage, drip irrigation, integrated pest management) can also reduce the impact of food production on resource degradation. Perhaps technology alone will not solve the world food problem, but increased production will be an integral part of any solution.

Finally, we should emphasize that although redistribution alone is unlikely to solve the world food problem, food distribution programs that target the poor and hungry are an absolutely critical part of any strategy to reduce undernutrition. We noted in Chapter 16 that providing adequate nutrition to the average undernourished person requires a very small investment—the average hungry person is a peanut butter sandwich a day away from adequate nutrition. The lack of political will in developed countries and in the world community to undertake the necessary investments can be explained in large part by the belief that the efforts will not in fact put food into the mouths of the needy, but will instead enrich the already well-fed who have learned how to use the programs to their own advantage.

21

Policies That Raise Prices Paid to Farmers: Direct Subsidies and Elimination of Urban Bias

> The most important class conflict in the poor countries of the world today is not between labor and capital. Nor is it between foreign and national interests. It is between the rural classes and the urban classes.
>
> —Michael Lipton (1977:13)

Chapter 20 showed that increasing production is almost certainly a necessary part of any solution to the world food problem. This conclusion is buttressed further by the findings in Table 21.1. This table shows the experiences of eighty-three countries during the period 1969–1971 to 2003–2005. All of the countries represented in the table began this period as "high hunger" countries—countries in which at least 26 percent of the population was undernourished. The countries are split into three groups according to how fast cereal yields grew between 1969–1971 and 2003–2005. "Low yield growth" is defined as growth of less than 20 percent; "high yield growth" is defined as growth of greater than 70 percent; and "medium yield growth" is between 20 percent and 70 percent. In 1969–1971, prevalence of undernutrition was similar in the three groups. But by the 2003–2005 period, the prevalence of undernutrition had declined sharply in the high-yield group, declined modestly in the medium-yield group, and actually increased in the low-yield group. This suggests that increasing agricultural yields and output may be an important component in any effort to reduce undernutrition.

Figure 21.1 illustrates two different ways of increasing the quantity of food produced. In this chapter we focus on the left-hand side of Figure 21.1. What kinds of policies can increase prices received by farmers? Here we consider two general approaches: increasing farm prices through subsidies, and increasing farm prices through removal of programs that impose an implicit tax on farm output.

Table 21.1 Yield Growth and the Prevalence of Undernutrition

Yield Growth	Percentage Growth in Cereal Yields Between 1969–1971 and 2003–2005	Number of Countries	Average Percentage of Population Undernourished in 1969–1971	Average Percentage of Population Undernourished in 2003–2005
Low	< 20	30	37.0	40.3
Medium	20–70	24	30.6	21.8
High	> 70	43	39.1	13.9

Source: Calculated from FAOSTAT data.

Figure 21.1 Difference Between an Increase in Quantity Supplied and an Increase (or Outward Shift) in the Supply Curve

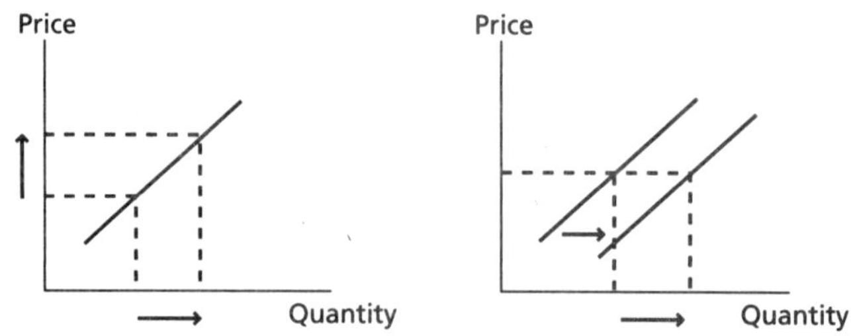

An increase in prices causes an increase in quantity supplied, moving along a stable supply curve

An outward shift in the supply curve causes an increase in quantity supplied, at a constant price

Direct Subsidies

We saw an example of a direct subsidy in Chapter 19 (Figure 19.2). Although the discussion there was couched in terms of a subsidy on consumption, we saw that the per unit subsidy created a wedge between the price consumers paid and the price farmers received. Direct price subsidies to farmers have exactly the same effect: the price farmers receive is the price paid by consumers plus the subsidy amount, so the program has the impact of increasing quantity, reducing consumer price paid, and increasing producer price received.

Historically, many agricultural subsidies have been administered as "target price–deficiency payment" programs. Under these programs the government announces the target price that they will guarantee for farmers. Then, once the crop is produced and sold at the market price, the government makes a payment to farmers (a "deficiency payment") that makes up the difference between the market price (the price paid by consumers) and the target price (the price received by farmers). As described in Chapter 20, in some cases these subsidies are implemented by having the government purchase farm output at a specified price and then act as the marketing agent to consumers.

The difficulties with these programs are the same as the difficulties with consumption subsidies: the programs are expensive, and they are economically inefficient (recall the discussion of Figure 19.1). In addition, they often fail to provide farmers with incentives to produce high-quality output (see Box 21.1 for an example).

Another method commonly used to keep farm prices high is through price support and purchase programs. These programs set a support price, and then promise to buy from farmers at that price; the market price can never fall

Box 21.1 Removing Farm Subsidies in Iraq

Under the regime of Saddam Hussein, agriculture in Iraq was dominated by government bureaucracy. The government provided seeds, fertilizer, chemicals, and machinery to farmers at below-market prices. It leased land to farmers at concessionary rates. It bought all wheat and barley at a fixed price, regardless of quality. Then government mills distributed flour for free to consumers.

When the Hussein government fell, the US-led occupation authority was in charge. It discovered that some of the grain produced by Iraqi farmers was of such low quality that it threatened to gum up the flour-milling equipment. To avoid upsetting the Iraqi farmers, the occupation authority continued to buy grain as under the old regime. But it fed much of the crop to cattle, and destroyed some by burning. Wheat was imported from the United States and Australia to provide flour for Iraqi consumers.

One Iraqi official told a *Washington Post* reporter (Cha 2004) that he blamed the subsidy system: "'People were making so much money that the incentive to work harder, to increase production—it wasn't there,' said Salam Iskender, the new head of the agriculture section for the Wasit [provincial] governorate. . . . As a result, the yield in some regions plummeted from one ton of wheat per [unit of land] to a third of that and, particularly in the last two years, a large percentage of the crop came up 'black,' meaning that it couldn't be eaten.

Western advisers to the provisional authority are certain that a more market-oriented approach will succeed in increasing output by 20 percent, and the upcoming crop year will see an end to government provision of inputs (though the output price subsidy will remain, and will even increase). Iraqi farmers are not so sure: 'We are afraid of the free economy. We don't understand it. If we grow crops, who will help us and who will buy it?' asks one farmer. 'We are like a child which stage by stage needs to grow up. . . . We need time,' says another" (Cha 2004).

below the support price because farmers always have the option of selling to the government at that price. Box 21.2 describes how this kind of program works in India. A basic problem with support price and purchase programs is that excess stocks accumulate in government warehouses. These can be dispensed by selling them at below-market prices to domestic consumers, or by selling them to foreign buyers with an export subsidy. In other instances, the government attempts to minimize surplus stocks by imposing limits on farm production. Of course, selling surplus production abroad, or limiting production, does not reduce domestic undernutrition.

Agricultural Subsidies and the World Trade Organization

As part of the WTO's negotiations to reduce impediments to freer international trade, countries have been negotiating reductions in their agricultural subsidies. In these negotiations, an important distinction has emerged about the difference between "distortionary" versus "nondistortionary" subsidies.

Distortionary subsidies are subsidies that work through the price mechanism, and give a farmer a subsidy *per unit of production.* This means that the more the farmer produces, the larger the subsidy received. Because of this, the subsidy program influences or "distorts" farmer decisions (the farmer moves up the supply

Box 21.2 Farm Subsidies in India

India has an extensive program of farm subsidies (see Landes 2004). In order to encourage self-sufficiency in wheat and rice, India supports the farm price of those two commodities. The support price is calculated as the amount a farmer would need to cover the full cost of production. Calculating support prices in this way has protected Indian farmers from the pressure of competition; as a result, crop yields and agricultural productivity remain low in India compared to other countries.

The grain purchased by the government at the support price goes into a ration program for storage and distribution (described in Chapter 19). Beginning in the late 1990s, support prices resulted in surpluses that were larger than the ration system could absorb, and government-held stocks soared. As a practical matter, market grain prices are determined by government decisions about how much grain to release from storage, rather than by the underlying supply-and-demand conditions. In order to reduce some of the government-held stocks, the Indian government began selling grain at a loss to foreign buyers for export.

As India's urban population becomes richer, there is growing demand for a broader range of food. But the agricultural support program encourages farmers to keep their land planted with wheat and rice, so consumers look to purchase imported food while Indian-grown wheat and rice sit unbought in government storehouses.

curve as shown in the left-hand side of Figure 21.1). In the context of international trade, a program that encourages increased production domestically interferes with free trade, because it reduces the market for imports, or because it creates additional exports that must compete with other countries. The deficiency payments described above are an example of a distortionary subsidy. Under WTO rules, distortionary subsidies are limited in size, and future negotiations are likely to limit them still further.

However, under WTO rules there are few restrictions on nondistortionary subsidies. Nondistortionary subsidies are subsidies that do not work through the price mechanism and therefore are not tied to the quantity produced by a farmer. For example, if we send every farmer a check for $1,000, this does not change a farmer's decision about what crops to plant or how much to produce. During the past two decades, agricultural subsidies in the United States and Europe have evolved from distortionary to nondistortionary subsidies. A farmer in the United States receives a payment based on how many acres he farmed at some point in the past. If he increases his acreage, that does not increase his subsidy payment. If he increases his yield per acre, that does not increase his subsidy payment. If he shifts crops from wheat to soybeans, that does not change his subsidy payment. Since the subsidy does not influence the farmer's decisions about what and how much to produce, it is regarded as nondistortionary, and is not limited by WTO agreements.

A few caveats are in order before leaving this subject. First, the description of US subsidies as nondistortionary applies to the bulk of US farm payments, but there are some exceptions. For example, in May 2004, the WTO determined that certain payments under the US Cotton Program violated WTO rules. Second, even these lump-sum subsidy payments can be distortionary in the longer run, for two reasons. First, the subsidies can influence (or distort) a farmer's decision about whether to quit or retire from farming. Low incomes from low prices may be compensated for by the government subsidy payment. Second, the subsidy system creates a guaranteed source of liquidity that may make it easier for many farmers to obtain loans and therefore to make investments that increase their future output.

In its 2008 *World Development Report,* the World Bank (2007) devoted a chapter to the current status of agricultural subsidies. Among its findings:

- Consistent with their commitments under the Uruguay Round agreement, developed countries have reduced overall agricultural subsidies by about 20 percent between 1986 and 2005, and have reduced distortionary subsidies by about 35 percent. In many developing countries, government programs that interfere in agricultural markets are being modified or eliminated, and the more market-oriented policies have often improved agricultural productivity.
- Although barriers to agricultural trade have been reduced over the past fifteen years, remaining barriers impose an efficiency cost of $67–200 billion

worldwide, with $20–60 billion of that cost being borne by developing countries. There are considerable differences among developing countries as to the extent to which their economies are harmed by trade restrictions; the cost is less than 1 percent of GDP for developing countries as a group, but for some countries (Vietnam, Thailand), the cost is in the 3–5 percent range.

- Most of the welfare costs cited above are from restrictions on trade, not from distortions caused by domestic agricultural subsidies. "Developed country agricultural policies cost developing countries about $17 billion per year—a cost equivalent to about five times the current levels of overseas development assistance to agriculture" (p. 103).
- A further reduction in trade distortions is likely to increase prices in world markets, especially for cotton, oilseeds, dairy products, and cereals. "Because many of the poorest countries spend a large part of their incomes on cereal imports, they may incur an overall welfare loss despite gains from price increases in non-food commodities such as cotton" (p. 106).
- Further reductions in trade distortions are projected to increase agricultural output in Latin America, and reduce agricultural output in developed countries. Poverty would decline in most countries, but not all.

Elimination of Urban Bias

Another way to increase farm prices is to remove policies that have the effect of depressing prices. Figure 19.1 illustrated a policy that reduced both producer and consumer prices in a country by increasing imports. Many developing countries have adopted policies that reduce food prices in the economy as a whole—both the prices paid by consumers and the prices received by farmers.

As pointed out in Box 21.3, one of the surprising anomalies of the world food problem is that developed countries (where agriculture is already highly productive and food supplies are abundant) have generally stimulated farm production by engaging in agricultural policies that result in high farm prices, whereas developing countries (where agricultural production is often marginal and food supplies are scarce) generally have discouraged farm production by engaging in agricultural policies that result in low farm prices. A central idea underlying developing-country policies that result in these low farm prices is that they represent an easy way to transfer income. Popular as they are, such pricing policies are not an efficient way of transferring income to the poor.

Recall our discussion about policies that lower consumer prices, illustrated in Figure 19.1. A government policy that also reduces producer prices, has the additional impact of discouraging domestic production. Because the domestic price of wheat is depressed below the world market price, farmers produce less than they would if they were paid the world market price. This loss of production is represented by triangle 2 in Figure 19.1.

Box 21.3 Urban Bias and Agriculture

Michael Lipton, in his 1977 book *Why Poor People Stay Poor: Urban Bias in World Development,* laid out a compelling case that the most important "class struggle" in developing countries was the competition between the rural population and the urban population for control of the policymaking apparatus. The urban class, he argued, was winning this competition because they had overwhelming advantages: lower poverty, better education, better capacity to organize and communicate.

As a result, urban-dominated ruling classes directed resources toward low-return projects that helped city dwellers rather than toward rural investments that had much higher payoffs. "Scarce investment, instead of going into water-pumps to grow rice, is wasted on urban super-highways. Scarce human skills design and administer not village wells and agricultural extension services, but world boxing championships" (p. 13).

Not only is urban bias inefficient, but it is also inequitable. The people who benefit most from government programs in developing countries are the comparatively rich urbanites, rather than the rural poor. Lipton concludes, "A shift of resources to the rural sector and within it to the efficient rural poor . . . is often, perhaps usually, the *overriding* developmental task" (p. 18).

Anandarup Ray (1986) pointed out that in the 1970s and 1980s, this urban bias could also be seen in agricultural pricing policies of developing countries. Whereas rich countries such as the United States, Canada, Europe, and Japan heavily subsidize their agricultural sectors, "developing countries tend to tax agriculture—even those low-income countries that depend critically on agriculture for their economic growth. Some pay their producers no more than half the world price for grains and then spend scarce foreign exchange to import food. Many subsidize consumption to help the poor, but end up reducing the incomes of farmers who are much poorer than many of the urban consumers who benefit from the subsidies" (p. 2).

Implicit Subsidies to Consumers, Implicit Taxes on Farmers

Government activities that result in low (that is, below–world market) farm prices include noncompetitive procurement of grain from farmers; below-market food prices set by law; foreign trade controls; support of an overvalued domestic currency; and limits on cash-cropping. All these activities are carried out in the name of lower food prices, and all of them, in turn, amount to implicit food subsidies to consumers, and implicit taxes on farmers.

Implicit food subsidies are almost always paid for by the farm sector through the below-market food prices. The difference between the depressed price the producer gets for his food (depressed because of an implicit consumer subsidy program in operation) and the international price for that food, is a hidden cost to the farmer. The farmer's contribution to the food subsidy represents an income transfer from the farmer to the recipient of the subsidized food.

In this way, farmers have sometimes paid the lion's share of a food subsidy. A study of food subsidies in Sri Lanka, India, and Pakistan in the 1970s found that farmers bore about half the costs of those subsidies in the form of reduced farm-level prices (Scandizzo and Tsakok 1985).

Noncompetitive procurement and administered prices. A number of countries have used compulsory procurement to obtain grain from farmers at below-market prices. This amounts to a tax on the growers of the commodities procured. Not only does it discourage farmers from producing more food, but it can also be a bigger burden on the poorest farmers, who have fewer opportunities to switch to production of nontaxed alternative crops, such as vegetables and fruits. The larger, less poor farmers who do make the switch increase the supply and reduce the price of these alternative crops, which chiefly benefits the high-income consumer.

The Indian food subsidy program allowed the government to force farmers to sell grain at government-set prices. In Tanzania, where, until the early 1990s, the government controlled most aspects of agricultural marketing, government-controlled farm prices were lowered between 1970 and 1984 so that the average of official producer prices declined by 46 percent. Rising export taxes and the costs of the government marketing program reduced the farmers' share of the final sales value of export crops to 41 percent in 1980. Output of some export crops (cashews, cotton, and pyrethrum) fell drastically in the 1970s. By 1984 the tonnage of export crops moving through the government marketing boards was 30 percent less than it had been in 1970 (World Bank 1986:74–75, *World Development Report*). The implicit tax on agriculture was a substantial disincentive to agricultural production.

Another aspect of administered prices has been pan-territorial and pan-seasonal pricing—that is, the practice of maintaining identical prices across time and place within the economy. This policy discourages private traders from storing food just after harvest and shifts the burden of storage, together with its costs, to the government. In Jamaica, prior to 1980, the government placed a ceiling on the retail price of wheat flour, most of which is imported. Stores near the port could just barely cover their costs at the ceiling price, but retailers further from the port stopped selling flour. In time, the only flour available in many remote locations was black-market flour, which sold at a considerable premium. Thus the rural poor ended up paying more for their flour than they would have paid without the government policy, while the urban rich found flour available at reasonable prices in their supermarkets.

Export taxes. Many third world governments have placed taxes on the export of agricultural commodities. This not only generates revenue for the government but also lowers the domestic price of the commodity, because exporters can pay farmers only the world price minus the export tax they have to pay to the government. The lower price can be a substantial disincentive to production.

In Argentina, from 1940 to 1972, the government generally maintained a policy to keep agricultural prices low relative to the prices of nonagricultural goods. This was accomplished through a variety of measures that, in general, added up to a high tax on agricultural exports and a tariff on nonagricultural imports. This resulted in an implicit tax on agriculture estimated at 50 percent of total agricultural output during the period. Among the consequences of this policy were that employment in agriculture declined, agriculture lost resources to nonagriculture, and agricultural productivity grew more slowly. In fact, per capita agricultural production in the 1970s was less than it was before World War II (Cavallo and Mundlak 1982:13–14).

During the months of soaring food prices in 2008, Argentina renewed its interest in taxing food exports. Such taxes would discourage exports, keep more domestic production in the country, and keep food prices from rising so drastically. Farmers objected vociferously. Argentine president Christina Kirchner (and her husband, former president Nestor Kirchner) denounced the farmers as unpatriotic, saying farm profits from high prices should "belong to all Argentines" (see Barrionuevo 2008).

Overvalued domestic currency. For countries that export agricultural goods and import nonagricultural goods, an overvalued exchange rate can have the same impact as an import tax: foreign buyers find the agricultural exports more expensive (hurting rural farmers) and domestic buyers find the nonagricultural imports less expensive (helping urban consumers) (Schuh 1988).

Combining the implicit tax resulting from the overvalued domestic currency with an explicit export tax, and thus forcing farm prices well below the international market, and in addition, requiring farmers to pay more than the world price for their modern inputs, creates a recipe for substantial disincentive to agricultural production.

Trade policies biased against agriculture were common in developing countries prior to the 1990s. The World Bank (1986, *World Development Report*) showed that in many developing countries, the manufacturing sector had a higher level of tariff protection than did the agricultural sector.

Bale (1985:24) studied five developing countries from the point of view of the impact of overvalued domestic currency on agricultural production. He found that the currencies were overvalued by between 25 and 45 percent. Bautista (1987) found a similar result for the Philippines.

Cleaver (1985) studied thirty-one countries in sub-Saharan Africa, and found that in countries that were reducing the amount of currency overvaluaton, agricultural production was growing much faster than in other countries in the sample.

Limits on cash-cropping. A cash crop is one that is produced for sale. The commercial orientation of the crop (regardless of whether it is a food or a nonfood crop) identifies it as a cash crop. An export crop is, of course, a particular kind of cash crop: one that is ultimately exported from a country (von Braun and Kennedy 1986:1). In contrast to cash crops, those grown by farm families for their own consumption are called subsistence crops.

It is often argued that the growing of cash crops, and in particular, the growing of cash crops for export, limits the local food supply and therefore raises local food prices. So limits on growing cash crops for export from third world countries are often proposed as a means of forcing a shift in cultivation to food crops, thereby lowering the local price of food.

Lappe and Collins, proponents of this point of view, quote a Colombian government economist as estimating that, in Colombia, "one hectare planted with carnations brings in a million pesos a year; planted with wheat or corn, the same hectare would bring only 12,500." In other words, the gross returns from a field of carnations in Colombia are eighty times the gross returns from grain. These authors assume that growing carnations for export will automatically raise local food prices through limiting the local food supply, and observe, rather sarcastically, that "if the local peasants cannot afford chicken or eggs, perhaps they can brighten their shacks with cut flowers" (1977:266).

The argument that growing cut flowers in Colombia deprives the local peasants of their food supply misses a couple of important points. First, Colombians can purchase a lot more grain from the United States (the recipient of the cut flowers) in exchange for a field of carnations, than they can raise on that field themselves. And second, the cut-flower industry is highly labor-intensive. Regardless of who owns or manages the field of carnations, many more peasants will be employed to produce an acre of cut flowers than to produce an acre of grain.

What happens when third world farmers do expand their production of cash crops? Let us look at the evidence. Kennedy and Cogill (1987) studied Kenyan smallholders who were reducing their activities in subsistence agriculture and increasing their commercial sugarcane production. As sugarcane acreage expanded, it replaced maize acreage. However, the return to labor for sugar was three times the daily agricultural wage rate and significantly higher than the return to maize. Incomes of the farmers who had joined the cane-growing scheme were significantly higher than those of nonsugar farmers, and the increased income positively affected household calorie consumption. For each 1 percent increase in sugarcane income, household energy intake was found to increase by 24 calories (p. 9). The increase in household calorie intake translated into modest increases in calorie intake among the children (Kennedy 1989:54). The expansion of the sugar industry in the area also increased employment. Typically the sugar mill hired laborers and supplied them to the sugar farmer for such tasks as weeding, cutting the cane, and transporting the cut cane to the mill (Kennedy and Cogill 1987:9).

In a study that looked at the household-level effects of cash-cropping in rural Guatemala, von Braun and colleagues (1989) surveyed 400 households, about half of which had recently started raising nontraditional vegetable crops for export. The nontraditional export crops were substantially more profitable than traditional crops and were adopted by even the smallest farmers. Net returns per acre from one of the export crops, snow peas, averaged fifteen times those of maize, the most important traditional crop. Returns per unit of family labor for the new crops in general were about twice as high as for maize and 60 percent higher than those for traditional vegetables.

Because the export-crop producers achieved yields for their subsistence food crops that were about 30 percent higher than the non-export-crop yields of their neighbors, the export-crop producers usually had larger amounts of maize and beans available, per capita, for home consumption. Among the reasons for their higher yields was their purchase and use of fertilizer; thus, their increased incomes helped increase yields on their subsistence crops. Nontraditional export crops enhance local employment, not only on the farm but also, through backward and forward linkages, off the farm. The farmers purchased locally manufactured sticks and ropes for tying snow pea plants, for instance. And the marketing of the vegetables for export is labor-intensive, requiring such tasks as selection, grading, and packing of the produce (von Braun et al. 1989:11–12, 48).

In a statistical analysis of seventy-eight developing countries that devote at least part of their farmland to cash crops, von Braun and Kennedy (1986:2) did not find support for the hypothesis that the expansion of cash-cropping happens at the expense of producing staples. To the contrary, growth in areas allocated to cash crops positively correlated with growth in staple food production. Furthermore, growth in the share of cropland allocated to cash crops is generally positively associated with per capita staple food production.

Alternatives to Taxing Farm Commodities

Government must be financed. And the taxation system that finances government should not only be fair, but also be economically efficient. For efficiency and fairness, all sectors of the economy must bear a portion of the total tax burden, and the tax incidence should fall proportionally across all sectors. A World Bank report (1982) in the early 1980s found that agricultural goods were taxed at 12 percent by direct taxes and 24 percent by indirect taxes such as those imposed by overvalued exchange rates. As noted previously, implicit taxes of 50 percent have been estimated as the cost of food subsidies to farmers. Rates of export taxation on the order of 50 to 75 percent for farm products have not been unusual (World Bank 1986:64, *World Development Report*).

Taxing agricultural land or income may be preferable alternatives to the implicit taxes on agriculture. Because these explicit taxes are readily identifiable, they are more likely to be applied equitably relative to other sectors of the economy than are implicit taxes. And because they are not commodity-specific, they do not favor the production of one agricultural commodity over another. For both of these reasons they are more economically efficient than are implicit taxes.

The Costs of Urban Bias

When third world governments adopt policies that lower the price of farm outputs and raise the price of industrial products, they are demonstrating a preference for industrial development over agricultural development. This preference for industry, or more broadly, for the people and resources concentrated in the cities, is sometimes called "urban bias."

Societies pay a heavy price for urban bias. As governments encourage the substitution of locally made industrial goods for imported goods (the policy is sometimes called import substitution, for short), the growth rate of the entire economy is inhibited. In a worldwide study of the effects of such policies, Chenery, Robinson, and Syrquin found that "economies which pursued export led growth—as opposed to a strategy of import substitution—grew faster, industrialized sooner, had higher rates of total factor productivity growth, and tended to achieve the input-output structure of an advanced economy faster" (1986:356–358). These researchers showed that shifting away from a tariff-induced import substitution trade policy to a neutral trade policy can account for an increase of as much as 1 percentage point in annual rate of growth of the entire economy. They further found that export-led economies are more likely to attract capital inflows. This helps to explain the success of such export-oriented economies as Korea and Taiwan.

A common aspect of third world urban bias is the implicit taxation of grain, which is often the leading agricultural product. Taxing one commodity or set of commodities to the exclusion of others shifts resource use in the direction of the untaxed commodities. When the taxed commodity is grain (the diet of the poor), the tax encourages the production of nongrain foods such as livestock products (livestock can eat the grain before it is taxed), fruits, and vegetables—favorites of the rich. It is hard to make a case that there is any nutritional gain in lowering the price of foods for the rich while limiting the production of the chief foods of the poor.

But perhaps the ultimate problem with urban bias is that, as it slows the growth of the entire economy, it deprives the poorest not only of jobs but also of possible income transfers from the rich to programs that would improve the welfare of the poorest.

An End to Urban Bias

A five-volume World Bank study (Krueger, Schiff, and Valdés 1991) confirmed the policy bias against agriculture in great detail. Schiff and Valdés (1995) presented a synopsis of that study's finding. They concluded that (for the eighteen countries in their own study) agricultural prices, compared to nonagricultural prices, were 43 percent lower than they would have been in the absence of policies biased against the rural sector.

In the years since the World Bank study was conducted, a growing body of evidence that developing countries are turning away from urban bias has accumulated. Some of this evidence is summarized and discussed in the World Bank's 2008 *Word Development Report* (World Bank 2007), and includes the following:

- The extent of currency overvaluation fell from 140 percent in the 1960s, to 80 percent in the 1970s and 1980s, to about 9 percent in the early 1990s.
- Direct and indirect taxation of agriculture in sub-Saharan African countries declined from 28 percent in 1980–1984 to 10 percent in 2000–2004. Of eleven countries studied, agricultural tax rates fell in nine.

The tax rate on agricultural export goods declined from 46 percent in 1980–1984 to 19 percent in 2000–2004.

- In agriculturally based developing countries, the net tax on sugar and rice has been eliminated and the net tax on maize and wheat has been substantially reduced.
- "Both China and India have reduced their antiagricultural bias substantially over the past three decades, not only directly but also indirectly via cuts to manufacturing protection" (p. 102).

Jensen and Robinson (2002) also find a reduction in the degree of urban bias:

> Empirical studies from the 1980s . . . supported the view that policies in many developing countries imparted a major incentive bias against agriculture. Eliminating this bias was one of the goals of policy reform strategies, including structural adjustment programs, supported by the World Bank and others; and many countries undertook such reforms in the 1990s. . . . [Our] analysis indicates that, in the 1990s, the economy-wide system of indirect taxes, including tariffs and export taxes, significantly discriminated against agriculture in only one country, was largely neutral in five, provided a moderate subsidy to agriculture in four, and strongly favored agriculture in five.

In an article reviewing the literature on agricultural policies in developing countries, Binswanger and Deininger (1997) explore the conditions under which policy reforms are likely to be initiated. They observe that "a fiscal crisis is usually necessary for initiating reform"—the efficiency costs of the urban bias must become high enough to have an impact on the government budgeting process. They argue that the sustainability of the reform effort depends on: (1) whether people who benefit from the reform are organized into groups that can exercise some political clout; (2) whether the central bank and ministry of finance have the will and power to maintain budgetary discipline; (3) whether the reform effort is encouraged by international credit and advice; and (4) whether policy analysts can support the reforms without succumbing to political pressures. It appears that the reform process is well under way in developing countries.

22

Policies That Reduce the Price of Food by Increasing Supply

> While the most important reasons for inadequate agricultural output are difficult to ascertain, T. W. Schultz, in the first Elmhurst Memorial Lecture to the International Association of Agricultural Economists, left no doubt as to his ranking of the causes. He stated that the level of agricultural production depends not so much on technical considerations, but in large measure, "on what governments do to agriculture."
>
> —Malcolm Bale and Ernst Lutz (1981:8)

In Figure 21.1, we noted the distinction between increasing the supply of food and increasing the quantity of food supplied. An increase in the supply of food—or as economists call it, an outward shift in the supply curve—occurs when farmers are willing to produce more food at the same price, or are willing to produce the same amount of food at a lower price. This is the type of change that would permit the quantity of food to increase while the equilibrium price of food declines (see Figure 22.1). What would make a farmer willing to produce more at the same price, or produce the same amount at a lower price? A reduction in the farmer's costs of production. And how can we reduce food production costs? By reducing prices farmers pay for inputs, by maintaining or improving quality of resources, by encouraging investment, and by developing new technologies that increase farm productivity.

Subsidies for Purchased Inputs

The most direct and obvious way to reduce production costs is to subsidize the prices farmers pay for inputs. The government can do this by compensating farmers for each unit of inputs purchased, by subsidizing the production of inputs, or by producing or distributing the inputs themselves. In this section, we will briefly explore programs that subsidize fertilizer, mechanization, and irrigation.

Figure 22.1 Effect of Reduced Costs or Improved Technology on the Supply Curve

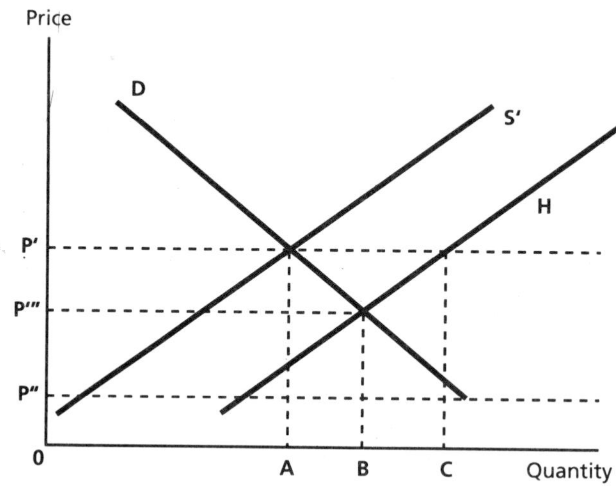

Note: When the supply curve shifts from S' to S", farmers are willing to furnish to the market: (1) an increased quantity (OC rather than OA) at the same, old price (P'); (2) the same, old quantity (OA) at a new price (P"); or (3) some other combination of quantity and price—for instance, the quantity and price represented by point H in the figure.

Fertilizer and Farm Chemicals

Fertilizer subsidies have been common in the developing world. During the 1980s, for example, urea sold at 56 percent below cost in Sri Lanka and at 60 percent below cost in Gambia. Arguments favoring subsidies to fertilizer include encouraging learning by doing, overcoming risk aversion and credit constraints, helping poor farmers, offsetting disincentives caused by taxing or pricing policies, and maintaining soil fertility (World Bank 1986:95, *World Development Report*). Let us look briefly at each of these arguments:

- When fertilizer was a new idea, it made sense to provide it to farmers at below cost as an incentive to try it. But knowledge of fertilizer is widespread now.
- Encouraging rural financial markets is the most appropriate way to deal with rural credit constraints, as discussed later.
- Fertilizer subsidies are an inefficient way to help poor farmers. Farmers with large operations, and those on better land, are likely to reap more benefits than the poor farmers. For example, data collected by Leclercq (1988) on size distribution of Brazilian soybean farms showed that the smallest 64

percent of farmers produce only 20 percent of the soybeans, while the largest 2 percent of farms produce over 30 percent of the soybeans.

- The best way to cope with production disincentives caused by antifarm taxation or pricing policies is to eliminate urban bias (Chapter 21). Subsidizing the price of fertilizer is an inequitable way to transfer income to the farm sector because it provides the greatest subsidy to the biggest farmers.
- Subsidizing fertilizer use to maintain soil fertility is a questionable practice, especially when making fertilizer cheap encourages farmers to substitute it for naturally occurring organic fertilizers that have better moisture-retaining properties.

Similarly, making pesticides cheap through subsidization encourages farmers to use more of the chemicals than they would if they paid the full costs. The subsidies undermine efforts to promote integrated pest management—a method of pest control that stresses biological suppression of insects and weeds and minimal use of chemical pesticides (Repetto 1985; USDA 1989a).

As fertilizer use increases, scarcities often develop because government-subsidized distribution fails to keep up with demand. Furthermore, fertilizer subsidies are expensive. Government involvement in production, importation, and distribution of fertilizer inhibits the development of a private distribution system. But the biggest inefficiency caused by a fertilizer subsidy program is that it encourages farmers to use more fertilizer than is economically or socially optimal. (See Box 22.1 for a description of experiences in sub-Saharan Africa.)

Mechanization

Many developing countries have pursued a mix of policies that tend to accelerate mechanization beyond the pace appropriate for their labor force (Binswanger et al. 1987:1). They do this in a variety of ways.

Governments often give preferential tariff treatment for machinery, and often give especially low tariffs for agricultural machinery. Farmers are sometimes given a tax shelter through deductions for farm machinery set at levels greater than the cost of the machinery. Brazil, for instance, has allowed a deduction for farm machinery of six times the value of the machine in the first year of operation. Other farm investments are treated less favorably, and labor costs enjoy no preferential tax treatment at all (World Bank 1986:97, *World Development Report*). Also, as mentioned in the preceding chapter, overvalued exchange rates make imports—including imported farm machinery—cheaper.

The desirability of machinery subsidies has been questioned on several points. First, the benefits of machinery subsidies typically go to large farms. Thus they provide the wealthy farmers with a competitive advantage relative to their poorer neighbors. Second, mechanization does not necessarily increase

Box 22.1 Impact of Eliminating Fertilizer Subsidies in Sub-Saharan Africa

During the 1970s and 1980s, many sub-Saharan African countries began government programs to distribute fertilizer at below-market prices. In the 1990s, responding to pressure from international organizations, these programs were eliminated. The hope was that getting governments out of the fertilizer distribution business would create a vacuum that would be filled by more efficient, privately owned companies.

The early indications of the impact of this policy reform are mildly disappointing. For example, in Ethiopia, it was estimated that eliminating the subsidy caused an increase in fertilizer prices of 21–39 percent. Derneke, Said, and Jayne (1997) examined fifty-one farms that differed in geography and crop choice; they found that with the subsidy, fertilizer use was profitable in forty-two of these cases. However, without the subsidy (assuming no increased efficiency of delivery), fertilizer use would be profitable in only twenty of the fifty-one cases. If cost efficiencies could be realized in the fertilizer delivery system, fertilizer use would be profitable in twenty-eight of the fifty-one cases.

A general review by Reardon and colleagues concluded: "African fertilizer use is the lowest in the world and has even decreased over the past decade and a half, i.e., over the same period in which fertilizer and seed subsidies and cheap input financial services programs have been reduced or eliminated. . . . Case study evidence points to a connection between the reduction in fertilizer use and these rising input prices. . . . Moreover, there is growing evidence that private fertilizer and seed merchants have responded much less than was expected to the liberalization of input markets enacted through elimination of fertilizer and seed parastatals. . . . Fertilizer markets in Africa are plagued by a series of fundamental problems such as risk, seasonal demand, high transport costs, underdeveloped financial services markets, and cash constrained farmers. Small markets add to the problem by limiting economies of scale. . . . Moreover, economies of scale in fertilizer production make domestic production inefficient in most African economies, so domestic fertilizer prices are sensitive to macro trade and exchange rate policies, and to volatile international fertilizer prices. While fertilizer subsidies and domestic fertilizer production schemes have generally proved ineffective in Africa, it appears clear that private market conditions in rural Africa cannot presently support necessary fertilizer deliveries, so some role for government is inevitable in the short-to-medium term. Given the considerable costs of delivering fertilizer to farmers on time and the restricted physical availability of fertilizer to most farmers, investment in improved private marketing infrastructure seems one of the most promising roles for the state" (1999:381–382).

yields. Binswanger, in a careful review of the studies concerning the impact of tractorization on yields in South Asia, found that the surveys failed "to provide evidence that tractors are responsible for substantial increases in intensity, yields, timeliness, and gross returns on farms in India, Pakistan, and Nepal" (1978:73). It seems that, generally, an acre of land tilled by hand or animal power does not yield less than an acre of land tilled by a tractor (Campbell 1984:47).

From the point of view of improving third world nutrition, there appears to be no justification for a machinery subsidy. When machinery is profitable, farmers will buy it and society will benefit from it. If the machinery is available only in fairly large units, small farmers can benefit from using rented machinery.

Subsidizing agricultural mechanization denies funding for alternative investments that are at least as productive as the machinery and that do not reduce labor demand as much as the subsidized machinery (Binswanger et al. 1987:1).

Stevens and Jabara (1988:272–273) list four undesirable effects from premature acceleration of agricultural mechanization resulting from subsidies: (1) reduced employment; (2) greater income disparities; (3) attempts by those with tractors and other machines to increase farm size; and (4) increased incentives to inventors and manufacturers to develop and produce even more labor-saving agricultural machinery.

Irrigation

Publically sponsored irrigation systems date back some 6,000 years, to when vast irrigation works were developed for the flood plain of the Tigris and Euphrates Rivers in Mesopotamia. In modern times, huge dams thrown across major rivers throughout the world (the Nile, the Indus, and the Colorado, for example) have furnished low-cost irrigation water to millions of farmers. Cost-benefit analysis has shown substantial gains to society from such projects, which usually are promoted for their multiplicity of benefits—flood control, electricity generation, and irrigation water for agriculture.

There is a natural government role in managing water resources. It is difficult or impossible to assign ownership to a flowing river, or an underground reservoir of indeterminate size, and it is difficult or impossible to make sure that only people who pay for the water use it. Therefore, public works projects are undertaken to build dams and canals.

But the allocation of water from public projects is a thorny problem. Should every user pay the same price, or should farmers pay lower prices than urban users? Should every user be allocated the same quantity (or the same quantity per unit of farmland), or should farmers with particular land quality or management skills who can use the water most productively be given relatively more water? Should water be auctioned off to the highest bidder, or sold at a flat rate? If sold at a flat rate, should that rate be set at or below or above the level that would cover the costs of building and maintaining the water project?

In many countries, governments "subsidize" irrigation water—they sell water to farmers at prices that are lower than costs, or lower than other nonagricultural users are willing to pay. In Egypt, irrigation water is free. In the Philippines, during 1980–1981, the subsidy amounted to 90 percent of the marginal cost (Bale 1985:17). In Sacramento, California, farmers pay between $2 and $19 per acre foot, while urban and suburban customers pay $1,275 per acre foot (California Department of Water Resources 2005).

Once the dam is built, it may be illogical to charge even the marginal costs of water delivery to the farmer if the water supply is so abundant that no allocation problem exists, or if monitoring may be more expensive than the marginal value of the water. In most other situations, economic efficiency would be improved through charging farmers the full cost of irrigation water or the value of water in its highest alternative use.

Maintaining or Improving Resource Quality

The issue of irrigation subsidies raises a more general issue, that of resource quality. When we look at a supply-and-demand diagram like Figure 22.1, we are inclined to see it as the supply-and-demand conditions for the current growing season. But with a little imagination, we can see the two different supply curves as representing alternative futures. If quality of land and water resources are seriously degraded, the agricultural supply curve will be low; but if resource quality is high, the agricultural supply curve will be high.

Meinzen-Dick and Rosegrant (2001) review some policy choices to increase water availability and encourage water conservation. New technology—low-cost desalination or methods of long-distance transport—could increase water availability, and policies definitely have a role in promoting research and development of new technology. Pricing and water markets are widely discussed theoretical solutions: if water is costly to buy, consumers have an incentive to conserve; if water can be sold for profit, water owners have an incentive to develop and preserve the quality and sustainability of water resources. However, Meinzen-Dick and Rosegrant warn that this solution may be better in theory than in practice. Measuring use and billing can be difficult, as can defining and enforcing ownership rights. Among their recommendations: "Education, social marketing, and public awareness campaigns to change behavior deserve much greater attention." (For other broad reviews of agricultural water policies, see Boggess, Lacewell, and Zilberman 1993; Rosegrant, Scheleyer, and Yadav 1995; and Sampath 1992.)

Maintaining quality of land resources is a little easier: people own land, and landowners have an economic incentive to preserve and improve land quality. Therefore, aspects of land reform (as discussed in Chapter 17)—especially land titling and easy enforcement of property rights—can be important elements of resource policy. As we will see in the next section, clear and enforceable land titles also facilitate credit by allowing land to be pledged as collateral. *The Economist* (2006) discusses the challenges involved in making a policy change in China that will allow tradable land titles. De Janvry, Key, and

Sadoulet (1997) describe a program in Colombia that facilitated purchases of small plots of land by poor landless households, by providing grants for 70 percent of the purchase price, and loans for the other 30 percent.

More generally, policies to reduce environmental damage fall into two categories. Some policies are aimed at limiting consumption that leads to environmental damage (Sagoff 1997). Methods of limiting consumption include direct rationing, tradable use permits, and taxes on consumption. For example, one policy to address anthropogenic climate change would be a carbon tax. A second approach to environmental policy involves requiring or encouraging technology that limits or reverses environmental damage. For example, another policy to address climate change would be government-sponsored research into methods of removing greenhouse gases from the atmosphere. (Other examples of this will be found later in this chapter.) Finally, we note the argument, advanced most famously in a World Bank memo by economist Larry Summers (1991), that environmental standards appropriate for rich countries may be more stringent than the appropriate standards for poor countries.

Policies to Encourage Investment

Investment, or actions that improve the quality of the capital stock, also can increase supply in the future. As a rudimentary agricultural example, consider that a farmer with a well-sharpened scythe can harvest more hay than a farmer with a dull scythe. But consider that a farmer who owns his own scythe and knows he will use it day after day is more likely to keep it well-sharpened than will a farm-worker who is handed a different scythe each day. The act of keeping the scythe well-sharpened is an "investment"—it is a use of effort to increase *future* output rather than to increase current output. Policies that will encourage investment are of two types: policies that encourage the *desire* to invest, and policies that facilitate the *ability* to invest.

Policies to Encourage the Desire to Invest

Property rights and institutions. The preceding discussions of the importance of land titling, and the problems of assigning ownership rights to water, illustrate the importance that ownership has for encouraging investment. Recall the discussion in Chapter 17 on the importance of institutions to economic growth. A farmer is more likely to want to make investments in her farm when she knows that she will reap the benefits of the improved productivity brought on by that investment. This means that she wants to have confidence in her right to use the improved asset in the future (ownership or long-term enforceable contracts); she wants to have confidence that the legal system will enforce her ownership or contractual rights; she wants to have confidence that the political system is stable and democratic—that her ownership rights won't be taken away by a change in political leadership; she wants to have confidence that the legal and political system is not corrupt.

Subsidizing farm-to-market roads and rural infrastructure. Building a new road into a region that formerly was reachable only by human or animal transport can have an important impact on that region's economy. Because it dramatically reduces the cost of transporting farm products out of the region, it raises the farm-gate price of what farmers sell to country marketing agents who buy and transport food to the city. And because it simultaneously reduces the cost of transporting materials into the region, it reduces the cost of purchased farm inputs such as fertilizer. Higher farm-gate prices and lower input costs increase farm income, stimulate greater farm production, increase the demand for farm labor, and raise local wages.

In a study of the impact of new roads on forty-six Philippine rural communities, Santos-Villanueva (1966) found a decrease in transportation costs of 17 to 60 percent per kilometer and a substantial increase in the amount of agricultural products sold off the farm (see Table 22.1).

In a study of sixteen villages in rural Bangladesh, it was found that villages with a good infrastructure, including good roads (all-weather, hard-surfaced), used 92 percent more fertilizer per hectare than villages with poor infrastructure. They used 4 percent more labor per hectare, and they paid their agricultural laborers 12 percent more per day than did villages with poor infrastructure (Ahmed and Hossain 1990).

Higher local wage rates reduce the pressure for out-migration. In other words, good roads help keep people home. Good roads encourage private entrepreneurs to start bus services, often with very small vehicles ranging from large three-wheeled motorcycles to minibuses, and thus make it possible for rural people who live within reasonable commuting distances to work in town

Table 22.1 Average Increase in Sales Volume of Selected Agricultural Products After Construction of a Nearby Road from Farm to Market, Philippines

	Percentage Increase
Corn	104
Chicken	69
Swine	47
Coconuts	30
Rice	24
Bananas	12

Source: Santos-Villanueva 1966.

but continue to live at home. Good roads enable employment outside the home neighborhood.

As roads lower the cost of transportation to and from the countryside, the annual fluctuation in the price of food is reduced. Remote communities find it cheaper to export food in good crop years and to import food in poor crop years, thus reducing price swings between the bad and the good harvests. Reducing these price swings reduces the probability that low-income families will face undernutrition during the years of bad harvest.

As a good road network reduces the cost of transporting agricultural commodities around the country, increased agricultural specialization by region can take place. For instance, perishable fruits and vegetables can be grown farther from urban markets than previously, making possible a higher-valued use of land far from market and reducing the income disparity between remote areas and the major cities.

The same sorts of advantages that accrue with new roads into a formerly remote territory apply to improvements in old roads. Putting a hard surface on a dirt road, or mending potholes on a worn-out, older, all-weather road, creates benefits that are similar to, albeit less dramatic than, those obtained from putting in a new road.

In addition to roads, the parts of rural infrastructure that probably have the most bearing on agricultural production are the electrical system, the communications network, and the marketing system.

Rural electrification makes possible the powering of a host of time-saving and production-enhancing devices. Small irrigation pumps and power tools are efficiently driven by electricity. Electrification also makes communications easier in the countryside, for example by making a modern telephone switching system possible. Access to a telephone (even if there is only one per village) can save lives in a medical emergency. But efficient communications are also important in marketing farm products as well as in gaining access to purchased farm inputs.

An important function of a marketing system is to reflect back to producers the wants and preferences of consumers. As economic development brings changes in food demand patterns, a good marketing system will efficiently transmit this information to farmers, who can then reallocate their resources to take advantage of the new production opportunities.

And as the sophistication of agricultural production increases with the adoption of new technology and improved management, a good marketing system for agricultural inputs will reflect farmers' preferences back to the farm suppliers. Bureaucratic, government-sponsored fertilizer distribution schemes usually supply only a limited choice of plant nutrient mixes. But if allowed to function in an appropriate institutional framework, a private-enterprise marketing system will make available a wide variety of fertilizers, allowing farmers to choose the most efficient ones for a particular situation.

Government support can help radio stations and newspapers pass along price information of interest to farmers. This is especially important for small farmers who, because of lack of information about market prices, might otherwise be at a disadvantage in comparison with the larger farmers who can afford to seek out such information on their own.

Government-sponsored terminal markets, where buyers can assemble farm products from the countryside and distribute them to retailers in the city, can increase the efficiency of the marketing system. Increasing the efficiency of the marketing system lowers prices to consumers and increases prices to producers, thus improving nutrition and stimulating increases in production. Appropriate government subsidies are important to an efficiently functioning agricultural marketing system (see Box 22.2).

Price stabilization. When farmers make decisions about whether to invest in projects that will improve their long-term productivity, they base their decisions on beliefs about future prices. Economists have found evidence that farmers are *risk-averse:* other things being equal, they prefer sure things or

Box 22.2 Public Goods and Public Investment

Public investments are usually made because of a failure of the market to provide certain public goods. A small public park, a downtown sidewalk, national defense, and free public education are examples of public goods. Public goods have two critical properties: (1) it is impossible to exclude individuals from enjoying the benefits of these goods, and (2) it is undesirable to exclude individuals from enjoying the benefits of these goods, since such enjoyment does not detract from that of others (Stiglitz 1986:119). No one wants to produce a public good because, once produced, it is available to all. The producer cannot sell it and recapture the costs of production.

The distinction between a private good and a public good may be fuzzy. For instance, we can think of expenditure on the education of our children as a private good, since the children benefit directly; but education also has aspects of a public good, because educated children will be more productive later in life than uneducated children. The fact that education provides benefits (external benefits) to people other than the students themselves helps explain why governments are involved in paying for public schools.

Likewise, when a consortium of governments pays for public agricultural research to develop high-yielding varieties of rice, we may want to call it a public investment. All rice farmers can benefit from the new rice varieties, and of course the public benefits from the lower market price of rice after the higher-yielding rice varieties are put into use. We could also call the rice research and the resulting high-yielding varieties a public good. So it is with many other public expenditures for enhancing agricultural production, such as research on irrigation machinery or agricultural extension programs—they are pitched at the farm sector but they benefit the public.

low-risk investments to high-risk investments. When output prices are highly variable and unpredictable, farm investments are riskier, and farmers are less likely to make these investments. For this reason, some governments have instituted policies to stabilize prices of agricultural commodities (Newbery and Stiglitz 1981).

Governments have used four general types of policies to achieve more stable prices:

1. The most direct policy is that of *administered prices,* under which the government sets prices, either by administrative fiat, or by acting as the sole buyer of the commodity. The difficulty with this type of policy is that the government frequently knows even less than farmers about likely future price levels. Therefore, the prices established by the government do not reflect underlying supply-and-demand conditions. Black markets emerge and either the programs are ineffective or they become an enormous drain on the government treasury.

2. A second method is a *buffer stock* program, under which the government buys the commodity when supplies are plentiful and prices are relatively low, and sells the commodity when supplies are tight and prices are relatively high. The government storage acts to even out prices over time. In theory, this program should put a ceiling and a floor on agricultural prices: when prices threaten to drop below the floor, the government buys the commodity and bids the price back up; when prices threaten to shoot through the ceiling, the government sells the commodity and brings prices back down. In practice, governments have not been particularly adept at establishing the correct range of prices. In addition, the subsidized government storage tends to discourage storage in the private sector.

3. A *buffer fund* accomplishes much the same thing as a buffer stock program, but does so without requiring the government to actually buy and sell the commodity. Under a buffer fund, the government taxes sales of the commodity in periods when prices are high (and thereby reduces the price received by farmers), and subsidizes sales in periods when prices are low (effectively increasing the farm price). Buffer funds require extensive bookkeeping to verify which farmer sold what quantity at what price.

4. Finally, governments can encourage the development of private market institutions, such as *forward contracting* and *futures markets,* to stabilize prices. These mechanisms allow farmers to lock in a price at the beginning of a growing season, at least for a major proportion of their crop. These types of contracts work well in theory, but farmers have been reluctant to adopt them in the United States and other developed countries.

Policies to Facilitate the Ability to Invest

A farmer who *wants* to invest also needs to have the *ability* to invest. For most assets—buildings, breeding stock, machinery, irrigation equipment—the acquisition, improvement, or maintenance requires cash. Unlike major corporations that can raise money by issuing stock or selling bonds, small farmers have to either save the money themselves, or borrow the money. Therefore, policies that facilitate borrowing are policies that lead to increased agricultural investment.

Credit subsidies. A number of developing-country governments have adopted programs that directly subsidize credit. These programs are seen as a way to spur agricultural productivity, to transfer income to the poor, and to compensate low-income farmers for their losses attributable to urban bias (Adams et al. 1984). However, experience has shown that subsidized credit is not an effective instrument for transferring income to the poor. Furthermore, subsidized credit may have harmful side effects on financial institutions and other segments of the economy, particularly the poor.

There are two common ways in which governments provide low-interest loans. One is directly through a governmental or quasi-governmental bank that loans the money to preferred borrowers at below-market interest rates. For instance, in Jamaica during the 1970s, the parastatal development bank supplied loans at less than half the interest rate changed by commercial banks. The other common way of providing low-interest loans is for the government to require commercial banks to supply a given amount of money to preferred borrowers at below-market rates. In Nigeria, for example, banks must devote 8 percent of their loans to the agricultural sector at approximately half the current commercial interest rate (Bale 1985:17).

Beginning in 1965 in Brazil, the law required commercial banks to lend at least 50 percent of their demand deposits to farmers at 17 percent interest per year—less than the inflation rate at that time of 20 to 40 percent per year. In 1971 the mandated rural interest rate was lowered to 15 percent, even though high rates of inflation continued (Sicat 1983:381).

When the government bears the cost of the credit subsidy, high costs often lead to deficit financing and high inflation, which discourage private-sector savings. In addition, the resources devoted to providing cheap credit might be better used in other government programs—such as to increase agricultural research, build better rural roads, or improve educational services.

When the cost of the credit subsidy is pushed onto the commercial banking system, by forcing banks to make low-interest loans to a preferred set of borrowers, then other, nonpreferred borrowers end up paying the cost in the form of higher interest rates, or reduced loan availability. If the group of preferred borrowers is large enough, the requirement that banks make low-interest loans will reduce the interest paid by the banks, discouraging some people from saving and motivating others to send their savings (and thus their capital) out of the country in search of better returns.

From the point of view of income distribution, the worst aspect of subsidized credit is that it discriminates against the poor. The subsidized credit almost always goes to those in the community who are better prepared to receive it.

In rural areas this means the rural elite: those who are well connected politically. Seldom does it go to the poorest of the poor, and the size of the subsidy is directly proportional to the size of the loan. The larger farmers get larger loans. Medium and small farmers get proportionally smaller subsidies. The smallest farmers and landless laborers get no subsidies at all (González-Vega 1983:371).

Other initiatives to encourage credit. The government can take other initiatives to facilitate the availability of credit. A government-operated or privately owned credit information system (such as the credit-scoring companies in the United States) gives lenders a source of information about a potential borrower's credit history, job history, income sources, and the like. By making this information available at a low cost, such a system reduces the risks of lending.

Government can also influence the ability of a borrower to pledge collateral for a loan. Collateral is an asset that the borrower promises to turn over to the lender in the event that the borrower is unable to make required repayments. For example, in the United States, car loans are secured (or "collateralized") by the cars themselves. If a borrower fails to make his car payments, his car will be repossessed by the lender. Collateral serves several important functions: it provides something of value to the lender even if the loan is not repaid; the threat of losing the collateral provides an incentive for the borrower to make repayments; and the possibility of losing collateral may discourage certain high-risk borrowers from taking out a loan in the first place.

As mentioned earlier in this chapter, creating transferable ownership rights to land allows land to be pledged as collateral, and thus improves the functioning of credit markets. A government-operated "collateral registry" allows a lender to see whether a certain asset has been pledged as collateral on any other loans. Government policies can also influence the ease with which collateral claims can be adjudicated and enforced.

Microcredit: an alternative to subsidized credit. During the 1970s and 1980s, economists began to piece together an alternative to subsidized credit. Muhammad Yunus, an economist from Bangladesh, observed that very small loans could make a huge difference to very poor households. He began to work on a method of providing small-scale loans to these households. Initially, Yunus himself provided the funding. But for the idea to work on a larger scale in impoverished rural areas, there would have to be a local source of loanable funds. Dale Adams (1983) found that the average rural household in five Asian countries *did* have positive savings, and that savings increased when interest rates rose.

Yunus's plan is now known as "microcredit" and is widely seen as a promising way to make credit available to small farmers (see Box 22.3 for an example of how microcredit works). Yunus founded the Grameen Bank of Bangladesh, which promoted the concept of making small loans to poor families who put

Box 22.3 Microcredit in Practice

An actual example of microcredit illustrates how it works. A farmer in the Philippines wanted to borrow $52 to buy two piglets. He planned to feed the piglets with table scraps and spoiled food from the family table, so there would be little cost other than the initial investment. He was able to convince the fellow members of his microcredit group that this was a reasonable investment, and they knew that the farmer had a reputation for honesty and hard work. The farmer received the loan and promised to repay $2.30 a week for twenty-six weeks (a total repayment of about $60). At the end of six months, the farmer was able to sell the fattened pigs for over $200. The money he repaid (plus interest) was made available to finance projects by other group members.

up no collateral other than a pledge to join and meet regularly with a support group. The sense of obligation to the group was a sufficiently strong incentive to ensure that borrowers paid back their loans (Hossain 1988a:25–26).

The use of small community groups to funnel credit to rural lenders has a number of advantages. The group members know the reputations of the borrowers—who is industrious, who is lazy, who is honest—and can take that into account in making loan decisions. The group members have similar backgrounds and experiences, and so are able to analyze the likelihood that a project will succeed in generating income sufficient to repay the loan. The group members can fairly easily determine whether difficulty in repaying is due to a borrower's bad luck or due to a borrower's poor performance. Finally, borrowers want to repay the loans in order to preserve their social standing in the community.

According to one estimate (Global Development Research Center 2003), there are about 13 million microcredit borrowers, and the average loan size is about $500. Repayment rates are reported to be above 95 percent.

However, Boudreaux and Cowen (2008) argue that mirocredit, although it can be beneficial, fails to live up to its promise in many respects. "[M]icrocredit is mostly a good thing. Very often it helps keep borrowers from even greater catastrophes, but only rarely does it enable them to climb out of poverty." But interest rates are high (50–100 percent annually), and many loans are for consumption (a doctor visit, school fees) rather than for investment.

Policies to Promote Technological Improvement

Promoting Agricultural Research

Technological change is one of the primary driving forces behind increasing production. Improved technology (e.g., a higher-yielding variety of rice) can increase the productivity of every item of the set of resources that a farmer uses—land, labor, management, and capital. Yet the agricultural researcher intent on developing new technology for agriculture faces an extraordinarily intricate range of scientific challenges. One analysis of technological change (Lele, Kinsey, and Obeya 1989:42) listed the following areas as important to researching improved crop varieties in third world agriculture:

- Yield potential and responsiveness to available chemical fertilizers and pesticides.
- Adaptation to the growing period and drought tolerance.
- Disease and pest resistance.
- Improvements in quality, palatability, and consumer acceptance.
- Storage, transport, and other handling methods (including processing) with available technology.
- Changes in production and processing labor requirements as a result of the available mechanical technology, in view of other requirements for household labor and incentives for labor use.
- Compatibility with other social, cultural, and economic norms.

Not only is the range of challenge complex, but the disciplines brought to bear on agricultural research are varied. Advancing agricultural technology involves research in biology, chemistry, and engineering, as well as in the social sciences, which are crucial to the appropriate integration of the new technology into the production system.

Despite the challenging nature of agricultural research, the payoff has been nothing short of spectacular. As described in Chapter 14, agricultural research has been instrumental in dramatically increasing crop yields per acre as well as livestock productivity. And when the costs of agricultural research are compared to the benefits to society, agricultural research turns out to be a real bargain.

Table 22.2 lists studies on payoffs from agricultural research projects conducted at experiment stations around the world. The last column, average annual internal rate of return, is of particular interest. It represents the average earning power of the resources during the project period, and is equivalent to the annual interest rate one would have to obtain from a bank savings account in order to receive the same return that the public received from the agricultural research.

In one of the earliest studies of this kind, returns to research on hybrid corn (maize) in the United States were calculated by economist Zvi Griliches (1958) (the corn-breeding research was carried out by many scientists). The internal rate of return to hybrid corn research in the United States was calculated to be 35 to 40 percent. This means that for every dollar the US public paid for agricultural research on hybrid corn prior to 1955, it collected 35 to 40 cents every year in benefits (mostly through lower prices for corn). It would be hard to find another set of investments that pay off as well as agricultural

Table 22.2 Studies on Rates of Return to Agricultural Research

Commodity	Number of Studies	Lowest Annual Internal Rate of Return (%)	Highest Annual Internal Rate of Return (%)	Average Annual Internal Rate of Return (%)
Multicommodity	436	–1	1,219	80
All agriculture	342	–1	1,219	76
Crops and livestock	80	17	562	106
Unspecified	14	16	69	42
Field crops	916	–100	1,720	74
Maize	170	–100	1,720	135
Wheat	155	–48	290	50
Rice	81	11	466	75
Livestock	233	3	5,645	121
Tree crops	108	1	1,736	88
All studies	1,772	–100	5,645	81

Source: Alston et al. 2000.

research. Swindale (1997) cites an example of research in sub-Saharan Africa into methods of controlling an insect pest—the cassava mealybug. The research cost $27 million, but the benefits from the research exceeded $4 billion.

If agricultural research has such a spectacular payoff, why don't farmers themselves pay for it? Or why don't private companies undertake the research and sell the results to farmers? (See Tables 22.3 and 22.4.) The two reasons why it is not appropriate to ask farmers to pay are that first, most farmers have far too small an operation to sponsor and benefit from agricultural research, and second, given that the elasticity of demand for most farm products is less than 1.0, the majority of the benefits from agricultural research go to consumers. Farmers generally lose revenue when a new technology is widely adopted, because they see their farm-gate prices fall faster than they can increase production. Long-term data from the United States, which has a history of public sponsorship of agricultural research dating back to the 1870s, are illustrative:

> The decline in the real price of food has been dramatic. Available data for the period 1888 to 1891 indicate that consumers spent an average of about 40 percent of their income for food. From 1930 to 1960, the food expenditure proportion of consumer incomes ranged from 20 to 24 percent. In the seventies, the proportion of total disposable personal income spent for food dropped to a range of 1–17 percent. By the mid-eighties, that proportion for the average family had dropped to a record low of 15 percent. (Lee and Taylor 1986: 20–21)

Patent laws protect mechanical and chemical innovations more effectively than biological innovations. For this reason, some agroindustrial firms have

Table 22.3 Estimated Public and Private Investment in Agricultural Research and Development, 1995

	Expenditures ($ millions)			
	Public	Private	Total	Percentage Private
Developing countries	11,770	609	12,379	4.9
Developed countries	21,567	10,962	32,530	51.4

Source: Pardey and Beintema 2001.

Table 22.4 Public Expenditures on Agricultural Research and Development as Percentage of Agricultural GDP

	1976	1996
Developing countries	0.5	0.6
Developed countries	1.5	2.6

Source: Pardey and Beintema 2001.

been able to sponsor research in farm machinery or agricultural chemicals and capture the benefits from that research. Hybrid seeds are protectable by patents and, because they do not breed true, farmers must purchase new supplies each year. So after government-sponsored research led the way, hybrid seed companies set up their own research and began developing their own varieties. But generally, biological innovations in agriculture have to be paid for by government. Thus, animal breeding, animal nutrition, plant breeding, plant pathology, entomology, agronomy, soil science, and so on, are government-sponsored (Judd, Boyce, and Evenson 1987:7).

Some countries are too small or too poor to sponsor agricultural research. Their funds are best spent on adaptive agricultural research—finding out which of the innovations discovered elsewhere are most adaptable to their own situations. The Consultative Group on International Agricultural Research, through its experiment stations, is helping to fill in the research gap felt by the smaller countries. Fifteen international agricultural research centers belong to this group and are sponsored by a variety of sources. The group includes the International Rice Research Institute in the Philippines and the International Maize and Wheat Improvement Center in Mexico, as well as a number of other centers whose activities range from plant and animal breeding to food policy.

Low-income rural householders could benefit from research on hardy but efficient scavenging animals. High-yielding, disease-resistant breeds of chickens, ducks, goats, pigs, cattle, bees, or fish that can utilize garbage or other food that may be locally available but unfit for human consumption would be of considerable benefit to the third world's poor.

Africa poses a particular challenge to agricultural research. Its soils are more diverse, its climate more varied, its pest and disease hazards more pronounced, and its farming systems more complex than those of monsoon Asia, where the green revolution has been such a success (Lele and Goldsmith 1989; Lipton and Longhurst 1990). African agriculture is characterized by *mixed-cropping* (more than one crop at a time is grown in a field), which occupies over 90 percent of the cropped area in most countries on the continent.

A special challenge for the agricultural research community working in Africa is to intensify agricultural production within this complex farm management system while maintaining its flexibility and its proven sustainability (Dommen 1988). One production innovation that holds promise for sustainability in the semiarid tropics that cover so much of Africa is *agroforestry,* a system of strip-cropping rows of trees between narrow strips of crops. The trees help conserve water, provide a ready source of organic matter, and reduce erosion. Another example of an effective agricultural research project in Africa is described in Box 22.4.

Policies to encourage development of new technology include not only direct government sponsorship of research, but also laws and regulations that encourage private research and development. The FAO 2004 *State of Food and*

Box 22.4 Cassava Research

In parts of western Africa, cassava—a starchy root crop—is a significant source of calories. From the mid-1980s to the mid-1990s, the International Institute of Tropical Agriculture funded research to develop new varieties. As described in the FAO's 2000 *State of Food Insecurity* report: "These new varieties yielded up to 55 tonnes per hectare, compared with about 10 tonnes per hectare from traditional varieties; matured early; were highly resistant to disease; . . . developed a broad leaf canopy, thus optimizing both weed control and yield potential; had compact root shapes, facilitating harvest" (FAO 2000a:21).

News about the new varieties spread rapidly, and farmers switched more and more land into cassava production. In Ghana, cassava area nearly doubled between 1983 and 1998. In Nigeria, it nearly tripled. In Ghana, yields increased from 7 metric tons per hectare in 1988 to 12 metric tons per hectare in 1997. Cassava research is one of the factors that explains Ghana's astonishing progress in reducing the prevalence of undernutrition—in 1980, over 60 percent of its population was undernourished; by 1997, that figure had dropped to 10 percent.

In addition to Ghana and Nigeria, other areas in western and central Africa are suitable for growing cassava, and the new varieties may help address the undernutrition problem in those locations as well.

Agriculture (FAO 2004c:88) report concludes that countries and the international community need to do the following:

- Establish regulatory procedures that are transparent, predictable, and science-based; and harmonize those procedures, where appropriate, at regional and global levels.
- Establish appropriate protections for intellectual property rights to ensure that developers can earn adequate returns on investment.
- Strengthen national plant-breeding programs and seed systems.
- Develop efficient input and output markets for agriculture and reduce trade barriers on agricultural technologies.

Subsidizing Technology Diffusion and Adoption

Profitable technology will spread from farmer to farmer by itself, but the rate of adoption can be accelerated by government-sponsored educational programs (see Feder, Just, and Zilberman 1985 for a review of the literature on this topic). Such initiatives are often called *extension programs,* because they were originally conceived to extend the knowledge developed in the US land-grant college system directly to the farmer. Programs that provide education and advice to farmers are now in place in most countries that have a significant agricultural economy.

A classic study of Iowa illustrates the typical growth curve of knowledge that an agricultural community experiences as its farmers gradually become aware of a new technology, then try it out, and finally adopt it as part of their regular farming activities. In this case, a new chemical weedicide control called 2,4–D had come onto the market. From 1944, when only a few of Iowa's farmers (4 percent) had even heard of it, it took approximately eleven years for full adoption. Notice in Figure 22.2 how awareness precedes trial, which in turn precedes adoption. In 1949, for instance, about midway through the process, 74 percent of Iowa's farmers were aware of the existence of the new weedicide, but only 40 percent had adopted it. By 1955, all farmers in the area were using it. Without the vigorous extension program conducted by Iowa State University, the new technology undoubtedly would have spread, but whether it would have spread as fast is open to question. It is the function of extension programs to accelerate the adoption of new technologies, whether new agricultural chemicals, better plant and animal varieties, or better farm management practices.

Accelerating the spread of technology begins with government sponsorship of agricultural training institutions—places where technicians learn the skills necessary for backstopping agriculture in the field, such as through training in plant and animal sciences, and farm management and finance. Not only do government-sponsored farm advisers need to know agricultural technology,

Figure 22.2 Cumulative Percentage of Farm Operators at the Awareness, Trial, and Adoption Stages for 2,4–D Weed Spray, Iowa, 1944–1955

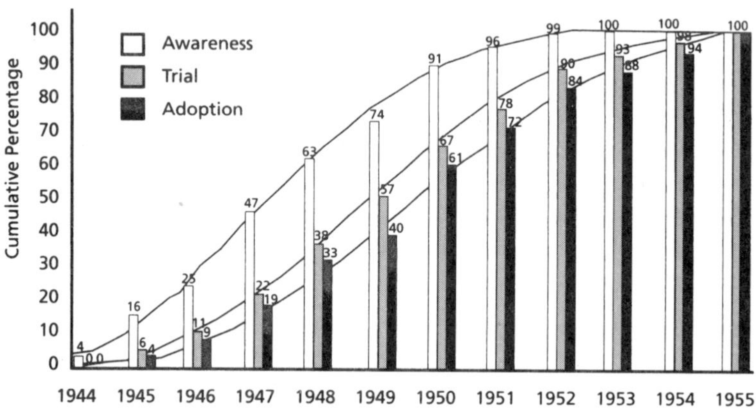

Source: Adapted from Beal and Rogers 1960:8.

but rural banks need farm appraisers, and rural tax collectors need to know about the economics of farming. Sponsoring the training of these technicians provides an important subsidy to agricultural production. The success of an agricultural extension program depends not only on the quality of the training that farm advisers receive at their local agricultural colleges, but also, and perhaps more important, on the quality of the technologies being extended. These technologies must be appropriate to the regions where the technical advisers are working if farmers are to adopt them.

Policies to Promote Sustainable Farming Methods

Closely related to policies that promote increased productivity in the farm sector through development and adoption of new technologies are policies that promote environmentally friendly (or sustainable) farming. Here too, government programs are directed at invention—or discovery—of environmentally appropriate practices, as well as at extension—teaching farmers about those methods. The kinds of policies discussed in the previous section apply here.

Earlier in this chapter, we noted the importance of maintaining or improving the quality of natural resources used in farming. Agricultural production puts a strain on the natural environment. If the strain is too great—if land and water resources become sufficiently degraded—future agricultural production will suffer. Technology policy has an important role here. Persuading farmers to adopt sustainable farming methods—methods that by definition do not entail resource degradation, which will in turn lead to future declines in production—will cause future food supply to be greater than it would be in the absence of these methods. One example of agricultural research that promotes soil quality is reported in Box 22.5.

Box 22.5 Research That Improves Soil Quality

Not all technological discoveries come out of government-funded laboratories using sophisticated equipment. Decades ago, the United Fruit Company began growing mucuna beans to provide cheap feed for mules on its banana plantations in Guatemala and Honduras. Local farmers noticed that the soil in fields where mucuna beans had been grown was richer than the soil in other fields.

With more systematic study (much of it funded by private aid groups), the advantages of this magic bean became better understood (see Pettifer 2001). When planted in land recently cleared and burned, mucuna beans grow amazingly fast, reducing the soil erosion that can occur on bare land. If the beans are left to rot on the soil, they create a natural fertilizer that makes use of commercial fertilizer unnecessary. Maize and mucuna can be grown together in fields, with mucuna replenishing the soil with nutrients used by the maize. The dense mucuna plants also suppress weeds and reduce or eliminate the need for commercial herbicides. Yields of maize double or triple. Mucuna can even be used to "create soil" (organic compost) on rocky hillsides.

23

World Food Supply and Demand for the Next Half Century: Alternative Scenarios

> And I beheld a black horse; and he that sat on him had a pair of balances in his hand. And I heard a voice . . . say, A measure of wheat for a penny, and three measures of barley for a penny. . . . And power was given unto them over the fourth part of the earth, to kill with the sword, and with hunger, and with death, and with the beast of the earth.
>
> —Revelation 6:5–8

Famine is one of the Four Horsemen of the Apocalypse. And when our visions of the future take an apocalyptic turn, the specter of widespread hunger is likely to appear. What does the future hold? Will the progress of the preceding decades continue? Or are we on the brink of a slide into catastrophe? Having boldly asked these questions, we less boldly reply, "It all depends . . ." This chapter will briefly review schools of thought about the future of the world food problem and present a framework for building scenarios about the future.

There are several points on which there is universal agreement. Of course, any prediction is likely to be wrong in certain fundamental ways. But the four Ps identified in Chapter 1—population, prosperity, pollution, and productivity—are almost certainly the major factors that shape the future. Growth in population and in per capita income will determine, to a considerable extent, demand for food. Growth in yields per hectare and limitations imposed by environmental quality will determine, to a considerable extent, supply of food. The interplay of these factors will ultimately be reflected in a fifth P—price of food. If demand grows more rapidly than supply, food prices will increase and (if prices increase faster than incomes) hunger will become more widespread. If supply grows more rapidly than demand, food prices will drop. The sixth P—policy—is the means by which humankind can hope to influence the future.

Two Views of the Future

It is tempting to characterize people's views about the future of world food supply and demand as "optimistic" or "pessimistic" (in fact, at several points in this book, such characterizations are made; see also McCalla 1998). But a more accurate division might be between the *establishment* view and the *anti-establishment* view. The establishment view is reflected in reports by the FAO (Alexandratos 1995), the World Bank (Mitchell, Ingco, and Duncan 1997), the International Food Price Research Institute (Pinstrup-Andersen, Pandya-Lorch, and Rosegrant 1998, and other "IFPRI 2020" reports), and the US Department of Agriculture (USDA 2008b). These reports embrace a common vision of the future in which world agricultural production continues to grow (with slight growth in agricultural area, and yield growth of 1.0 to 1.5 percent per year), world population continues to grow, but ever more slowly than the current rate of 1.17 percent, and income per capita continues to grow. Implicit in their vision is an assumption that there will be no catastrophic changes in environmental conditions.

The antiestablishment view, which puts a much greater emphasis on the possibility of environmental catastrophe (see Gilland 2002 for a careful exposition), is reflected in the works cited in Chapters 12–14 by David Pimentel of Cornell University and by Lester Brown and his colleagues at the Worldwatch Institute. The types of catastrophe that could occur include losses of agricultural land due to erosion or degradation, reductions in availability of usable water for agriculture due to overirrigation, and global warming, which could (in the worst cases projected by scientists) cause significant loss of land to rising sea levels, and significant yield reductions from changing climate patterns. The antiestablishment view is also notably less confident about the possibility of a "technological fix" for these future catastrophes if they do occur. As described in Chapters 13 and 14, this lack of confidence is based on the slowing of yield growth, and the belief that we are unlikely to see any significant breakthroughs in basic science that would provide a foundation for a new spurt of growth in yields.

Past predictions of global food shortages have been wrong largely because they underestimated the ability of technological progress to increase food output. The establishment view has confidence that the institutions and processes that generated yield growth in the past will continue to generate yield growth in the future. The antiestablishment view sees the yield growth of the past fifty years as a one-time stroke of good fortune that is unlikely to be repeated.

Both the establishment and antiestablishment views accept the premise that human actions—especially government policies—can influence the future state of the world. The establishment view is more sympathetic to incrementalism—small changes in policies to ensure that the world continues to make progress.

The antiestablishment view holds that major sweeping changes—especially in environmental and population policies—must be made to accommodate and ameliorate future problems.

The antiestablishment view tends to see current prosperity—and the fact that "only" 18 percent of the world's population is currently undernourished—as coming at the expense of future generations. In effect, as proponents of this view argue, we are "eating the seed"—satisfying our current hunger by guaranteeing even more severe and widespread problems in the future. The appropriate response, therefore, is a radical reduction in current levels of consumption of food and nonfood alike. At the extreme, it is argued, accomplishing this radical reduction may require radical restructuring of economic and political systems. The establishment view does not accept the vision of inherent conflict between present and future generations: current prosperity will not cause future poverty; future generations will be on average at least as prosperous as we are today. This attitude leads to a broad endorsement of existing institutions. This is not quite the same thing as saying that the establishment view endorses a "business as usual" policy. Rather, the kinds of policies endorsed by the establishment view—market-oriented pricing, increased government funding of research and development, appropriate macroeconomic policies to promote general economic growth, for example—tend to be policies that can be pursued within the existing institutional framework.

Predicting the Future: Principles for Scenario Building

Looking at different scenarios for the future allows us to see how intelligent, well-informed people can arrive at such different views about the future. In this section, we present a few alternative scenarios. The main purpose is not to prove that one or the other of these scenarios is "correct," but rather to provide a template from which readers can construct their own scenarios.

These scenarios make projections 42 years (2008–2050) into the future. This is further than most projections; but most projections are intended for policymakers, who recognize that today's projections will be replaced by a new set within a few years. This period has been chosen to give readers a view of world food supply and demand during their lifetimes.

All scenarios are based on assumptions about annual rates of growth: growth in population, growth in per capita income, growth in area harvested, and growth in yields per hectare. The rates of growth assumed are *average* rates for the entire period. Growth may be higher than average for some years, and lower than average for others. As laid out at the end of Chapter 10, projections about world food demand are based on assumptions about growth in population, growth in income per capita, and growth in biofuels. As suggested at the beginning of

Chapter 11, projections about world food supply are based on assumptions about growth in agricultural land area and growth in yields.

In developing the scenarios in this chapter, we use a set of assumptions that are internally consistent. The complexity of interconnections among the factors has been illustrated throughout the book. Agricultural productivity, and the concomitant prosperity in the farm sector, contribute to economy-wide prosperity and growth in per capita income. Higher per capita income is associated with better healthcare and sanitation, and therefore lower infant mortality; this in turn leads to reduced fertility and ultimately to lower population growth rates. Environmental catastrophes, should they occur, are likely to be associated with lower per capita incomes, and increased mortality—and therefore possibly lower population growth.

An "Establishment View" Scenario

At the end of Chapter 10, we presented some sample scenarios for growth in total food demand. We saw a number of different assumptions that were consistent with a doubling of demand for food over the next fifty years. Here, we revisit that hypothesis, and expand it to include the supply side. The basic demand-side assumption is that demand will increase by 95 percent by 2050. We also consider assumptions and conclusions that are summarized as scenario E (for "establishment") in Table 23.1.

Population and income. As reflected in Table 10.5 (scenario 1), demand growth of about 95 percent is consistent with annual population growth of 0.8 percent and annual growth of per capita income of 2 percent. To put these numbers into perspective, the USDA's ten-year projection (USDA 2008b) assumes population growth of 1.1 percent per year and per capita income growth of 2.4 percent per year. We also construct scenarios with total demand growth of 115 percent (if

Table 23.1 Outcomes of Various Scenarios to Address Food Supply and Demand, 2050

	Percentage Growth Between 2008 and 2050				
Scenario	Demand Growth	Land Growth	Yield Growth	Supply Growth	Price Increase
E	95	11	70	89	5
E-1	65	11	70	89	−19
E-2	115	11	70	89	19
E-3	95	11	40	55	34
AE	95	−15	9	−7	120
AE-1	95	0	40	40	51
AE-2	115	−15	9	−7	137

biofuel demand grows faster, for example) and with total demand growth of 75 percent.

Increase in agricultural land. Chapter 11 discusses the potential for increasing agricultural land. Here we assume that agricultural land increases by 2.5 percent per decade (not per year), or 11 percent by 2050. To put this into perspective, worldwide agricultural land grew by 11 percent in the forty years between 1961 and 2001.

Increases in agricultural yields. As mentioned, the establishment view projects increases of 1.0 to 1.5 percent per year in yields. The graphs in Chapter 14 show that this range is consistent with historical trends and experience. Here we take the midpoint of that range (1.25 percent per year) as our assumption. Under this assumption, yields grow to 170 percent of current levels by 2050. To put our assumptions into perspective, the USDA's ten-year projection assume that US yields will grow at a little over 1 percent for wheat, corn, and soybeans (USDA 2008b).

Total increase in supply. Under these assumptions, total food supply (at constant prices) would increase to 188.7 percent (1.11 × 1.70) of current levels.

Effect on price. If food demand grows to 195 percent of current levels while supply grows to 188.7 percent of current levels, we see a slight upward pressure on prices. Assuming a demand elasticity of 0.2 and a supply elasticity of 0.5, food price would increase by about 4.7 percent from current levels.

Extent of undernutrition. Of course, we cannot estimate what happens to the extent of undernutrition with any precision by using only these worldwide aggregate numbers. But ample evidence shows that the extent of undernutrition will decline substantially under this scenario. Calories per capita per day are projected to increase from 2,800 currently to 3,360 by 2050 (about the level in the United States during the mid-1980s). In addition, notice that while the price of food (in this scenario) is essentially unchanged at the end of 42 years, average income per capita has more than doubled. This should lead to a substantial reduction in the extent of undernutrition. In fact, this is the prediction of the establishment view's published projections. For example, the FAO projects that the number of food-insecure people should drop from 840 million (about 20 percent of the developing world's population) in the early 1990s to 680 million in 2010 (about 12 percent of the population). The International Food Policy Research Institute projects that the number of malnourished children will decrease by 20 percent from 1993 to 2020; during the same period, total population will increase by more than 30 percent.

"Sensitivity Analysis": What Happens If We Change Assumptions?

The base scenario permits us to modify assumptions one at a time and see how sensitive the results are to changes in the assumptions.

Scenario E-1: lower demand growth. Scenario E-1 in Table 23.1 assumes that demand only grows by 65 percent by 2050. This is consistent with slower population growth (such as in the UN's low variant), or slower income growth (1 percent per year, rather than 2 percent per year). If we hold all other assumptions constant, this would mean supply growing much faster than demand, and substantial price decreases (a decline of about 18.6 percent). Although the supply curve shifts out by 88.7 percent, the price decline causes a movement along the supply curve of 9 percent (based on a 0.5 supply elasticity). Thus the total increase in quantity of food supplied is about 80 percent. As in scenario E, income grows and prices decline, so calories per capita increase and prevalence of undernutrition declines.

Scenario E-2: higher demand growth. If demand for biofuels grows much more than expected in the base scenario, total demand growth could be 115 percent rather than 95 percent. If we again maintain the supply-side assumptions of scenario E (land growth of 11 percent total, and yield growth of 1.25 percent per year), then demand growth outstrips supply growth and prices increase by about 19.1 percent. Even in this scenario, the price increase is smaller than the income growth. The undernutrition picture improves, though by less than under scenarios E and E-1.

Scenario E-3: lower yield growth. If average yield growth is 0.8 percent per year, rather than 1.25 percent in the base scenario (scenario E), then yields in 2050 are 40 percent higher than current levels. Food supply increases much less rapidly than food demand (95 percent increase in this scenario); therefore, prices rise substantially, although, again, less rapidly than the increase in incomes. Per capita calorie intake increases modestly.

An "Antiestablishment View" Scenario

A scenario that is consistent with the antiestablishment view (scenario AE in Table 23.1) is likely to reflect the following assumptions.

Population and income. The antiestablishment view is more pessimistic about income growth, so we use a growth rate of 1 percent per year. The antiestablishment view does not take a strong stand on population projections; but we assume higher population growth to be consistent (according to the theory of demographic transition) with our assumption of lower income growth.

Total shift in demand. Given the above assumptions—see scenario 4 in Table 10.5—total demand for food in 2050 will be about 95 percent higher than current demand.

Increase in agricultural land. As noted in Chapter 10, the antiestablishment view believes there may be declines in agricultural land due to degradation and a rise in sea level from global warming. In this scenario, we assume that agricultural area decreases by 15 percent by 2050.

Increases in agricultural yields. The antiestablishment view is doubtful that historical rates of growth can be continued. It may be reasonable (from the standpoint of this view) to assume that yields do not increase at all, or even that they decline as irrigation and land quality decline. Here we assume that yields do continue to grow, but at only a 0.2 percent annual rate. Under this assumption, yields in 2050 are 9 percent higher than current yields.

Total increase in supply. Under these assumptions, total food supply (at constant prices) would decrease to about 92.6 percent (0.85 × 1.09) of current levels.

Effect on price. If demand grows by 95 percent and supply declines by 7 percent, price must increase. Using the 0.2 demand elasticity and 0.5 supply elasticity, this 102 percent difference between quantity demanded and quantity supplied at current prices means that price must rise by 120 percent to return supply and demand to equilibrium.

Extent of undernutrition. The antiestablishment conclusion is that the extent of undernutrition increases substantially. The decline in average purchasing power supports this conclusion—average incomes increase by about 50 percent while prices increase by 120 percent. Additional evidence of food shortage is found in the fact that population is projected to increase by 60 percent while food production increases by less than 50 percent. (Although the supply curve declines under this scenario, we see an increase in quantity supplied in response to the much higher prices. See the appendix to this chapter for details.) Calories per capita decline from 2,800 per day currently to under 2,600 by 2050. This is the level of calories per capita per day that prevailed in the early 1980s, when 28 percent of the developing world suffered from undernutrition.

Alternative Antiestablishment Scenarios

Scenario AE-1: higher yield growth and less land degradation. Suppose things are not quite as bad as projected under scenario AE. Suppose that the quantity of agricultural land does not decline, but stays constant at current levels. And suppose that yield growth is not 0.2 percent per year, but is 0.8 percent per year. Under these assumptions (and maintaining the demand-side assumptions of scenario AE), supply grows by 40 percent, more than under scenario AE, but still less than the demand growth of 95 percent. Prices increase about 50 percent, about the same as per capita income growth. Calories per capita per day increase slightly, and the prevalence of undernutrition remains at close to current levels.

Scenario AE-2: higher biofuel demand. Suppose we start with the basic AE scenario, but add in an assumption of increased biofuel demand (as in scenario 5 of Table 10.5), so that total demand increases by 115 percent. In this case, prices increase by 137 percent, compared to income growth of about 50 percent. Under these assumptions, the prevalence of undernutrition is even worse than it was under scenario AE.

* * *

The reader is invited to create other alternative scenarios. Again, we stress the importance of consistency. For example, if we construct a scenario involving high income growth, we may want to include an assumption of higher yield growth. Economy-wide prosperity is likely to be linked to increased agricultural productivity: if the economy is prosperous, more money can be invested in agricultural productivity; in developing countries where agriculture is a large sector of the economy, agricultural productivity is a prerequisite for high rates of economic growth.

Other projections for the future can be found in USDA 2008b, FAO 2006, and IFPRI 2005. The USDA's projection is only for a ten-year period, as detailed previously. The FAO projects that by 2050, food consumption will exceed 3,000 calories per capita per day worldwide. The IFPRI projects that maize prices will increase by a small amount under the "policy failure" scenario, and will fall by 5 percent or so under the "progressive policy actions" scenario. These are broadly consistent with our "establishment view" scenario. The IFPRI warns, however, that maize prices could increase by 60 percent under the "technology and natural resource management failure" scenario—much more consistent with our "antiestablishment view" scenario.

Policies

Scenarios such as those presented here may be useful in evaluating the likely impacts of various policies. Policy action or inaction will influence the future course of events and the severity of the worldwide hunger problem. Experts do not agree on all details about what comprises a "best policy," nor do they agree on the likely impacts of any policy option. However, it seems clear that policy initiatives should focus two basic objectives:

- *Invest in improved agricultural productivity.* Direct government investment in agricultural research and extension, and improved access to rural credit and markets will spur agriculturual production.
- *Encourage economic growth among the poorest.* Appropriate macroeconomic policies, reliance on competitive markets, and investment in human capital will increase food security by making food more affordable to the world's poorest households.

These two basic policy objectives have impacts on other basic elements of the world food problem. Income growth creates incentives that lead to reductions in fertility and population growth. Agricultural research not only discovers new ways to increase yields, but also discovers ways to preserve quality of natural resources. Some initiatives—such as improving institutional arrangements to enforce property rights—can contribute to both of these basic policy objectives.

Appendix: Mathematics Used in Making Projections

Using Growth Rates to Project the Future

To understand the calculations behind the scenarios presented in this chapter, it is necessary to know some elementary mathematics of growth. A simple example explains the basic point. Suppose a city starts the year with a population of 100,000, and suppose the population grows by 10 percent during the year. At the end of the year (year 1), the city's population will be 110,000. We get this by multiplying the population at the beginning of the year (100,000) by the sum of 1 plus the annual growth rate (10 percent or 0.10):

$$110{,}000 = 100{,}000 \times (1 + 0.10)$$

To conserve verbiage, we assign symbols to these words: V_1 is the value at the end of year 1, V_0 is the value at the beginning of year 1 (or at the end of year 0), and r is the growth rate:

$$V_1 = V_0\,(1 + r) \qquad (1)$$

Now suppose that the population grows by 10 percent for the second year. We can use equation 1 to calculate the population at the end of year 2:

$$110{,}000 \times (1 + 0.10) = 121{,}000$$

Notice that we can also write this as:

$$(100{,}000 \times [1 + 0.10]) \times (1 + 0.10) = 121{,}000$$

Similarly, at the end of year 3 (if growth continues at 10 percent), the population would be:

$$\{[100{,}000 \times (1 + 0.10)] \times (1 + 0.10)\} \times (1 + 0.10)$$
$$= 121{,}000 \times (1 + 0.10) = 133{,}100$$

The last two calculations show us a pattern:

$$V_1 = V_0\,(1 + r)^1$$
$$V_2 = V_0\,(1 + r)^2$$
$$V_3 = V_0\,(1 + r)^3$$

Thus the general rule used to calculate values at the end of T periods is:

$$V_T = V_0\,(1 + r)^T \qquad (2)$$

To see how equation 2 is applied, look at the assumption about yield growth in scenario E. Yield is assumed to grow at an average rate of 1.25 percent (or 0.0125) over the 42-year period between 2008 and 2050. Applying equation 2 gives us:

$$V_{2050} = V_{2008}\,(1 + 0.0125)^{42} = V_{2008} \times 1.685$$

Rounding 1.685 to 1.7, this means that yield in 2050 is 1.7 times (or 170 percent of) yield in 2008; in other words, yield has increased by 70 percent. Thus "70" is entered into the yield column of scenario E in Table 23.1.

Calculating the Total Impact When Two Multiplicative Factors Are Growing

Total supply is land times yield. In our scenarios, both of these factors (land and yield) change. As explained in Chapter 10, when this happens, growth rate for total supply is calculated as follows. Taking scenario E in Table 23.1 as an example, if land grows by 11 percent and if yield grows by 70 percent over the period, growth in total supply is calculated like this:

$$(1 + 0.11) \times (1 + 0.70) = 1.887$$

Total supply in 2050 is 1.887 times (or 189 percent of) supply at the beginning of the period; in other words, supply has increased by 89 percent.

Calculating the Price Change

The changes in demand and supply in our scenarios assume constant prices—we are projecting the degree to which demand and supply curves shift in the future. (Those with more training in economics will notice that our demand factors—population and income—are traditional demand shifters in economic theory; however, our supply factors—area and yield—are not consistent with theoretical economics of supply. We analyze growth in supply using these supply factors because they make the supply side of the equation easier to understand.) If demand grows faster than supply, equilibrium price will increase from current levels. If supply grows faster than demand, equilibrium price will decrease.

How do we calculate the size of the price change? Scenario E in Table 23.1 illustrates the calculation. Supply is projected to grow by 88.7 percent and demand is projected to grow by 95 percent. Thus, at constant prices, quantity demanded would exceed quantity supplied. To see that a 4.7 percent increase in prices brings us back to equilibrium—where quantity supplied equals quantity demanded—do the following calculations. We start out-of-equilibrium, where quantity demanded is at 195 and quantity supplied is at 188.7. A 4.7 percent increase in price generates an increase in quantity supplied of 2.35 percent (since supply elasticity is assumed to be 0.5), or $188.7 \times 1.0235 = 193.13$. A 4.7 percent increase in price generates a 0.94 percent decrease in quantity demanded (since demand elasticity is assumed to be 0.2), or $195 \times (1 - 0.0094) = 193.16$. Supply demanded is equal to quantity supplied and equilibrium has been restored.

References

Aalangdong, O. I., J. M. Kombiok, and A. Z. Salifu. 1999. Assessment of non-burning and organic manuring practices. *ILEIA Newsletter,* September. Amersfoort, Netherlands: Centre for Information on Low External Input and Sustainable Agriculture.

Abbot, Patrick. 2003. Ireland's great famine, 1845–1849. http://www.irelandstory.com/past/famine.

Adams, Dale W. 1983. Mobilizing household savings through rural financial markets. In *Rural financial markets in developing countries: Their use and abuse,* ed. J. D. Von Pischke et al., 399–407. Baltimore: Johns Hopkins University Press.

Adams, Dale W., and Douglas H. Graham. 1981. A critique of traditional agricultural credit projects and policies. *Journal of Development Economics* 8:347–366.

Adams, Dale W., et al., eds. 1984. *Undermining rural development with cheap credit.* Boulder: Westview.

Adamu, H. 2000. We'll feed our people as we see fit. *Washington Post,* September 11, p. A23.

Adelman, I., and C. T. Morris. 1973. *Economic growth and social equity in developing countries.* Stanford: Stanford University Press.

African Development Fund. 2004. *Democratic Republic of Congo: Agricultural and rural sector rehabilitation support project appraisal report.* Abidjan, Côte d'Ivoire. http://www.afdb.org/pls/portal/docs/page/adb_admin_pg/documents/operations information/adf_bd_wp_2004_35_e.pdf.

Ahluwalia, Montek S. 1976a. Income distribution and development: Some stylized facts. *American Economic Review* 66 (May):128–135.

———. 1976b. Inequality, poverty, and development. *Journal of Development Economics* 3 (September):307–342.

Ahluwalia, Montek S., N. Carter, and H. Chenery. 1979. Growth and poverty in developing countries. In *Structural change and development policy,* ed. H. Chenery. Oxford: Oxford University Press. Also available in *Journal of Development Economics* 6 (September):299–341.

Ahmed, Raisuddin. 1988. Structure, costs, and benefits of food subsidies in Bangladesh. In *Food subsidies in developing countries,* ed. Per Pinstrup-Andersen, 219–228. Baltimore: Johns Hopkins University Press.

———. 1989. Making rural infrastructure a priority. *IFPRI Report* 11, no. 1:1, 4. Washington, D.C.: International Food Policy Research Institute.

Ahmed, R., and F. Goletti. 1997. Food policy reform in Bangladesh. In *1997 annual report of the International Food Policy Research Institute.* Washington, D.C.: International Food Policy Research Institute.

Ahmed, Raisuddin, and Mahabub Hossain. 1990. *Developmental impact of rural infrastructures: Bangladesh.* Research Report no. 83. Washington, D.C.: International Food Policy Research Institute.

Alberts, Tom. 1983. *Agrarian reform and rural poverty: A case study of Peru.* Boulder: Westview.

Alderman, Harold. 1986. *The effect of food price and income changes on the acquisition of food by low-income households.* Washington, D.C.: International Food Policy Research Institute.

Alderman, H., J. Hoddinott, and B. Kinsey. 2003. Long-term consequences of early childhood malnutrition. Paper presented to the International Conference on Chronic Poverty and Development Policy, April 7–9, IDPM, University of Manchester. http://idpm.man.ac.uk/cprc/conference/conferencepapers/alderman.22.01.pdf.

Alderman, H., and K. Lindert. 1998. The potential and limitations of self-targeted food subsidies. *The World Bank Research Observer* 13, no. 2:213–229.

Alderman, Harold, and Joachim von Braun. 1984. *The effects of the Egyptian food ration and subsidy system on income distribution and consumption.* Research Report no. 45. Washington, D.C.: International Food Policy Research Institute.

Alexandratos, N., ed. 1995. *World agriculture: Towards 2010, an FAO study.* London: Wiley.

Allison, Graham T. 1971. *Essence of decision: Explaining the Cuban missile crisis.* Boston: Little, Brown.

Alston, J., et al. 2000. *A meta-analysis of rates of return to agricultural R&D: Ex pede herculem.* Research Report no. 113. Washington, D.C.: International Food Policy Research Institute.

Alston, Philip. 1997. Recognition of the right to food. *World Food Summit fact sheet.* Rome: United Nations and Food and Agriculture Organization. http://www.fao.org./wfs/fs/e/img/right-e.pdf.

Anderson, J. R., and J. A. Roumasset. 1985. Microeconomics of food insecurity: The stochastic side of poverty. Unpublished paper available through the Department of Economics, University of Hawaii, Manoa.

Anderson, Jock, et al. 1985. *International agricultural research centers: A study of achievements and potential—summary.* Washington, D.C.: World Bank, Consultative Group on International Agricultural Research.

Anderson, Mary Ann, et al. 1981. *Nutrition intervention in developing countries, study I: Supplementary feeding.* Cambridge, Mass.: Oelgeschlager, Gunn, and Hain.

Angé, A. L. 1993. *Trends of plant nutrient management in developing countries.* Rome: Food and Agriculture Organization.

Angel, J. Lawrence. 1975. Paleoecology, paleodemography, and health. In *Population, ecology, and social evolution,* ed. S. Polgar, 167–190. The Hague: Mouton.

———. 1984. Health as a crucial factor in the changes from hunting to developed farming in the Eastern Mediterranean. In *Paleopathology at the origins of agriculture,* ed. M. N. Cohen and G. J. Armelagos, 51–73. New York: Academic Press.

Anonymous. 1974. How hunger kills. *Time,* November 11, p. 68.

Anonymous. 1988. Women and development: Education and fertility. *Finance and Development* 43 (September).

Applebaum, Anne. 2008. When China starved. *Washington Post,* August 12, p. A13. http://www.washingtonpost.com/wp-dyn/content/article/2008/08/11/ar2008081102015.html?sub=ar.

Arnold, Jesse C., R. W. Engel, D. B. Aguillon, and M. Caedo. 1981. Utilization of family characteristics in nutritional classification of preschool children. *American Journal of Clinical Nutrition* 34 (November):2546–2550.

Aron, Robert, et al. 1962. *Les origines de la guerre d'Algerie.* Paris: Fayard.

Askari, Hossein, and John T. Cummings. 1976. *Agricultural supply response: A survey of the econometric evidence.* New York: Praeger.

Associated Press. 2000. UN: Nutrition improving in North Korea. November 4.

Astawa, I. B. 1979. Using the local community: Bali, Indonesia. In *Birth control: An international assessment,* ed. M. Potts and P. Bhiwandiwala, 55–70. Baltimore: University Park Press.

Baden, John. 2003. Move over culture war: Here's the aquaculture war. Tech Central Station, August 29. http://www.techcentralstation.com/082903d.html.

Bale, Malcolm D. 1984. Opening of the discussion on plenary paper 5. In *Proceedings of the fourth congress of the E.A.A.E.: Agricultural markets and prices.* Available in *European Review of Agricultural Economics* 12:82–83.

———. 1985. *Agricultural trade and food policy: The experience of five developing countries.* Working Paper no. 724. Washington, D.C.: World Bank.

Bale, Malcolm D., and Ernst Lutz. 1981. Price distortions in agriculture and their effects: An international comparison. *American Journal of Agricultural Economics* 63 (February):8–22.

Baliunas, Sallie. 1999. Why so hot? Don't blame man, blame the sun. *Wall Street Journal,* August 5.

Banerjee, A., and E. Duflo. 2007. The economic lives of the poor. *Journal of Economic Perspectives* 21, no. 1:141–167.

Barrionuevo, A. 2008. Argentina export tax sets off political furor. *New York Times,* July 5.

Bautista, Romeo M. 1987. *Production incentives in Philippine agriculture: Effects of trade and exchange rate policies.* Research Report no. 59. Washington, D.C.: International Food Policy Research Institute.

Bayliss, K., and D. Hall. 2001. A PSIRU response to the World Bank's 'private sector development strategy': Issues and options. University of Greenwich, Public Services International Research Unit. http://www.psiru.org/reports/2001-10-u-wb-psd.doc.

Baylor, K. 1996. Biochemical studies on the toxicity of isocyanates. Abstract from a PhD thesis submitted to University College, Cork, Ireland, May. http://www.connect.net/dreggie/methyl%20i.htm.

Beal, George M., and D. J. Hobbs. 1969. *Social action: The process in community and area development.* Ames: Iowa State University Cooperative Extension Service, August.

Beal, George M., and Everett M. Rogers. 1960. *The adoption of two farm practices in a central Iowa community.* Special Report no. 26. Ames: Iowa State University, Agricultural and Home Economics Experiment Station.

Bearak, B. 2003. Why people still starve. *New York Times Sunday Magazine,* August 10, p. 33.

Beaton, George H., and Hossein Ghassemi. 1982. Supplementary feeding programs for young children in developing countries. *American Journal of Clinical Nutrition* 35 (April):864–916.

Becker, Gary. 1975. *Human capital.* New York: Columbia University Press.

Becker, K. 1989. Chief of Basic Data Unit, Statistics Division, Food and Agriculture Organization, Rome. Letter to author. January 16.

Behrman, J., and A. Deolalikar. 1987. Will developing country nutrition improve with income? A case study for rural south India. *Journal of Political Economy* 95:108–138.

Belmont, Lillian, and Francis A. Marolla. 1973. Birth order, family size, and intelligence: A study of a total population of 19-year-old men born in the Netherlands. *Science* 182:1096–1101.

Bengoa, J. M. 1972. Nutritional significance of mortality statistics. In *Proceedings of the third western hemisphere nutrition congress.* New York: Futura.

Bennett, M. K. 1941. International contrasts in food consumption. *Geographical Review* 31:365–374.

Berg, Alan. 1973. *The nutrition factor: Its role in national development.* Washington, D.C.: Brookings Institution.

———. 1987. *Malnutrition: What can be done?* Baltimore: Johns Hopkins University Press.

Berry, A. R., and W. R. Cline. 1979. *Agrarian structure and productivity in developing countries.* Baltimore: Johns Hopkins University Press.

Bettany, G. T. 1890. Introduction. Essay published with the 1890 Ward edition of *Essay on population,* by T. R. Malthus.

Bezuneh, Mesfin, Brady J. Deaton, and George W. Norton. 1988. Food aid impacts in rural Kenya. *American Journal of Agricultural Economics* 70 (February):181–191.

Bhargava, Alok. 1996. Econometric analysis of psychometric data: A model for Kenyan schools. Washington, D.C.: World Bank, Policy Research Unit.

Bilger, Burkhard. 2004. The height gap. *New Yorker,* April 5.

Binswanger, Hans. 1978. *The economics of tractors in South Asia: An analytical review.* New York: Agricultural Development Council.

Binswanger, Hans, and Klaus Deininger. 1997. Explaining agricultural and agrarian policies in developing countries. *Journal of Economic Literature* 35 (December): 1958–2005.

Binswanger, Hans, et al. 1987. *Agricultural mechanization: Issues and options.* Washington, D.C.: World Bank, Policy Research Unit.

Birdsall, Nancy. 1984. Population growth: Its magnitude and implications for development. *Finance and Development* 21 (September):10–13.

Blake, Judith. 1989. *Family size and achievement.* Berkeley: University of California Press.

Bleichrodt, Nico, and Marise P. H. Born. 1994. A metaanalysis of research on iodine and its relationship to cognitive development. In *The damaged brain of iodine deficiency: Cognitive, behavioral, neuromotor, educative aspects,* ed. John Stanbury. Port Washington, N.Y.: Cognizant Communication.

Bliss, C. J., and N. Y. Stern. 1982. *Palanpur: The economy of an Indian village.* Oxford: Clarendon.

Blow, L., A. Leicester, and Z. Smith. 2003. London's congestion charge. Institute for Fiscal Studies, Briefing Note no. 31. http://www.ifs.org.uk/consume/bn31.pdf.

Boediono. 1978. Elastisitas permintaan untuk berbagai barang di Indonesia: Penerapan metode Frisch. *Ekonomi dan Keuangan Indonesia* 26 (September):362.

Boggess, W., R. Lacewell, and D. Zilberman. 1993. Economics of water use in agriculture. In *Agricultural and environmental resource economics,* ed. G. Carlson, D. Zilberman, and J. Miranowski, 319–392. Oxford: Oxford University Press.

Bongaarts, John. 1982. The fertility-inhibiting effects of the intermediate fertility variables. *Studies in Family Planning* 13:179–189.

———. 1994. Population policy options in the developing world. *Science* 263:771–776.

Boserup, Ester. 1981. *Population and technological change: A study of long-term trends.* Chicago: University of Chicago Press.

Boudreaux, K., and T. Cowen. 2008. The micromagic of microcredit. *Wilson Quarterly* (Winter).

Bouis, Howarth E. 1991. The changing focus of economic research on nutrition. *IFPRI Report* 13, no. 2:1, 4. Washington, D.C.: International Food Policy Research Institute.

Bouis, Howarth E., and Lawrence Haddad. 1990. *The effects of agricultural commercialization on land tenure, household resource allocation, and nutrition in the*

Philippines. Research Report no. 79. Washington, D.C.: International Food Policy Research Institute.

———. 1992. Are estimates of calorie-income elasticities too high? *Journal of Development Economics* 39:333–364.

Boulding, Kenneth. 1964. *The meaning of the 20th century*. New York: Harper and Row.

Bowles, S., and H. Gintis. 2002. The inheritance of inequality. *Journal of Economic Perspectives* 16, no. 3:3–30.

Bread for the World. 1997. *Hunger in the global economy: Hunger 1998*. Silver Spring, Md.: Bread for the World Institute.

Briscoe, J. 1979. The qualitative effect of infection of the use of food by young children in poor countries. *American Journal of Clinical Nutrition* 32 (March):648–676.

Bromley, Daniel. 1981. The role of land reform in economic development: Policies and politics—discussion. *American Journal of Agricultural Economics* 63 (May):399–400.

Brooker, S., P. J. Hotez, and D. Bundy. 2008. Hookworm-related anaemia among pregnant women: A systematic review. *PLoS Negl Trop Dis* 2, no. 9:e291. http://www.plosntds.org/article/info%3adoi%2f10.1371%2fjournal.pntd.0000291.

Brown, Lester R. 1970. *Seeds of change: The green revolution and development in the 1970s*. New York: Praeger.

———. 1974. *In the human interest: A strategy to stabilize world population*. New York: Norton.

———. 1983. *Population policies for a new economic era*. Paper no. 53. Washington, D.C.: Worldwatch Institute.

———. 1988. *The changing world food prospect: The nineties and beyond*. Paper no. 85. Washington, D.C.: Worldwatch Institute.

Brown, L., and B. Halweil. 1998. China's water shortage could shake world food security. *Worldwatch* (July–August).

Brown, Lester, and Hal Kane. 1994. *Full house: Reassessing the earth's population carrying capacity*. Washington, D.C.: Worldwatch Institute.

Brown, Lynn. 1997. The potential impact of AIDS on population and economic growth rates. *2020 Brief* no. 43 (June). Washington, D.C.: International Food Policy Research Institute. http://www.cgiar.org/ifpri/2020/briefs/2br43.htm.

Bulatao, Rodolfo A. 1984a. Fertility control at the community level: A review of research and community programs. In *Rural Development and Human Fertility*, ed. W. Schutjer and C. Stokes, 269–290. New York: Macmillan.

———. 1984b. *Reducing fertility in developing countries: A review of determinants and policy servers*. Population and Development Series no. 5, Working Paper no. 680. Washington, D.C.: World Bank.

Bulatao-Jayme, J., D. Dela Paz, and C. Gervacio. 1971. Recommended height and weight standards for Filipinos. *Philippine Journal of Nutrition* 24:161–178.

Bumb, Balu L., and Carlos A. Baanante. 1996. World trends in fertilizer use and projections to 2020. *2020 Brief* no. 38 (October). Washington, D.C.: International Food Policy Research Institute.

Burki S., and Robert Ayres. 1986. A fresh look at development aid. *Finance and Development* 23 (March):6.

Byerlee, D., and P. Moya. 1993. *Impacts of international wheat breeding research in the developing world, 1966–90*. Mexico City: Centro Internacional de Mejoramiento de Maize y Trigo.

Caldwell, John C. 1983. Direct economic costs and benefits of children. In *Determinants of fertility in developing countries*, vol. 1, *Supply and demand for children*, ed. Rodolfo A. Bulatao et al., 458–493. New York: Academic Press.

Calegar, Geraldo M., and G. Edward Schuh. 1988. *The Brazilian wheat policy: Its costs, benefits, and effects on food consumption.* Research Report no. 66. Washington, D.C.: International Food Policy Research Institute.

California Department of Water Resources. 2005. *California water plan, 2005.* Vol. 4. Sacramento. http://www.waterplan.water.ca.gov/previous/cwpu2005/index.cfm.

Campbell, Joseph K. 1984. Machines and food production. In *World food issues,* ed. Matthew Drosdoff, 47–50. Ithaca: Cornell University, College of Agriculture.

Carner, George. 1984. Survival, interdependence, and competition among the Philippine rural poor. In *People-centered development,* ed. David Korten and Rudi Klauss, 133–145. West Hartford, Conn.: Kumarian.

Carson, Rachel. 1962. *Silent spring.* Greenwich, Conn.: Fawcett.

Carter, Michael. 1989. *US farm exports and third-world agricultural development.* Economic Issues no. 111. Madison: University of Wisconsin, Department of Agricultural Economics.

Casey, M. 2007. Experts target rice as climate culprit. *Boston Globe,* May 1. http://www.boston.com/news/world/asia/articles/2007/05/01/experts_target_rice_as_climate_culprit?mode=pf.

Cassidy, Claire. 1980. Benign neglect and toddler malnutrition. In *Social and biological predictors of nutritional status, physical growth, and neurological development,* ed. Lawrence S. Greene, 109–139. New York: Academic Press.

———. 1987. World-view conflict and toddler malnutrition: Change agent dilemmas. In *Child survival: Anthropological perspectives on the treatment and maltreatment of children,* ed. Nancy Scheper-Hughes, 293–324. Norwell, Mass.: Reidel.

CAST (Council on Agricultural Science and Technology). 2002. Comparative environmental impacts of biotechnology-derived and traditional soybean, corn, and cotton crops. Washington, D.C. http://www.cast-science.org/cast/biotech/pubs/biotechcrops benefit.pdf.

———. 2003. *Biotechnology in animal agriculture.* Washington, D.C.

Cavallo, Domingo, and Yair Mundlak. 1982. *Agriculture and economic growth in an open economy: The case of Argentina.* Research Report 36. Washington, D.C.: International Food Policy Research Institute.

CDC (Centers for Disease Control). 2002. Investigation of human health affects associated with potential exposure to genetically modified corn. Atlanta. http://www.cdc.gov/nceh/ehhe/cry9creport/complete.htm.

CGIAR (Consultative Group on International Agricultural Research). 1995. Wheat is doing well in Syria. *CGIAR Newsletter* (October). Washington, D.C.: World Bank. http://www.worldbank.org/html/cgiar/newsletter/oct95/3syria.htm.

———. 1997. *25 years of food and agriculture improvement in developing countries.* Washington, D.C.: World Bank. http://www.worldbank.org/html/cgiar/25years/25 cover.html.

Cha, A. 2004. Iraqis face tough transition to market-based agriculture. *Washington Post,* January 22, p. A1.

Chambers, Robert, Richard Longhurst, David Bradley, and Richard Feacham. 1979. *Seasonal dimensions to rural poverty: Analysis and practical implications.* Discussion Paper no. 142. Brighton: University of Sussex, Institute of Development Studies.

Champakam, S., S. C. Srikantia, and C. Gopalan. 1968. Kwashiorkor and mental development. *American Journal of Clinical Nutrition* 21 (August):844–852.

Chandra, Ranjit K. 1980. Immunocompetence in undernutrition and overnutrition. *Nutrition Review* 39:225–231.

———. 1988. Nutritional regulation of immunity: An introduction. In *Nutrition and immunology,* ed. Ranjit Chandra, 1–7. New York: Alan R. Liss.

Chaney, J. 2008. As Chinese wealth rises, pets take a higher place. *International Herald Tribune,* March 17. http://www.iht.com/articles/2008/03/17/business/pet.php.

Chang, Andrew. 2001. Bitter pill: Is the chocolate that you eat the product of child slavery? May 4. http://abcnews.go.com/sections/world/dailynews/cotedivoire010504_choco.html.

Chavez, Adolfo, and Celia Martinez. 1982. Growing up in a developing community: A bio-ecological study of the development of children from poor peasant families in Mexico. Mexico City: Instituto Nacional de la Nutricion. (Translated from the Spanish.)

Chen, P. C. 1981. China's birth planning program. In *National Research Council Committee on Population and Demography: Research on the population in China, proceedings of a workshop,* 78–90. Washington, D.C.: National Academy Press.

Chen, P. C., and A. Kols. 1982. Population and birth planning in the People's Republic of China. *Population Reports* series J, no. 25 (January–February), vol. 10, no. 1: 577–618.

Chen, R. S. 1990. Global agriculture, environment, and hunger: Past, present, and future links. *Environmental Impact Assessment Review* 10, no. 4:335–358.

Chen, Y., and L. Zhou. 2007. The long-term health and economic consequences of the 1959–1961 famine in China. *Journal of Health Economics* 26, no. 4:659–681.

Chenery, Hollis B. 1971. Growth and structural change. *Finance and Development Quarterly* 3:16–27.

Chenery, Hollis, Sherman Robinson, and Moshe Syrquin. 1986. *Industrialization and growth: A comparative study.* New York: Oxford University Press.

Chesapeake Bay Foundation. 2008. "Save the bay: Animals" website. http://www.cbf.org/site/pageserver?pagename=exp_sub_watershed_animals.

Chevalier, P. 1995. Zinc and duration of treatment of severe malnutrition. *Lancet* 345, no. 8956 (April 22):1046–1047.

Chhibber, Ajay. 1988. Raising agricultural output: Price and nonprice factors. *Finance and Development* (June):44–47.

Chidambaram, G. 1989. *Tamil Nadu integrated nutrition project: Terminal evaluation.* Madras, India: State Planning Commission.

Chisholm, Anthony H., and Rodney Tyers, eds. 1982. *Food security: Theory, policy, and perspectives from Asia and the Pacific rim.* Lexington, Mass.: Lexington Books.

Chu, Yung-Peng. 1982. Growth and distribution in a small, open economy. PhD dissertation, University of Maryland.

CIMMYT (Centro Internacional de Mejoramiento de Maize y Trigo). 1989. *Towards the 21st century: CIMMYT's strategy.* Mexico City.

———. 2003. Innovation for development: Annual report 2002–2003. Mexico City. http://www.cimmyt.org/english/docs/ann_report/recent/pdf/ar03_reducing.pdf.

Clarendon Press. 1959. *The shorter Oxford economic atlas of the world.* 2nd ed. Oxford: Oxford University Press.

Clark, Colin G. 1973. More people, more dynamism. *CERES* (November–December). Rome: Food and Agriculture Organization.

Clay, Jason W., and Bonnie K. Holcomb. 1986. *Politics and the Ethiopian famine, 1984–1985.* Cambridge, Mass.: Cultural Survival.

Cleaver, Kevin M. 1985. *The impact of price and exchange rate policies on agriculture in sub-Saharan Africa.* Working Paper no. 728. Washington, D.C.: World Bank.

Coale, Ansley J., and Edgar M. Hoover. 1958. *Population growth and economic development in low-income countries: A case study of India's prospects.* Princeton: Princeton University Press.

Cogill, B. 2003. *Anthropometric indicators measurement guide: 2003 revision.* Washington, D.C.: Food and Nutrition Technical Assistance.

Cohen, Joel. 1996a. *How many can the earth support?* New York: Norton.

———. 1996b. Maximum occupancy. *American Demographics* (February).

Cohen, Mark N. 1984. An introduction to the symposium. In *Paleopathology at the origins of agriculture,* ed. Mark Cohen and George Armelagos, 1–11. New York: Academic Press.

Cohen, Mark N., and George Armelagos, eds. 1984. *Paleopathology at the origins of agriculture.* New York: Academic Press.

Collier, P. 2007. *The bottom billion: Why the poorest countries are failing and what can be done about it.* Oxford: Oxford University Press.

———. 2008. The politics of hunger: How illusion and greed fan the food crisis. *Foreign Affairs* 87, no. 6 (November–December).

Collier, P., and D. Dollar. 2002. *Globalization, growth, and poverty: Building an inclusive world economy.* Oxford: Oxford University Press.

Conquest, Robert. 1986. *The harvest of sorrow: Soviet collectivization and the terror-famine.* New York: Oxford University Press.

Cook, Robert C. 1962. How many people have ever lived on earth? *Population Bulletin* 18 (February).

Cooke, G. W. 1967. *The control of soil fertility.* London: Crosby Lockwood.

Costa, D. L., and R. H. Steckel. 1997. Long-term trends in health, welfare, and economic growth in the United States. In *Health and welfare during industrialization.* Chicago: University of Chicago Press.

Court, J., and T. Yanagihara. N.d. Asia and Africa into the global economy. http://www.unu.edu/hq/academic/pg_area4/august-intro.html.

Coutsoudis, A., and N. Rollins. 2003. Breast-feeding and HIV transmission: The jury is still out. *Journal of Pediatrics, Gastroenterology, and Nutrition* 36:434–442.

Cowell, F. A. 1977. *Measuring inequality: Techniques for the social sciences.* New York: Wiley.

Cowen, T. 2006. *Good and plenty: The creative successes of American arts funding.* Princeton: Princeton University Press.

Cox, W. M., and R. Alm. 2008 You are what you spend. *New York Times,* February 10, p. WK14.

Craig, B. J., P. Pardey, and J. Roseboom. 1994. International agricultural productivity patterns. Working Paper no. WP 94-1. St. Paul: Center for International Food and Agricultural Policy, University of Minnesota. http://agecon.lib.umn.edu/cgi-bin/pdf_view.pl?paperid=1746&ftype=.pdf.

———. 1997. International productivity patterns: Accounting for input quality, infrastructure, and research. *American Journal of Agricultural Economics* 79, no. 4 (November):1064–1076.

Crosson, Pierre. 1996a. Resource degradation? *Perspectives on the Long-Term Global Food Situation* (Summer). http://www.fas.org/food/issue2.html.

———. 1996b. Who will feed China? *Perspectives on the Long-Term Global Food Situation* 2 (Spring). http://www.fas.org/food/issue2.html.

Cunningham, A. S., D. B. Jelliffe, and E. F. P. Jelliffe. 1991. Breastfeeding and health in the 1980s: A global epidemiological review. *Journal of Pediatrics* 15:659–668.

Daberkow, S., K. Isherwood, J. Poulisse, and H. Vroomen. 1999. Fertilizer requirements in 2015 and 2030. IFA Agricultural Conference, Barcelona.

Dagum, Camilo. 1987. Gini ratio. In *The new Palgrave: A dictionary of economics,* vol. 2, ed. John Eatwell et al., 529–532. New York: Stockton.

Dalrymple, Dana G. 1964. The Soviet famine of 1932–1934. *Soviet Studies* 15 (January): 250–284.

———. 1979. The adoption of high-yielding grain varieties in developing countries. *Agricultural History* 53 (October):704–726.

———. 1985. The development and adoption of high-yielding varieties of wheat and rice in developing countries. *American Journal of Agricultural Economics* 67 (December):1067–1073.

———. 1986a. See US Department of State 1986a.

———. 1986b. See US Department of State 1986b.

Dam, Marjory. 1989. *Report of world health.* Geneva: World Health Organization, September.

Dao, James. 2003. US to resume food aid to North Korea after 2-month halt. *New York Times,* February 25.

Das, J., J. Hammer, and K. Leonard. 2008. The quality of medical advice in low-income countries. *Journal of Economic Perspectives* 22, no. 2:93–114.

Das Gupta, Monica. 1988. Selective discrimination against female children in rural Punjab, India. *Population and Development Review* 13:77–100.

Datta, S. K., S. H. Ghosh, and C. N. Bairagya. 1988. Growth and yield of wet season rice with tilapia fish. *International Rice Research Newsletter* 13 (August):46.

Dawkins, K. 2003. *Gene wars: The politics of biotechnology.* New York: Seven Stories.

De Janvry, Alain. 1981. The role of land reform in economic development: Policies and politics. *American Journal of Agricultural Economics* 63:384–392.

De Janvry, A., N. Key, and E. Sadoulet. 1997. *Agricultural and rural development policy in Latin America: New directions and new challenges.* Rome: Food and Agriculture Organization. http://www.fao.org/docrep/w7441e/w7441e00.htm#contents.

De Soto, H. 2001. *The mystery of capital: Why capitalism triumphs in the west and fails everywhere else.* New York: Basic.

De Zoysa, Isabelle, et al. 1985. *Focus on diarrhoea.* London: Ross Institute, London School of Hygiene and Tropical Medicine.

Deaton, Angus, and John Muellbauer. 1980. *Economics and consumer behavior.* Cambridge: Cambridge University Press.

Deininger, K., and P. Olinto. 2000. Asset distribution, inequality, and growth. Working Paper no. 2375. Washington, D.C.: World Bank.

Deininger, Klaus, and Lyn Squire. 1997. Economic growth and income inequality: Reexamining the links. *Finance and Development* (March):38–41.

———. 1998. New ways of looking at old issues: Inequality and growth. *Journal of Development Economics* 52, no. 2:259–287.

Delgado, C. 2003. Rising consumption of meat and milk in developing countries has created a new food revolution. *Journal of Nutrition* 133:3907S–3910S.

DeLong, G. R., et al. 1996. Effect of iodination of irrigation water on crop and animal production in Long Ru, Hotien County, Xinjiang. In *Mineral problems in sheep in northern China and other regions of Asia: Proceedings of a work-shop held in Beijing, People's Republic of China, 25–30 September 1995,* 49–51. Canberra: Australian Centre for International Agricultural Research.

den Biggelaar, C., L. Rattan, K. Wiebe, H. Eswaran, V. Breneman, and P. Reich. 2004. The global impact of soil erosion on productivity II: Effect on crop yields and production over time. *Advances in Agronomy* 81:49–95.

Derneke, M., A. Said, and T. Jayne. 1997. Relationships between fertilizer use and grain sector performance. Working Paper no. 5. Addis Ababa: Grain Market Research Project, Ministry of Economic Development and Cooperation.

Deschenes, O., and M. Greenstone. 2007. The economic impacts of climate change: Evidence from agricultural output and random fluctuations in weather. *American Economic Review* 97, no. 1 (March):354–385.

DeVer, James R. 1983. Determinants of nutritional status in a North Indian village: An economic analysis. Master's thesis, University of Maryland.

DFID (Department for International Development). 2002. Better livelihoods for poor people: The role of agriculture. Glasgow. http://www.tradeobservatory.org/library/uploadedfiles/better_livelihoods_for_poor_people_the_role_of.htm.

Diamond, J. 1987. The worst mistake in the history of the human race. *Discover Magazine,* May. http://www.scribd.com/doc/2100251/jared-diamond-the-worst-mistake-in-the-history-of-the-human-race.

Dickens, Charles. 1958 [1859]. *A tale of two cities.* London: Oxford University Press.

Diro Pusat Statik. 1981. Statistik harga yan diterima dan yan dibayar pentani untuk biaya produksi pertanian dan Kebutuhan Rumah Tangga Tani; Jawa: Madura dan beberapa Propinsi Luar Jawa. Jakarta, Indonesia, December.

Dixon, John A. 1982. *Food consumption patterns and related demand parameters in Indonesia: A review of available evidence.* Working Paper no. 6. Washington, D.C.: International Food Policy Research Institute, International Fertilizer Development Center, and International Rice Research Institute.

Dixon, Robyn. 2008. White farmer's ordeal in Zimbabwe. *Los Angeles Times.* August 7. http://www.latimes.com/news/nationworld/world/la-fg-farmer7-2008aug07,0,1115417.story.

Doctors Without Borders. 2008a. *Malnutrition.* http://www.doctorswithoutborders.org/news/issue.cfm?id=2396.

———. 2008b. *Starved for attention: The neglected crisis of childhood malnutrition—a symposium.* New York, September 11. http://www.doctorswithoutborders.org/events/symposiums/2008/nutrition.

Dollar, D., and A. Kraay. 2002. Growth is good for the poor. *Journal of Economic Growth* 7, no. 3:195–225.

Dolot, Miron. 1985. *Execution by hunger: The hidden holocaust.* New York: Norton.

Dommen, Arthur J. 1988. *Innovation in African agriculture.* Boulder: Westview.

Donnelly, James. 2001. *The great Irish potato famine.* Phoenix Mill, Gloucestershire: Sutton.

Dover, Michael, and Lee M. Talbot. 1987. *To feed the earth: Agro-ecology for sustainable development.* Washington, D.C.: World Resources Institute.

Dreze, J. 2001. Starving the poor. *The Hindu,* February 26.

———. 1989. *Hunger and public action.* Oxford: Oxford University Press.

Dreze, Jean, and Amartya Sen, eds. 1990. *The political economy of hunger.* 3 vols. Oxford: Oxford University Press.

Duke, Lynne. 1998. Land reform plan divides Zimbabweans. *Washington Post,* February 15, p. A27.

Durand, C. H., and J. P. Pigney. 1963. Revue de 410 cas de diarrhees aqueuses infectieuses chez le nourisson et l'enfant de moins de deux ans, traites pendant quatre ans dan les meme service hospitalier. *Annals of Pediatrics* 39:1386.

Easterlin, R. A. 1973. Does money buy happiness? *Public Interest* 30 (1973):3–10.

———. 1974. Does economic growth improve the human lot? Some empirical evidence. In *Nations and households in economic growth: Essays in honor of Moses Abramowitz,* ed. P. A. David and M. W. Reder. New York: Academic Press.

Easterly, W. 2005. Review of Sach's *The end of poverty. Washington Post Book World,* March 27.

———. 2006. Why doesn't aid work? *Cato Unbound,* April 3. Washington, D.C.: Cato Institute.

Easterly, W., and R. Levine. 2003. Tropics, germs, and crops: How endowments influence economic development. *Journal of Monetary Economics* 50, no. 1 (January): 3–39.

The Economist. 2001. No title. March 29.

———. 2002. The road to hell is not paved. December 19.

———. 2006. How to make China even richer. March 23.

———. 2008. Cereal offenders. March 27.

Edirisinghe, Neville. 1987. *The food stamp scheme in Sri Lanka: Costs, benefits, and options for modification.* Washington, D.C.: International Food Policy Research Institute.

Edirisinghe, Neville, and Thomas T. Poleman. 1983. *Behavioral thresholds as indicators of perceived dietary adequacy or inadequacy.* International Agricultural Economics Study no. 17 (July). Ithaca: Cornell University Press.

Edwards, Clark. 1988. Real prices received by farmers keep falling. *Choices* (Fourth Quarter):22–23.

Elias, Victor J. 1985. *Government expenditures on agriculture and agricultural growth in Latin America.* Research Report no. 50. Washington, D.C.: International Food Policy Research Institute.

Elliott, Kathleen. 1978. Editorial. *Lancet* 2:300.

EPA (Environmental Protection Agency). 2008. *Climate Change: Agriculture and Food Supply.* Washington, D.C. http://www.epa.gov/climatechange/effects/agriculture .html.

Erickson, J. D. 2002. Folic acid and prevention of spina bifida and anencephaly. *Morbidity and Mortality Weekly Report,* September 13. Centers for Disease Control. http://www.cdc.gov/mmwr/preview/mmwrhtml/rr5113a1.htm.

Erlich, P., and A. Erlich. 1991. *Healing the planet.* Reading, Mass.: Addison-Wesley.

Evelth, P. G., and J. M. Tanner. 1967. *Worldwide variation in human growth.* Cambridge: Cambridge University Press.

Evenson, Robert E. 1981. Benefits and obstacles to appropriate agricultural technology. *Annals of the American Academy of Political and Social Science* 458:54–67. As quoted in *Agricultural development in the third world,* ed. Carl K. Eicher and John M. Staatz, 348–361. Baltimore: Johns Hopkins University Press, 1984.

Evenson, R. E., and P. M. Flores. 1978. Social returns to rice research. In *Economic consequences of the new rice technology,* ed. R. Barker and Y. Hayami, 243–265. Los Banos, Philippines: International Rice Research Institute.

Evenson, R., and D. Gollen, eds. 2003a. Assessing the impact of the green revolution, 1960 to 2000. *Science* 300:758–762.

———. 2003b. *Crop variety improvement and its effect on productivity: The impact of international research.* Wallingford, UK: CAB International.

Evenson, R., and M. Rosegrant. 2003. The economic consequences of crop genetic improvement programs. In *Crop variety improvement and its effect on productivity: The impact of international research,* ed. R. Evenson and D. Gollen, 473–497. Wallingford, UK: CAB International.

Ewen, S., and A. Pusztai. 1999. Effects of diets containing genetically modified potatoes expressing *Galanthus nivalis* lectin on rat small intestine. *Lancet* 354, no. 9187.

Falconer, J. 1990. Hungry season food from forests, unasylva 41. http://www.fao.org/docrep/t7750e/t7750e00.htm#contents.

Family Health International. 1997. http://www.fhi.org/fp/fpfaq/index.html.

Fan, S., P. Hazell, and S. Thorat. 2000. Government spending, growth, and poverty in rural India. *American Journal of Agricultural Economics* 82, no. 4:1038–1051.

FAO (Food and Agriculture Organization). 1974. *FAO/WHO handbook on human nutritional requirements.* Nutritional Studies no. 28. Rome.

———. 1989. *Food outlook.* Rome, May.

———. 1991. *Food balance sheets, 1984–86 average.* Rome.

———. 1994. *Body mass index: A measure of chronic energy deficiency in adults.* Rome.

———. 1996a. Backgrounder on plant genetic resources and plant breeding. Rome. http://www.fao.org/focus/e/96/06/02-e.htm.

———. 1996b. *Fact sheet on water and food security.* Rome. http://www.fao.org/wfs/fs/e/watirr-e.htm.
———. 1996c. *Sixth world food survey.* Rome.
———. 1996d. Special feature: The cereals sector outlook to 2010 seen from mid-1996. *Food Outlook* 5–6 (May–June). http://www.fao.org/waicent/faoinfo/economic/giews/english/fo/fo9606/fo960607.htm.
———. 1996e. World Food Summit (WFS) technical background papers, nos. 1–15. Rome. http://www.fao.org/wfs/final/e/list-e.htm.
———. 1997a. Contribution of greenhouse gases to global warming. http://www.fao.org/news/factfile/ff9715-e.htm.
———. 1997b. Hungry season in the Sahel. Famine Early Warning System (FEWS) Special Report no. 97-5. Rome. http://www.fews.org/fb970825/fb97sr5.html#hungry.
———. 2000a. *State of food insecurity in the world, 2000.* Rome.
———. 2000b. World agriculture: Towards 2015/2030. Rome. http://www.fao.org/docrep/004/y3557e/y3557e00.htm#topofpage.
———. 2002. Rural livelihoods devastated by epidemic. Press Release, July 5, 2002, FAO, Rome. http://www.fao.org/english/newsroom/news/2002/7302-en.html.
———. 2004a. FAOSTAT website. http://www.fao.org/waicent/portal/statistics_en.asp.
———. 2004b. Food balance sheet. http://www.fao.org/waicent/portal/statistics_en.asp.
———. 2004c. *State of food and agriculture, 2004.* Rome.
———. 2006. *World agriculture: Towards 2030/2050.* Rome.
———. 2008. *State of food insecurity, 2008.* Rome. http://www.fao.org/docrep/009/a0750e/a0750e00.htm.
———. N.d. Programme Against African Trypanosomiasis website. http://www.fao.org/ag/againfo/programmes/en/paat/home.html.
———. N.d. World Food Summit background technical documents. http://www.fao.org/wfs/index_en.htm.
———. Various years. *State of food insecurity in the world.* http://www.fao.org/sof/sofi/index_en.htm.
FAOSTAT. 2008a. FAOSTAT website. www.fao.org/waicent/portal/statistics_en.asp.
———. 2008b. FAOSTAT Food Balance Sheets. www.fao.org/waicent/portal/statistics_en.asp.
———. 2008c. FAOSTAT ProdSTAT. www.fao.org/waicent/portal/statistics_en.asp.
———. 2008d. FAOSTAT ResourceSTAT. www.fao.org/waicent/portal/statistics_en.asp.
Fass, Simon M. 1982. Water and politics: The process of meeting a basic need in Haiti. *Development and Change* 13:347–364.
Feacham, R. G., and M. A. Koblinsky. 1983. Interventions for the control of diarrhoeal diseases among young children: Measles immunization. *Bulletin of the World Health Organization* 61, no. 4:641–652.
———. 1984. Interventions for the control of diarrhoeal diseases among young children: Promotion of breast-feeding. *Bulletin of the World Health Organization* 62, no. 2:271–291.
Feder, G., R. Just, and D. Zilberman. 1985. Adoption of agricultural innovations in developing countries: A survey. *Economic Development and Cultural Change* 33: 255–294.
Federico, G. 2005. *Feeding the world: An economic history of agriculture, 1800–2000.* Princeton: Princeton University Press.
Fei, John C. H., and Gustav Ranis. 1964. *Development of the labor surplus economy: Theory and policy.* New Haven: Yale University Press.
FEMA (Federal Emergency Management Agency). 1991. *Projected impact of sea-level rise on the national flood insurance program.* Report to Congress. Washington, D.C.: Federal Insurance Administration.

FIAN (FoodFirst Information and Action Network). 1997. *Twelve misconceptions about the right to food.* http://www.fian.org/miscon.htm.

Filmer, D., J. S. Hammer, and L. H. Pritchett. 2000. Weak links in the chain: A diagnosis of health policy in poor countries. *World Bank Research Observer* 15, no. 2 (August):199–224.

Finch, V. C., and O. E. Baker. 1917. See USDA 1917.

Fishstein, Paul. 1985. Pre and post green revolution income distribution in a North Indian village. Master's thesis, University of Maryland.

Fitzhugh, H. 1998. Competition between livestock and mankind for nutrients: Let ruminants eat grass. In *Feeding a world population of more than eight billion people: A challenge to science,* ed. J. C. Waterlow et al., 223–231. Oxford: Oxford University Press.

Fogel, Robert W. 1994. Economic growth, population theory, and physiology: The bearing of long-term processes on the making of economic policy. *American Economic Review* 84, no. 3 (June):369–395.

———. 2004. *The escape from hunger and premature death, 1700–2100.* Cambridge: Cambridge University Press.

Food for the Hungry. 1998. Website. http://www.fh.org.

Foster, Phillips. 1972. *Introduction to environmental science.* Homewood, Ill.: Irwin.

———. 1978. See US Department of State 1978.

———. 1991. Malnutrition, starvation, and death. In *Horrendous death, health, and well-being,* ed. Dan Leviton, 205–218. New York: Hemisphere.

Foster, Phillips, and Herbert Steiner. 1964. *The structure of Algerian socialized agriculture.* College Park: University of Maryland, Agricultural Experimental Station.

Frejka, Thomas. 1973. The prospects for a stationary world population. *Scientific American,* March, p. 15.

Frisancho, A. Roberto. 1981. New norms of upper limb fat and muscle areas for the assessment of nutritional status. *American Journal of Clinical Nutrition* 34:2540–2545.

———. 1989. *Anthropometric standards for the evaluation of nutritional status of children and adults.* Ann Arbor: University of Michigan Press.

Gabbert, S., and H. P. Weikard. 2001. How widespread is undernourishment? A critique of measurement methods and new empirical methods. *Food Policy* 26, no. 3: 209–228.

Galler, Janina R. 1986. Malnutrition: A neglected cause of learning failure. *Journal of Postgraduate Medicine* 80 (October):225–230.

Galway, Katrina, et. al. 1987. *Child survival: Risks and the road to health.* Columbia, Md.: Westinghouse Institute for Resource Development, Demographic Data for Development Project.

Garcia, Marito, and Per Pinstrup-Andersen. 1987. *The pilot food price subsidy scheme in the Philippines: Its impact on income, food consumption, and nutritional status.* Research Report no. 61. Washington, D.C.: International Food Policy Research Institute.

Gardner, Bruce L. 1979. *Optimal stockpiling of grain.* Lexington, Mass.: Heath.

———. 1987. *The economics of agricultural policies.* New York: Macmillan.

Gardner, Gary. 1996. *Shrinking fields: Cropland loss in a world of eight billion.* Washington, D.C.: Worldwatch Institute.

General Accounting Office (GAO). 2008. Genetically engineered crops. GAO Report 09-60. November 2008. GAO, Washington D.C. http://www.gao.gov/new.items/d0960.pdf.

George, P. S. 1988. Costs and benefits of food subsidies in India. In *Food subsidies in developing countries,* ed. Per Pinstrup-Andersen, 229–241. Baltimore: Johns Hopkins University Press.

Gershwin, M. Eric, et al. 1985. *Nutrition and immunity.* New York: Academic Press.

Gifford, R. C. 1992. Agricultural engineering in development: Mechanization strategy formulation, vol. 1, Concepts and principles. *FAO Agricultural Services Bulletin* 99, no. 1:74.

Gilland, B. 2002. World population and food supply: Can food production keep pace with population growth in the next half century? *Food Policy* 27:47–63.

Gillis, J. 2003. Debate grows over biotech food; Efforts to ease famine in Africa hurt by US, European dispute. *Washington Post,* November 30, p. A1.

Gilmore, Richard, and Barbara Huddleston. 1983. The food security challenge. *Food Policy* 8 (February):31–45.

Glauber, J. 2008a. Statement before the Joint Economic Committee on Recent Developments in Food Prices. Washington, D.C.: US Department of Agriculture, May 1. http://www.usda.gov/oce/newsroom/archives/testimony/2008/foodpricetestimony.pdf.

———. 2008b. Statement before the Committee on Energy and Natural Resources. Washington, D.C.: US Department of Agriculture, June 12. http://www.usda.gov/oce/newsroom/archives/testimony/2008/glaubersenate061208.pdf.

Glewwe, Paul, Hanan Jacoby, and Elizabeth King. 1996. *An economic model of nutrition and learning: Evidence from longitudinal data.* Washington, D.C.: World Bank, Policy Research Unit.

Global Development Research Center. 2003. Microfacts: Data snapshots on microfinance. http://www.gdrc.org/icm/data/d-snapshot.html.

Godwin, William. 1793. Political justice. In *A reprint of the essay on "Property,"* ed. H. S. Salt. London: Allen and Unwin, 1949.

Goldenberg, R. L., et al. 1995. The effect of zinc supplementation on pregnancy outcome. *Journal of the American Medical Association* 274, no. 6 (August 9):463–468.

Gómez, F., R. Galvan, S. Frank, R. Chavez, and J. Vazquez. 1956. Mortality in third degree malnutrition. *Journal of Tropical Pediatrics* 2:77.

Gommes, R., A. Bakum, and G. Farmer. 1998. An el nino primer. *SD Dimensions.* Rome: Sustainable Development Department of the Food and Agriculture Organization. http://www.fao.org/sd/eidirect/EIan0008.htm#topofpage.

González-Vega, Claudio. 1983. Arguments for interest rate reform. In *Rural financial markets in developing countries: Their use and abuse,* ed. J. D. von Pischke et al., 365–372. Baltimore: Johns Hopkins University Press.

Goodall, Roger M. 1984. CDD Information Papers nos. 1–2. New York: United Nations Children's Fund, June.

Gopalan, C. 1970. Some recent studies in the nutrition research laboratories: Hyderabad. *Journal of Clinical Nutrition* (January):35–53.

———. 1986. Vitamin A deficiency and child mortality. *Nutrition Foundation of India Bulletin* 7, no. 3.

Gopalan, C., and K. S. Rao. 1979. Nutrient needs. In *Human nutrition: A comprehensive treatise,* vol. 2, *Nutrition and growth,* ed. D. Jelliffe and E. Jelliffe. New York: Plenum Press.

Graham, K. K., et al. 1994. Pharmacologic evaluation of megestrol acetate oral suspension in cachectic AIDS patient. *Journal of Acquired Immune Deficiency Syndromes* 7, no. 6:580–585.

Graham, S. 2002. Rice paddy methane emissions depend on crops' success. *Scientific American,* August 20. http://www.sciam.com/article.cfm?articleid=00005E42-5440-1D61-90FB809EC5880000.

Grantham-McGregor, S., L. Fernald, and K. Sethuraman. 1999a. Effects of health and nutrition on cognitive and behavioural development in children in the first 3 years of life, pt 1., Low birthweight, breast-feeding, and protein-energy malnutrition. *Food and Nutrition Bulletin* 20:53–75.

———. 1999b. Effects of health and nutrition on cognitive and behavioural development in children in the first 3 years of life, pt. 2, Infections and micronutrient deficiencies: Iodine, iron, and zinc. *Food and Nutrition Bulletin* 20:76–99.

Gray, Cheryl W. 1982. *Food consumption parameters for Brazil and their application to food policy.* Research Report no. 32. Washington, D.C.: International Food Policy Research Institute.

Greenland, D. J., P. J. Gregory, and P. H. Nye. 1998. Land resources and constraints to crop production. In *Feeding a world population of more than eight billion people: A challenge to science,* ed. J. C. Waterlow et al. Oxford: Oxford University Press.

Griffiths, Marcia. 1985. *Growth monitoring of preschool children: Practical considerations for primary health care projects.* Geneva: World Federation of Public Health Associations.

Grigg, D. 1993. *The world food problem.* 2nd ed. Cambridge: Blackwell.

Griliches, Zvi. 1958. Research costs and social returns: Hybrid corn and related innovations. *Journal of Political Economy* 66:419–431.

Guggenheim, Karl Y. 1981. *Nutrition and nutritional diseases: The evolution of concepts.* Lexington, Mass.: Heath.

Gupta, Arun, and Jon E. Rohide. 1993. Economic value of breast-feeding in India. *Economic and Political Weekly* 28, no. 26:1390.

Haag, A. L. 2007. Pond powered biofuels: Turning algae into America's new energy. *Popular Mechanics,* March 29. http://www.popularmechanics.com/science/earth/4213775.html.

Haggblade, Steven, and Peter Hazell. 1989. Agricultural technology and farm-nonfarm growth linkages. *Agricultural Economics: The Journal of the International Association of Agricultural Economists* 3:345–364.

Hancock, G. 1985. *Ethiopia: The challenge of hunger.* London: Victor Gollancz.

Harberger, Arnold. 1983. Basic needs versus distributional weights in social cost-benefit analysis. *Economic Development and Cultural Change* 32, no. 3:455–474.

Hardin, G. 1974. Lifeboat ethics: "The case against helping the poor." *Psychology Today,* September 8, pp. 38–43.

Harlan, Jack R. 1975. *Crops and man.* Madison, Wis.: American Society of Agronomy.

Harris, M. 1974. *Cows, pigs, wars, and witches: The riddles of culture.* New York: Vintage.

———. 1977. *Cannibals and kings: The origin of cultures.* New York: Random House.

Hartini, T., A. Winkvist, L. Lindholm, H. Stenlund, V. Persson, D. Nurdiati, and A. Surjono. 2003. Nutrient intake and iron status of urban poor and rural poor without access to rice fields are affected by the emerging economic crisis: The case of pregnant Indonesian women. *European Journal of Clinical Nutrition* 57:654–666.

Haslberger, A. G. 2003. GM food: The risk assessment of immune hypersensitivity reactions covers more than allergenicity. *Food, Agriculture, and Environment* 1:42–45. http://www.biotech-info.net/hypersensitivity.html.

Haub, Carl. 1987. Understanding population projections. *Population Bulletin* 42, no. 4.

Haupt, A., and T. Kane. 2004. *Population Reference Bureau's population handbook.* 5th ed. Washington, D.C.: Population Reference Bureau.

Hayami, Yujiro, and Robert Herdt. 1977. Market price effects of technological change on income distribution in semisubsistence agriculture. *American Journal of Agricultural Economics* 69:245–256.

Hayward, S. 2006. *Index of leading environmental indicators.* Washington, D.C.: American Enterprise Institute. http://www.aei.org/docLib/20060413_2006index.pdf.
Hearts and Minds. 2003. "Socially responsible food" website. http://www.change.net/articles/foodiss.htm.
Heilbroner, Robert L. 1953. *The worldly philosophers.* New York: Simon and Schuster.
Hendricks, L. 2002. How important is human capital for development? Evidence from immigrant earnings. *American Economic Review* 92, no. 1:198–219.
Herbert, Sandra. 1971. Darwin, Malthus, and selection. *Journal of History of Biology* 4:209–217.
Herdt, Robert W. 1970. A disaggregate approach to aggregate supply. *American Journal of Agricultural Economics* 52:512–520.
———. 1983. Mechanization of rice production in developing Asian countries. In *Consequences of small-farm mechanization,* 1–13. Manila: International Rice Research Institute.
Herdt, Robert W., and Jock R. Anderson. 1987. The contribution of the CGIAR Centers to world agricultural research. In *Policy for agricultural research,* ed. Vernon W. Ruttan and Carl E. Pray, 39–64. Boulder: Westview.
Herdt, Robert W., and John W. Mellor. 1964. The contrasting response of rice to nitrogen: India and United States. *Journal of Farm Economics* 46:150–160.
Herring, Ronald J. 1983. *Land to the tiller: The political economy of agrarian reform in South Asia.* New Haven: Yale University Press.
Hicks, L. E., R. A. Langham, and J. Takenaka. 1992. Cognitive and social measures following early nutritional supplementation: A sibling study. *American Journal of Public Health* 72.
Ho, M., T. Traavik, O. Olsvik, B. Tappeser, C. V. Howard, C. von Weizsacker, and G. C. McGavin. 1998. Gene technology and gene ecology of infectious diseases. *Microbial Ecology in Health and Disease* 10:33–59.
Ho, T. J. 1984. *Economic status and nutrition in East Java.* Washington, D.C.: World Bank. Unpublished data.
Hoddinott, J., M. Cohen, and M. Bos. 2003. Redefining the role of food aid. Washington, D.C.: International Food Policy Research Institute.
Hoffman, V. 2009. Intrahousehold allocation of free and purchased mosquito nets. *American Economic Review* 99, no. 2 (May):236–241.
Holick, M. F. 2004. Vitamin D: Importance in the prevention of cancers, type 1 diabetes, heart disease, and osteoporosis. *American Journal of Clinical Nutrition* 79, no. 3 (March):362–371. http://www.ajcn.org/cgi/content/full/79/3/362.
Holmes, S., and C. Sunstein. 1999. *The cost of rights: Why liberty depends on taxes.* New York: Norton.
Hopkins, Raymond F. 1988. Political calculations in subsidizing food. In *Food subsidies in developing countries,* ed. Per Pinstrup-Andersen, 107–125. Baltimore: Johns Hopkins University Press.
Hossain, Mahabub. 1988a. *Credit for alleviation of rural poverty: The Grameen Bank in Bangladesh.* Research Report no. 65. Washington, D.C.: International Food Policy Research Institute.
———. 1988b. *Nature and impact of the green revolution in Bangladesh.* Research Report no. 67. Washington, D.C.: International Food Policy Research Institute.
Houser, Daniel, and Barbara Sands. 2000. How centrally planned was China's great leap forward demographic disaster? University of Arizona, Department of Economics, June. http://w3.arizona.edu/~econ/working_papers/china629.pdf.
Huang, Kuo W. 1985. *US demand for food: A complete system of price and income effects.* Technical Bulletin no. 1714. Washington, D.C.: US Department of Agriculture.
Huddleston, Barbara. 1984a. *Briefs.* New York: CARE.

———. 1984b. *Closing the cereals gap with trade and food aid.* Research Report no. 43. Washington, D.C.: International Food Policy Research Institute.

Hull, T. H. 1978. Where credit is due: Policy implications of the recent rapid fertility decline in Bali. Paper presented at the annual meeting of the Population Association of America, Atlanta.

Hull, T. H., et al. 1977. Indonesia's family planning story: Success and challenge. *Population Bulletin* 32, no. 6.

Human Rights Watch. 1996. The small hands of slavery: Bonded child labor in India. http://www.hrw.org/reports/1996/india3.htm.

Hutabarat, Pos M. 1990. Proyeksi distribusi konsumsi kalorie menurut kelompok-kelompok pendapatan di Indonesia tahun 1990 [Projections of the distribution of caloric consumption by income groups in Indonesia in 1990]. Master's thesis, Bogor Agricultural University, Agricultural School.

Ibe, A. C., and L. F. Awosika. 1991. Sea level rise impact on African coastal zones. In *A change in the weather: African perspectives on climate change,* ed. S. H. Omide and C. Juma, 105–112. Nairobi: African Centre for Technology Studies.

IFPRI (International Food Policy Research Institute). 2005. New risks and opportunities for food security: Scenario analyses for 2015 and 2050. *2020 Brief* no. 73 (February). Washington, D.C.: International Food Policy Research Institute.

———. 2008. High food prices: The what, who, and how of proposed policy actions. IFPRI Policy Brief, May 2008. Washington, D.C.: IFPRI. http://www.ifpri.org/PUBS/ib/FoodPricesPolicyAction.pdf.

ILO (International Labour Organization). 1987. *Yearbook of Labor Statistics, 1987.* Geneva.

———. 1988. *I.L.O. Information* 16, no. 3 (August). Geneva.

———. n.d. Facts on HIV/AIDS and the world of work. ILO, New York. http://www.ilo.org/wcmsp5/groups/public/dgreports/dcomm/documents/publication/wcms_067561.pdf.

Imam, Izzedin I. 1979. *Peasant perceptions: Famine.* Dahka, Bangladesh: Bangladesh Rural Advancement Committee, July. In *People centered development: Contributions toward theory and planning frameworks,* ed. David C. Korten and Rudi Klauss, 152–155. West Hartford, Conn.: Kumarian, 1984.

Indonesia Oleh Direktorat Gizi Departemen Kesehatan R.I. 1979. *Daftar Komposisi Bahan Makanan.* Jakarta: Bhratara Karya Askara.

IPCC (Intergovernmental Panel on Climate Change). 2007a. *Climate change 2007 synthesis report: Contribution of Working Groups I, II and III to the fourth assessment report of the Intergovernmental Panel on Climate Change.* Geneva. http://www.ipcc.ch/ipccreports/ar4-syr.htm.

———. 2007b. *Report of Working Group II on climate change impacts, adaptation, and vulnerability.* Geneva. http://www.ipcc-wg2.org/index.html.

IRRI (International Rice Research Institute). 1989. Azolla helps organic farmer earn more. *IRRI Reporter* (June). Manila: International Food Policy Research Institute.

———. 2002. Project summary and highlights, project 3, Genetic enhancement for yield, grain quality, and stress resistance. Manila. http://www.irri.org/science/progsum/pdfs/dgreport2002/project%203.pdf.

ISRIC (International Soil Reference and Information Center) and UNEP (United Nations Environment Programme). 1991. World map of the status of human-induced soil degradation, by L. R. Oldeman, R. T. A. Hakkeling, and W. G. Sombroek. In *Global assessment of soil degradation,* 2nd ed. Nairobi: Wageningen.

Jackson, R., A. Ramsay, C. Christensen, S. Beaton, D. Hall, and I. Ramshaw. 2001. Expression of mouse interleukin-4 by a recombinant ectromelia virus suppresses cytolytic lymphocyte responses and overcomes genetic resistance to mousepox. *Journal of Virology* 75:1205–1210.

Jackson, Tony, with Deborah Eade. 1982. *Against the grain: The dilemma of project food aid.* Oxford: Oxfam.

James, Clive. 2002. Preview: Status of commercialized transgenic crops, 2002. Brief no. 27. Ithaca: International Service for the Acquisition of Agri-Biotech Applications.

———. 2007. *ISAAA Report on global status of biotech/GM crops.* Ithaca: International Service for the Acquisition of Agri-Biotech Applications. http://www.isaaa.org/resources/publications/briefs/37/pptslides/brief37slides.pdf.

James, W. P. T., and E. C. Schofield. 1990. *Human energy requirements: A manual for planners and nutritionists.* Oxford: Oxford University Press.

Jamison, Dean T., et al., eds. 1993. *Disease control priorities in developing countries.* New York: Oxford University Press.

Jayne, T. S., L. Rubey, M. Chisvo, and M. T. Weber. 1996. Zimbabwe food security success story. *Policy Synthesis* no. 18 (April).

Jayne, T., L. Rubey, D. Tschirley, M. Mukumbu, M. Chisvo, A. Santos, M. Weber, and P. Diskin. 1995. *Effects of market reform on access to food by low-income households: Evidence from four countries in eastern and southern Africa.* International Development Paper no. 19. East Lansing: Michigan State University Press.

Jelliffe, D. B. 1966. *The assessment of the nutritional status of the community.* Monograph no. 53. Geneva: World Health Organization.

Jensen, H., and S. Robinson. 2002. *General equilibrium measures of agricultural policy bias in fifteen developing countries.* Discussion Paper no. 105. Washington, D.C.: International Food Policy Research Institute.

Jensen, R. T., and N. H. Miller. 2008. Giffen behavior and subsistence consumption. *American Economic Review* 98, no. 4:1533–1577.

Johnson, D. G. 1975. *World food problems and prospects.* Washington, D.C.: American Enterprise Institute.

Johnson, Stanley R., Zuhair A. Hassan, and Richard D. Green. 1984. *Demand systems estimation methods and applications.* Ames: Iowa State University Press.

Joy, Leonard. 1973. Food and nutrition planning. *Journal of Agricultural Economics* 24:166–197.

Judd, M. Ann, James K. Boyce, and Robert E. Evenson. 1987. Investment in agricultural research and extension. In *Policy for agricultural research,* ed. Vernon Ruttan and Carl E. Pray, 7–38. Boulder: Westview.

Kahkonen, S., and H. Leathers. 1997. *Is there life after liberalization? Transaction costs analysis of maize and cotton marketing in Zambia and Tanzania.* College Park: University of Maryland Press, IRIS Center.

Kaiser, J. 2004. Wounding Earth's fragile skin. *Science* 304:1616–1618.

Kakturskaya, M. 2003. Why aren't Russians having babies? *Argumenty I Fakty,* July 23. Reproduced in *World Press Review,* October.

Kakwani, Nanak 1987. Lorenz curve. In *The new Palgrave: A dictionary of economics,* vol. 3., ed. John Eatwell et al., 243–244. New York: Stockton.

Kamrin, M. N.d. Environmental "hormones" pesticide information project. Michigan State University. http://ace.ace.orst.edu/info/extoxnet/tics/env-horm.txt.

Kantor, L. S., K. Lipton, A. Manchester, and V. Oliveira. 1997. Estimating and addressing America's food losses. *Food Review,* January, pp. 2–12. USDA, ERS. Washington, D.C. http://www.ers.usda.gov/publications/foodreview/jan1997/jan97a.pdf.

Karim, Rezaul, Manjur Majid, and F. James Levinson. 1984. The Bangladesh sorghum experiment. *Food Policy* 5:61–63.

Kates, Robert. 1996. Ending hunger: Current status and future prospects. *Consequences* 2, no. 2. http://www.gcrio.org/consequences/vol2no2/article1.html.

Kates, Robert W., et al. 1988. *The hunger report, 1988.* Providence, R.I.: Brown University Press, World Hunger Program.

Keilmann, A. A., and C. McCord. 1978. Weight-for-age as an index of death in children. *Lancet* (June):1247–1250.

Kekic, L. 2007. The Economist Intelligence Unit's index of democracy. *The Economist.* http://www.economist.com/media/pdf/democracy_index_2007_v3.pdf.

Kendall, H. W., and D. Pimentel. 1994. Constraints on the expansion of the global food supply. *Ambio* 23:198–205.

Kendall, H. W., et al. 1997. Bioengineering of crops: Report of the World Bank panel on transgenic crops. Washington, D.C.: World Bank.

Kennedy, Eileen T. 1989. *The effects of sugar cane production on food security, health, and nutrition in Kenya: A longitudinal study.* Research Report no. 78. Washington, D.C.: International Food Policy Research Institute.

Kennedy, Eileen T., and Bruce Cogill. 1987. *Income and nutritional effects of the commercialization of agriculture in southwestern Kenya.* Research Report no. 63. Washington, D.C.: International Food Policy Research Institute.

Kennedy, Eileen T., and Odin Knudsen. 1985. A review of supplementary feeding programmes and recommendations on their design. In *Nutrition and development,* ed. Margaret Biswas and Per Pinstrup-Andersen, 77–96. Oxford: Oxford University Press.

Kennedy, Eileen T., et al. 1983. *Nutrition-related policies and programs: Past performance and research needs.* Washington, D.C.: International Food Policy Research Institute, February.

Kennickell, A. 2006. Currents and undercurrents: Changes in the distribution of wealth, 1989–2004. Washington, D.C.: Federal Reserve Bank. http://www.federalreserve.gov/pubs/feds/2006/200613/200613pap.pdf.

Kenya Ministry of Planning and National Development. 1984. *Kenya contraceptive prevalence survey.* Nairobi: Central Bureau of Statistics.

Keys, Ancel, et al. 1950. *The biology of human starvation.* Minneapolis: University of Minnesota Press.

Kimbrell, Andrew. 2002. Seven deadly myths of industrial agriculture. In *Fatal harvest: The tragedy of industrial agriculture,* ed. Andrew Kimbrell. Washington, D.C.: Island.

Kinealy, Christine. 2002. *The great Irish famine.* New York: Palgrave.

Kinzer, S. 2007. After so many deaths, too many births. *New York Times,* February 11.

Kirchick, J. 2007. Killing them softly: The other African genocide. *New Republic,* March 8. http://www.tnr.com/docprint.mhtml?i=w070305&s=kirchick030807.

Kluender, S. 2003. The Peace Corps in Zambia. Personal website of a Peace Corps volunteer. http://peacecorpsonline.org/messages/messages/467/2018900.html.

Komlos, J. 1989. *Nutrition and economic development in the eighteenth-century Habsburg monarchy: An anthropometric history.* Princeton: Princeton University Press.

Kostermans, Kees. 1994. *Assessing the quality of anthropometric data: Background and illustrated guidelines for survey managers.* Washington, D.C.: World Bank.

Kremer, M. 1993. Population growth and technological change: One million B.C. to 1990. *Quarterly Journal of Economics* 108, no. 3:681–716.

Krick, Jackie. 1988. Using the Z score as a descriptor of discrete changes in growth. *Nutritional Support Services* 6, no. 8 (August).

Krueger, Anne, Maurice Schiff, and Alberto Valdés. 1991. *The political economy of agricultural pricing policy.* Baltimore: Johns Hopkins University Press.

Kuznets, Simon. 1955. Economic growth and income inequality. *American Economic Review* 65:1–28.

Lakshmi, R. 2004. Opening files, Indians find scams. *Washington Post,* March 9, p. A17.

Lancet. 1995. Editorial: Health effects of sanctions on Iraq. December 2, p. 1439.

Landes, M. 2004. The elephant is jogging: New pressures for agricultural reform in India. *Amber Waves,* February. http://www.ers.usda.gov/amberwaves/february04/features/elephantjogs.htm.

Landman, Lynn. 1983. China's one-child families: Girls need not apply. *RF Illustrated,* December, pp. 8–9. New York: Rockefeller Foundation.

Lappe, Frances Moore. 1971. *Diet for a small planet.* New York: Ballantine.

———. 1992. *Diet for a small planet twentieth anniversary edition.* New York: Ballantine.

Lappe, Francis Moore, and Joseph Collins. 1977. *Food first.* Boston: Houghton Mifflin.

Lashof, D., and D. Tirpak. 1990. Policy options for stabilizing global climate. US Environmental Protection Agency report. New York: Hemisphere.

Latham, Michael C. 1984. International nutrition problems and policies. In *World food issues,* ed. Matthew Drosdoff, 55–64. Ithaca: Cornell University Press, Center for the Analysis of World Food Issues, Program in International Agriculture.

Leclercq, Vincent. 1988. *Conditions et limites de l'insertion du Bresil dans les echanges mondiaux du soja.* Montpellier, France: INRA.

Lee, John E., and Gary C. Taylor. 1986. Agricultural research: Who pays and who benefits? In *Research for tomorrow: 1986 yearbook of agriculture,* 14–21. Washington, D.C.: US Department of Agriculture.

Lee, R. 1968a. Problems in the study of hunter gathers. In *Man the hunter,* ed. R. Lee and I. Devore, 3–12. Chicago: Aldine.

———. 1968b. What hunters do for a living, or how to make out on scarce resources. In *Symposium on man the hunter,* ed. R. B. Lee and Irven DeVore, 30–48. Chicago: Aldine.

———. 1969. !Kung bushmen subsistence: An input-output analysis. In *Environment and cultural behavior,* ed. A. Vayda, 47–49. Garden City, N.J.: Natural History Press.

———. 1972. Population growth and the beginnings of sedentary life among the !Kung bushmen. In *Population growth: Anthropological implications,* ed. B. Spooner, 329–342. Cambridge: Massachusetts Institute of Technology Press.

Leggett, J. 2007. *Climate change: Science and policy implications.* Report no. RL33849. Washington, D.C.: Congressional Research Service.

Lele, Uma. J. N.d. Overall flows of official development assistance to the MADIA countries. In *Aid to African agriculture: Lessons from two decades of donor experience,* ed. Uma Lele. World Bank discussion paper.

Lele, Uma J., and Arthur Goldsmith. 1989. The development of national agricultural research capacity: India's experience with the Rockefeller Foundation and its significance for Africa. *Economic Development and Cultural Change* 37:305–343.

Lele, Uma J., Bill H. Kinsey, and Antonia O. Obeya. 1989. Building agricultural research capacity in Africa: Policy lessons from the Madia countries. Unpublished working paper presented for the Joint TAC/CGIAR Center Directors Meeting, Rome.

Lemieux, T. 2006. Increasing residual wage inequality: Composition effects, noisy data, or rising demand for skill? *American Economic Review* 96, no. 3:461–498.

Lemons, J., R. Heredia, D. Jamieson, and C. Spash. 1995. Climate change and sustainable development. In *Sustainable development: Science, ethics, and public policy,* ed. Lemons and Brown. Amsterdam: Kluwer, pp. 110–157.

Levine, R. E., et al. 1990. *Breastfeeding saves lives: An estimate of breastfeeding related infant survival.* Bethesda, Md.: Center to Prevent Childhood Malnutrition.

Levinger, Beryl. 1994. *Nutrition, health, and education for all.* New York: United Nations Development Programme.

———. 1995. Critical transitions: Human capacity development across the lifespan. http://www.edc.org/int/hcd.

Lewis, W. Arthur. 1954. Economic development with unlimited supplies of labor. *Manchester School of Economic and Social Studies* (May):139–191.

Li, R., et al. 1994. Functional consequences of iron supplementation in iron-deficient female cotton mill workers in Beijing, China. *American Journal of Clinical Nutrition* 59, no. 4 (April):908–913.

Lin, Justin Yifu. 1990. Collectivization and China's agricultural crisis in 1959–1961. *Journal of Political Economy* (December):1228–1252.

Lin, Justin Yifu, and Dennis Tao Chang. 2000. Food availability, entitlements, and the Chinese famine of 1959–61. *Economic Journal* (January):136–158.

Lipton, Michael. 1977. *Why poor people stay poor: Urban bias in world development.* Cambridge: Harvard University Press.

Lipton, Michael, and Richard Longhurst. 1990. *New seeds and poor people.* Baltimore: Johns Hopkins University Press.

Lobine, E. 2000. Waterwar in Cochabamba, Bolvia. University of Greenwich, Public Services International Research Unit. http://www.psiru.org/reports/cochabamba.doc.

López, R. E. 1980. The structure of production and the derived demand for inputs in Canadian agriculture. *American Journal of Agricultural Economics* 62:38–45.

Lorenz, Max C. 1905. Methods of measuring the concentration of wealth. *Publications of the American Statistical Association* 9:209–219.

Luo, Z., R. Mu, and X. Zhang. 2006. Famine and overweight in China. *Review of Agricultural Economics* 28, no. 3 (Fall):296–304.

Lutz, W., W. Sanderson, and S. Sherbov. 2001. The end of world population growth. *Nature* 412:543–545.

Lynch, Colum. 2008. U.N. warns of impending food crisis in Zimbabwe. *Washington Post,* June 19, p. A13.

Lynn, R., and T. Vanhanen. 2002. *I.Q. and the wealth of nations.* New York: Praeger.

Ma, T., D. Wang, and Z. P. Chen. 1994. Mental retardation other than typical cretinism in IDD endemias in China. In Stanbury (ed.). New York: Cognizant Communication Corporation, pp. 265–272.

Mabbs-Zeno, C. C. 1987. *Where, if anywhere, is famine becoming more likely?* College Park, Md.: World Academy of Development and Cooperation.

Mace, James. 1984. Historical introduction. In *Human life in Russia,* ed. E. Ammende. Cleveland: Zubal.

Maddison, A. 1995. *Monitoring the world economy, 1820–1992.* Washington, D.C.: Organization for Economic Cooperation and Development.

———. 2001. *The world economy: A millennial perspective.* Paris: Organization for Economic Cooperation and Development. http://www.j-bradford-delong.net/articles_of_the_month/maddison-millennial.html.

Malthus, T. R. 1803–1826. *An essay on the principle of population, or a view of its past and present effects on human happiness with an inquiry into our prospects respecting the future removal or mitigation of the evils which it occasions.* [1st ed. 1803, 6th and last 1826.] London: Ward, 1890.

Mamarbachi, D., et al. 1980. Observations on nutritional marasmus in a newly rich nation. *Ecology of Food and Nutrition* 9:43–54.

Mann, Charles. 1997. Reseeding the green revolution. *Science* 277:1038–1043.

Martin, A. 2008. Spam turns serious and Hormel turns out more. *New York Times,* November 14.

Martorell, Reynaldo. 1980. The impact of ordinary illnesses on the dietary intakes of malnourished children. *American Journal of Clinical Nutrition* 33:345–350.

———. 1988. Seminar at University of Maryland, Department of Nutrition. December 12.

———. 1989. Body size, adaptation, and function. *Human Organization* 48:15–20.

———. 2001. Obesity. In *Health and nutrition: Emerging issues in developing countries.* Brief no. 7 (February). Washington, D.C.: International Food Policy Research Institute. http://www.ifpri.org/2020/focus/focus05/focus05_07.htm.

Masoro, E. J., B. P. Yu, and H. A. Bertrand. 1982. Action of food restriction in delaying the aging process (longevity/metabolic rate/lifetime caloric expenditure/life prolongation). *Proceedings: National Academy of Science* 79:4239–4241.

Mataev, Olga. 2001. The time of trouble: As seen by Dutch merchant Isaak Abrahamsz Massa. *Olga's Gallery,* October 1. http://www.abcgallery.com/list/2001oct01.html.

Mauro, P. 1997. The effects of corruption on growth, investment, and government expenditure: A cross-country analysis. In *Corruption and the global economy,* ed. K. A. Elliott. Washington, D.C.: Peterson Institute.

Maxwell, Bill. 2002. Slavery alive in Florida agricultural industry. *St. Petersburg Times,* July 3. http://www.sptimes.com/2002/07/03/columns/slavery_alive_in_flor.shtml.

Maxwell, Simon J. 1978a. *Food aid, food for work, and public works.* Discussion Paper no. 127 (March). Brighton: University of Sussex, Institute of Development Studies.

———. 1978b. Food aid for supplementary feeding programmes: An analysis. *Food Policy* 3:289–298.

Maxwell, Simon J., and H. W. Singer. 1979. Food aid to developing countries: A survey. *World Development* 7:225–247.

Mayer, Jean. 1976. The dimensions of human hunger. In *Scientific American* 235, no. 3 (September):40–49.

Mazumdar, D. 1965. Size of farm and productivity: A problem of Indian peasant agriculture. *Economica* 32 (May):161–173.

———. 1975. The theory of sharecropping with labor market dualism. *Economica* 42 (August):261–271.

McCalla, A. 1998. Agriculture and food needs to 2025. In *International agricultural development in the third world,* ed. Carl K. Eicher and John M. Staatz, 39–54. Baltimore: Johns Hopkins University Press.

McDevitt, Thomas M. 1996. Trends in adolescent fertility and contraceptive use in the developing world. Washington, D.C.: Bureau of the Census.

McFarland, William E., et al. 1974. *Demos, demographic-economic models of society: A computerized learning system.* Santa Barbara, Calif.: General Electric Tempo.

McGuire, Judy S. 1988. *Malnutrition: Opportunities and challenges for A.I.D.* Washington, D.C.: Resources for the Future, November.

McKigney, John, and Hamish Munro, eds. 1976. *Nutrient requirements in adolescence.* Cambridge: Massachusetts Institute of Technology Press.

McLaughlin, M. 1984. Interfaith action for economic justice. N.p.

Meier, G. 1979. Family planning in the banjars of Bali. *International Family Planning Perspectives* 5:63–66.

Meinzen-Dick, R., and M. Rosegrant. 2001. Overcoming water scarcity and quality constraints. *2020 Brief* no. 9 (October). Washington, D.C.: International Food Policy Research Institute.

Mellor, John W. 1984. Food price policy and income distribution in low-income countries. In *Agricultural development in the third world,* ed. Carl K. Eicher and John M. Staatz. Baltimore: Johns Hopkins University Press.

———. 1985a. *Agricultural change and rural poverty.* Food Policy Statement no. 3. Washington, D.C.: International Food Policy Research Institute.

———. 1985b. *The role of government and new agricultural technologies.* Food Policy Statement no. 4. Washington, D.C.: International Food Policy Research Institute.

———. 1986a. Dealing with the uncertainty of growing food imbalances: International structures and national policies. In *Proceedings of the nineteenth international conference of agricultural economists,* 191–198. Brookfield, Vt.: Grower.

———. 1986b. *The new global context for agricultural research: Implications for policy.* Food Policy Statement no. 6. Washington, D.C.: International Food Policy Research Institute.

———. 1988. Global food balances and food security. *World Development* 16:997–1011.

Mellor, John W., and Bruce F. Johnston. 1984. The world food equation: Interrelations among development, employment, and food consumption. *Journal of Economic Literature* 22:531–574.

Merrick, Thomas W., et al. 1986. World population in transition. *Population Bulletin* 41.

Meteorological Service of Canada. 2006. A warmer Canada. In *Understanding atmospheric change.* Ottawa: Environment Canada. http://www.msc-smc.ec.gc.ca/saib/climate/climatechange/soe_95-2/sections/9_e.html#fig27.

Micronutrient Initiative. 2008. Website. http://www.micronutrient.org/home.asp.

Miller, Gay Y., Joseph Rosenblatt, and Leroy Hushak. 1988. The effects of supply shifts on producer's surplus. *American Journal of Agricultural Economics* 70:886–891.

Mincer, Jacob. 1976. Unemployment effects of minimum wages. *Journal of Political Economy* 84, no. 4, pt. 2 (August):87–104.

Mintz, Sidney W. 1989. Food and culture: An anthropological view. In *Completing the food chain: Strategies for combating hunger and malnutrition,* ed. Paula M. Hirschoff and Neil G. Kolter, 114–121. Washington, D.C.: Smithsonian.

Mitchell, Donald O., Merlinda D. Ingco, and Ronald C. Duncan. 1997. *The world food outlook.* Cambridge: Cambridge University Press.

Mittal, Anuradha. 2002. The growing epidemic of hunger in a world of plenty. In *Fatal harvest: The tragedy of industrial agriculture,* ed. Andrew Kimbrell. Washington, D.C.: Island.

Mokyr, Joel. 1985. *Why Ireland starved: A quantitative and analytical history of the Irish economy, 1800–1850.* 2nd ed. London: Allen and Unwin.

Mokyr, Joel, and Cormac Ó Gráda. 1999. Famine disease and famine mortality: Lessons from Ireland, 1845–1850. Northwestern University. http://www.faculty.econ.northwestern.edu/faculty/mokyr/mogbeag.pdf.

Monto, A. S., and J. W. Koopman. 1980. The Tecumseh Study XI: Occurrence of acute enteric illness in the community. *American Journal of Epidemiology* 112:323–333.

Moore, M. 2006. As Europe grows grayer, France devises a baby boom. *Washington Post,* October 18, p. A1.

Moses, S., F. Plummer, E. Ngugi, N. Nagelkerke, A. Anzalat, and J. Ndinya-Achola. 1991. Controlling HIV in Africa: Effectiveness and cost of an intervention in a high frequency STD transmitter core group. *AIDS: Official Journal of the International AIDS Society* 5, no. 4:407–412.

Mudimu, G. 2003. Zimbabwe food security issues paper. In *Forum for food security in southern Africa.* http://www.odi.org.uk/food-security-forum/docs/zimbabwecipfinal.pdf.

Mundlak, Yair, Donald Larson, and Al Crego. 1996. *Agricultural development: Issues, evidence, and consequences.* Washington, D.C.: World Bank, International Economics Department.

Mwanaumo, A., P. Preckel, and P. Farris. 1994. Motivation for marketing system reform for the Zambian maize market. *Journal of International Food and Agribusiness Marketing* 5:29–49.

Myers, Robert G. 1988. *Programming for early child development and growth.* Paris: UNESCO-UNICEF Cooperative Program, June.

———. 1992. *The twelve who survive: Strengthening programmes of early childhood development in the third world.* London: Routledge.

Naiken, L. 1988. Comparison of the FAO and World Bank methodology for estimating the incidence of undernutrition. *FAO Quarterly Bulletin of Statistics* 1, no. 3:iii–v.

Nakajima, Hiroshi. 1989. *World health statistics annual.* Geneva: World Health Organization.

National Academy of Science. 1989. *Recommended dietary allowances.* 10th ed. Washington, D.C.

———. 2002. *Dietary reference intakes for energy, carbohydrates, fiber, fat, protein, and amino acids (macronutrients).* Washington, D.C.

———. 2003. *Dietary reference intakes: Applications in dietary planning.* Washington, D.C.

National Health Service (UK), Birmingham Community Nutrition and Dietetic Service. 2006. *Treatment of undernutrition in the community, including rationale for oral nutritional supplement (SIP) prescribing.* http://www.dietetics.bham.nhs.uk/docs/final55975%20ebham%20sip%20feed2.pdf.

National Research Council. 2000. *Beyond 6 billion: Forecasting the world's population.* Washington, D.C.: National Academy Press.

National Science and Technology Council. N.d. *Biotechnology for the 21st century: New horizons.* http://www.nal.usda.gov/bic/bio21/tablco.html.

Natsios, Andrew. 1999. The politics of famine in North Korea. Washington, D.C.: US Institute of Peace, August. http://www.usip.org/pubs/specialreports/sr990802.html.

Neue, H. 1993. Methane emission from rice fields: Wetland rice fields may make a major contribution to global warming. *BioScience* 43, no. 7:466–473.

Newbery, D., and J. Stiglitz. 1981. *The theory of commodity price stabilization: A study in the economics of risk.* Oxford: Clarendon.

Nicol, Mark. 2003. Famine-struck North Koreans eating children. *Telegraph,* June 9.

Nin, A., C. Arndt, T. Hertel, and P. Preckel. 2003. Bridging the gap between partial and total factor productivity measures using directional distance functions. *American Journal of Agricultural Economics* 85:937–951.

Nord, M., M., Andrews, and S. Carlson. 2003. *Household food security in the United States, 2002.* Food Assistance and Nutrition Research Report no. 35. Washington, D.C.: US Department of Agriculture, Food and Rural Economics Division, Economic Research Service.

North, D. 2005. The Chinese menu for development. *Wall Street Journal,* April 7, p. A14.

Notestein, Frank W. 1945. Population: The long view. In *Food for the world,* ed. Theodore W. Schultz. Chicago: University of Chicago Press.

O'Brien, Kevin. 2003. AIDS and African armies. *Atlantic Monthly,* July–August. http://www.theatlantic.com/issues/2003/07/rand.htm.

OECD (Organization for Economic Cooperation and Development) and FAO (Food and Agriculture Organization). 2007. *Agricultural outlook, 2007–2016.* Paris. http://www.oecd.org/dataoecd/6/10/38893266.pdf.

Olmsted, A., and P. Rhode. 2002. The red queen and the hard reds: Productivity growth in American wheat, 1800–1940. *Journal of Economic History* 62, no. 4:929–966.

Oram, Peter. 1995. *The potential of technology to meet world food needs in 2020.* 2020 Briefing Paper no. 13. Washington, D.C.: International Food Policy Research Institute.

Orwin, A. 1999. The privatization of water and wastewater utilities: An international survey. http://www.environmentprobe.org/enviroprobe/pubs/ev542.html#south%20america.

Overpeck, M., H. Hoffman, and K. Prager. 1992. The lowest birth-weight infants and the US infant mortality rate: NCHS 1983 linked birth/infant death data. *American Journal of Public Health* 82, no. 3:441–444.

Overton, M. 1996. Agricultural revolution in England: The transformation of the agrarian economy, 1500–1850. Cambridge: Cambridge University Press.

Oxfam. 1997. Fighting famine in North Korea and Ethiopia. October. http://www.caa.org.au/aware/1997/october-1997.html.

Paarlberg, D. 1988. *Toward a well-fed world.* Ames: Iowa State University Press.

Paglin, M. 1974. The measurement and trend of inequality: A basic revision. *American Economic Review* 65:598–609.

Pardey, Philip G., and Julian M. Alston. 1995. *Revamping agricultural R&D.* 2020 Briefing Paper no. 24. Washington, D.C.: International Food Policy Research Institute.

Pardey, P., and S. Beintema. 2001. Slow magic: Agricultural R&D a century after Mendel. Washington, D.C.: International Food Policy Research Institute.

Parham, Walter. 2001. Degraded lands: South China's untapped resource. *FAS Public Interest Report* 54, no. 2 (March–April). http://www.fas.org/faspir/2001/v54n2/resource.htm.

Parizokova, Jana. 1977. *Body fat and physical fitness.* The Hague: Nijhoff.

Park, Robert Ezra. 1934. Forward. In *Shadow of the plantation,* ed. Charles Spurgen. Chicago: University of Chicago Press.

Parry, M. L., A. R. Magalhaes, and N. H. Nih. 1992. *The potential socio-economic effects of climate change: A summary of three regional assessments.* Nairobi: United Nations Environment Programme.

Parry, M., C. Rosenzweig, and M. Livermore. 2005. Climate change, global food supply, and risk of hunger. *Philosophical Tranactions of the Royal Society B* 360: 2125–2138.

Payne, Philip R. 1985. The nature of malnutrition. In *Nutrition and development,* ed. Margaret Biswas and P. Pinstrup-Andersen. Oxford: Oxford University Press.

Pearson, N. O. 2007. Meat, sugar scarce in Venezuela stores. Associated Press, February 8. http://news.yahoo.com/s/ap/20070208/ap_on_re_la_am_ca/venezuela_food_crunch.

Pelletier, D. L., E. A. Frongillo Jr., and J. P. Habicht. 1993. Epidemiologic evidence for a potentiating effect of malnutrition on child mortality. *American Journal of Public Health* 83 (August):1130–1133.

———. 1995. The effects of malnutrition on child mortality in developing countries. *Bulletin of the World Health Organization* 73, no. 4:443–448.

Pellett, Peter L. 1977. Marasmus in a newly rich urbanized society. *Ecology of Food and Nutrition* 6:53–56.

———. 1987. Problems and pitfalls in the assessment of nutritional status. In *Food and evolution: Toward a theory of food habits,* ed. Marvin Harris and Erick B. Ross, 163–179. Philadelphia: Temple University Press.

Pelto, Gretl H. 1987. Cognitive performance and intake in preschoolers. In *Cognitive performance and intake in preschoolers,* ed. Lindsay H. Allen, Adolfo Chavez, and Gretl H. Pelto. Mexico City: University of Connecticut and Instituto Nacional de la Nutricion.

Penning de Vries, F. W. T., H. Van Keulen, R. Rabbinge, and J. C. Luyten. 1995. Biophysical limits to global food production. *2020 Brief* no. 18 (May). Washington, D.C.: International Food Policy Research Institute.

Perisse, J., F. Sizaret, and P. Francoise. 1969. The effect of income on the structure of the diet. *FAO Nutrition Newsletter* 7 (July–September):2.

Perkins, Sid. 2008. Disaster goes global. *Science News,* August 30. http://www.sciencenews.org/view/feature/id/35245/title/disaster_goes_global.

Peterson, Willis L. 1979. International farm prices and the social cost of cheap food policies. *American Journal of Agricultural Economics* 61 (February):12–21.

Pettifer, J. 2001. The magic bean. *BBC News,* June 8. http://news.bbc.co.uk/1/hi/programmes/correspondent/1363320.stm.

Pfeifer, Karen. 1985. *Agrarian reform under state capitalism in Algeria.* Boulder: Westview.

Philippines Ministry of Agriculture. 1981a. Food consumption and nutrition. Memo to agricultural minister Tanco. Manila: National Agricultural Policy Staff, September 8.

———. 1981b. National agricultural policy staff memo. Manila: National Agricultural Policy Staff, September 8.

———. 1981c. *Seasonal price indices of selected agricultural commodities.* National Policy Paper no. 81-2. Manila: Ministry of Agriculture.

———. 1983. *National consumption patterns for major foods, 1977–1982.* Manila: National Food Authority, Economic Research and Statistics Directorate, Special Studies Division.

Philippines National Economic Development Authority. 1983. 1987–1988 integrated survey of households (ISH), as quoted in *1983 economic and social indicators,* p. 157. Manila: National Economic Development Authority.

Philippines National Science and Technology Authority. 1983. FNRI Publication no. 82-ET-10. Manila, February.

———. 1984. *Second nationwide nutrition survey: Philippines, 1982.* Manila: Food and Nutrition Research Institute, October.

Phillips, Marshall, and Albert Baetz, eds. 1980. *Diet and resistance to disease.* New York: Plenum.

Phipps, Tim T. 1984. Land prices and farm-based returns. *American Journal of Agricultural Economics* 66 (November):422–429.

Pike, Ruth L., and Myrtle Brown. 1984. *Nutrition: An integrated approach.* New York: Wiley.

Piketty, T., and E. Saez. 2003. Income inequality in the United States, 1913–1998. *Quarterly Journal of Economics* 118, no. 1:1–39.

Pimentel, D. 1993. Climate changes and food supply. *Forum for Applied Research and Public Policy* 8, no. 4:54–60. http://www.ciesin.org/docs/004-138/004-138.html.

Pimentel, David, and Mario Giampietro. 1994. *Food, land, population, and the US economy.* Washington, D.C.: Carrying Capacity Network.

Pimentel, D., et al. 1994. Natural resources and an optimum human population. *Population and Environment* 15:347–369.

———. 1995. Environmental and economic costs of soil erosion and conservation benefits. *Science* 267:1117–1123.

———. 1996. Impact of population growth on food supplies and environment. Presented at the annual meeting of the American Association for the Advancement of Science, Baltimore, February 9.

Pingali, P. 1987. *From hand tillage to animal traction: Causes and effects and the policy implications for sub-Saharan African agriculture.* African Livestock Policy Analysis Network Paper no. 15. Addis-Ababa: International Livestock Centre for Africa. http://www.ilri.cgiar.org/infoserv/webpub/fulldocs/x5509e/x5509e00.htm#contents.

Pingali, P., and P. Heisey. 1999. *Cereal crop productivity in developing countries: Past trends and future prospects.* Economics Working Paper no. 99-03. Mexico City: Centro Internacional de Mejoramiento de Maize y Trigo.

Pinstrup-Andersen, Per, and Elizabeth Caicedo. 1978. The potential impact of changes in income distribution on food demand and human nutrition. *American Journal of Agricultural Economics* 60 (August):402–415.

Pinstrup-Andersen, Per, and Peter Hazell. 1985. The impact of the green revolution and prospects for the future. *Food Reviews International* 1, no. 1:11.

Pinstrup-Andersen, Per, David Nygaard, and Annu Ratta. 1995. *The right to food: Widely acknowledged and poorly protected.* 2020 Briefing Paper no. 22. Washington, D.C.: International Food Policy Research Institute.

Pinstrup-Andersen, P., R. Pandya-Lorch, and M. Rosegrant. 1998. *The world food situation: Recent developments, emerging issues, and long-term prospects.* Washington, D.C.: International Food Policy Research Institute.

Pinstrup-Andersen, Per, et al. 1976. The impact of increasing food supply on human nutrition: Implications for commodity priorities in agricultural research and policy. *American Journal of Agricultural Economics* 58:137–138.

Pollitt, E., K. Gorman, P. Engle, R. Martorell, and J. Rivera. 1993. Early supplementary feeding and cognition. *Monographs of the Society for Research in Child Development* 58, no. 235:7.

Pope John Paul II. 1996. Message to World Food Summit. November 13. http://www.fao.org/wfs/index_en.htm.

Population Information Program. 1985. Fertility and family planning surveys. *Population Reports* series M, no. 8 (September–October). Baltimore: Johns Hopkins University.

Population Reference Bureau. 1970. *World population data sheet, 1965.* Washington, D.C.

———. 1987. *World population data sheet, 1987.* Washington, D.C.

Posner, R. A. 1986. *Economic analysis of the law.* 3rd ed. Boston: Little, Brown.

Postel, Sandra. 1997. Dividing the waters. *Technology Review,* April. http://web.mit.edu/techreview/www/articles/apr97/toc.html.

———. 2003. Water for food production: Will there be enough in 2025? In *Global environmental challenges of the twenty-first century: Resources, consumption, and sustainable solutions,* ed. D. Lorey. London: Rowman and Littlefield.

Postel, Sandra, G. C. Daily, and P. R. Erlich. 1996. Human appropriation of renewable fresh water. *Science* 271:785–788.

Prentice, A. M., G. R. Goldberg, and Ann Prentice. 1994. Body mass index and lactation performance. *European Journal of Clinical Nutrition* 48, supp. 3 (November):78.

Pullum, Thomas W. 1983. Correlates of family-size desires. In *Determinants of fertility in developing countries,* vol. 1, *Supply and demand for children,* ed. Rodolfo A. Bulatao et al., 334–386. New York: Academic Press.

Pustilnik, Lev, and Gregory Yom Din. 2003. Influence of solar activity on state of wheat market in medieval England. Proceedings of the International Cosmic Ray Conference, p. 4131.

Qaim, M., and A. de Janvry. 2003. Adoption of BT cotton in Argentina. *American Journal of Agricultural Economics* 85:814–828.

Qaim, M., and D. Zilberman. 2003. Yield effects of genetically modified crops in developing countries. *Science* 299:900–902.

Quandt, Sara A. 1987. Methods for determining dietary intake. In *Nutritional anthropology,* ed. Francis E. Johnson, 67–84. New York: Alan R. Liss.

Ramalingaswami, Vulimiri, Urban Jonsson, and John Rohde. 1996. Commentary: The Asian enigma. In *The progress of nations.* Washington, D.C.: United Nations Children's Fund.

Ranade, C. G., and R. W. Herdt. 1978. Shares of farm earnings from rice production. In *Economic consequences of the new rice technology,* ed. R. Barker and Y. Hayami, 87–104. Los Banos, Philippines: International Rice Research Institute.

Rand, W., R. Uauy, and N. Scrimshaw. 1984. Protein-energy-requirement studies in developing countries: Results of international research. *Food and Nutrition Bulletin Supplement* no. 10. Tokyo: United Nations University Press. http://www.unu.edu/

unupress/unupbooks/80481e/80481e03.htm#5.%20protein-energy%20interactions.

Rangarajan, C. 1982. *Agricultural growth and industrial performance in India.* Research Report no. 33. Washington, D.C.: International Food Policy Research Institute.

Rask, Norman. 1986. Economic development and the dynamics of food needs. Unpublished paper delivered at Global Development Conference, University of Maryland, September.

Ravallion, Martin. 1997. Famines and economics. *Journal of Economic Literature* 35 (September):1205–1242.

Ray, Anandarup. 1986. Trade and pricing policies in world agriculture. *Finance and Development* 23 (September):2–5.

Reardon, T., C. Barrett, V. Kelly, and S. Kimseyinga. 1999. Policy reforms and sustainable agricultural intensification in Africa. *Development Policy Review* 17, no. 4: 375–395.

Rejesus, R., P. Heisey, and M. Smale. 1999. *Sources of productivity growth in wheat: A review of recent performance and medium- to long-term prospects.* Economics Working Paper no. 99-05. Mexico City: Centro Internacional de Mejoramiento de Maize y Trigo.

Repetto, Robert. 1985. *Paying the price: Pesticide subsidies in developing countries.* Research Report no. 2. Washington, D.C.: World Resources Institute, December.

Reuters. 1996. Tension in Jordan's Karak after bread riots. August 16. http://www.nando.net/newsroom/ntn/world/081696/world7_23344.html.

Reutlinger, Shlomo, and Marcelo Selowsky. 1976. *Malnutrition and poverty: Magnitude and policy options.* Occasional Paper no. 23. Washington, D.C.: World Bank.

Reutlinger, Schlomo, et al. 1983. Policy implications of research on energy intake and activity levels with reference to the debate on the energy adequacy of existing diets in developing countries. Discussion Paper no. 7. Washington, D.C.: World Bank, Department of Agriculture and Rural Development, Research Unit.

———. 1985. Food security and poverty in LDCs. *Finance and Development* 22 (December):7–11.

———. 1986. *Poverty and hunger: Issues and options for food security in developing countries.* Washington, D.C.: World Bank.

Rich, S. 2007. Africa's village of dreams. *Wilson Quarterly* (Spring).

Rivera, Juan, and Reynaldo Martorell. 1988. Nutrition, infection, and growth. Pts. 1–2. *Clinical Nutrition* 7:156–167.

Roberts, D. F. 1953. Body weight, race, and climate. *American Journal of Physical Anthropology* 11:533–558.

Rodrik, D. 2008. Spence christens a new Washington Consensus. *Economists' Voice* 5, no. 3:1–3. http://www.bepress.com/ev/vol5/iss3.

Rogers, Beatrice Lorge. 1988a. Design and implementation considerations for consumer-oriented food subsidies. In *Food subsidies in developing countries,* ed. Per Pinstrup-Andersen, 127–146. Baltimore: Johns Hopkins University Press.

———. 1988b. Economic perspectives on combating hunger. Presentation at the second annual World Food Prize Celebration, Washington, D.C., September 30. An edited version of this paper is available in *Completing the food chain: Strategies for combating hunger and malnutrition,* ed. Paula M. Hirschoff and Neil G. Kolter, 122–126. Washington, D.C.: Smithsonian, 1989.

———. 1988c. Pakistan's ration system: Distribution of costs and benefits. In *Food subsidies in developing countries,* ed. Per Pinstrup-Andersen, 242–252. Baltimore: Johns Hopkins University Press.

Rogers, J. E. Thorold. 1887. *Agriculture and prices in England.* Vols. 1–8. Oxford: Clarendon. Reprint, Vaduz, Liechtenstein: Kraus, 1963.

Rose, D., P. Strasberg, J. Jefe, and D. Tschirley. 1999. Higher calorie intakes related to higher incomes in northern Mozambique. *Flash,* no. 17. Michigan State University. http://www.aec.msu.edu/agecon/fs2/mozambique/flash17e.pdf.

Rose, Stephen, and David Fasentast. 1988. *Family incomes in the 1980s.* Working Paper no. 103. Washington, D.C.: Economic Policy Institute, November.

Rosegrant, Mark W. 1986. Irrigation with equity in Southeast Asia. *IFPRI Report* no. 8 (January):1, 4. Washington, D.C.: International Food Policy Research Institute.

Rosegrant, M., R. Scheleyer, and S. Yadav. 1995. Water policy for efficient agricultural diversification: Market-based approaches. *Food Policy* 20:203–223.

Rosen, S. 1999. Potato paradoxes. *Journal of Political Economy* 107, no. 6: S294–S313.

Rosenzweig, Cynthia, and Daniel Hillel. 1995. Potential impacts of climate change on agriculture and food supply. *Consequences* (Summer).

Rosenzweig, C., and M. L. Parry. 1994. Potential impact of climate change on world food supply. *Nature* 367, no. 6459.

Rosenzweig, C., M. L. Parry, G. Fischer, and K. Frohberg. 1993. *Climate change and world food supply.* Research Report no. 3. Oxford: Oxford University, Environmental Change Unit.

Rosset, P., J. Collins, and F. M. Lappe. 2000. Lessons from the green revolution. *Tikkun Magazine,* March 1. http://www.foodfirst.org/media/printformat.php?id=148.

Rowley, R. J., J. C. Kostelnick, D. Braaten, X. Li, and J. Meisel. 2007. Risk of rising sea level to population and land area. *EOS* 88, no. 9 (February):105–116.

Royal Society. 1999. *Review of data on possible toxicity of GM potatoes.* London. http://www.royalsoc.ac.uk/files/statfiles/document-29.pdf.

Ruddiman, W. 2005. *Plows, plagues, and petroleum: How humans took control of climate.* Princeton: Princeton University Press.

Runge, C., B. Senauer, P. Pardey, and M. Rosegrant. 2003. *Ending hunger in our lifetime: Food security and globalization.* Baltimore: Johns Hopkins University Press.

Rustein, Shea O. 1984. Infant and child mortality: Levels, trends, and demographic differentials. In *World fertility survey comparative studies no. 45,* rev. ed., tab. 14. Voorburg, Netherlands: International Statistical Institute.

Ryan, James G. 1977. *Human nutritional needs and crop breeding objectives in the Indian semi-arid tropics.* Hyderabad, India: International Crops Research Institute for the Semi-Arid Tropics.

Sachs, J. 2005a. *The end of poverty.* New York: Penguin.

———. 2005b. A practical plan to end poverty. *Washington Post,* January 17, p. A17.

Sagoff, M. 1997. Do we consume too much? *Atlantic Monthly,* June.

Sahlins, Marshall. 1968. Notes on the original affluent society. In *Man the hunter,* ed. R. B. Lee and I. DeVore, 85–89. Chicago: Aldine.

Sahn, David E., ed. 1989. *Seasonal variability in third world agriculture: The consequences for food security.* Baltimore: Johns Hopkins University Press.

Sahn, David E., and Neville Edirisinghe. 1993. Politics of food policy in Sri Lanka: From basic human needs to an increased market orientation. In *The political economy of food and nutrition policy,* ed. Per Pinstrup-Andersen. Baltimore: Johns Hopkins University Press.

Salaff, Janet W., and Arline Wong. 1983. *Incentives and disincentives in population policies.* Report no. 12. Washington, D.C.: Draper World Population Fund, August.

Sala-i-Martin, X. 2002. *The world distribution of income (estimated from individual country distributions).* Working paper. New York: Columbia University, May 1. http://www.econ.upf.es/deehome/what/wpapers/postscripts/615.pdf.

Sampath, R. 1992. Issues in irrigation pricing in developing countries. *World Development* 20:967–977.

Samuels, B. 1986. Infant mortality and low birth weight among minority groups in the United States: A review of the literature. In *Report of the secretary's task force on black and minority health,* vol. 4, *Infant mortality and low birth weight.* Washington, D.C.: US Department of Health and Human Services.

Sancoucy, R. 1995. Livestock: A driving force for food security and sustainable development. In *World animal review,* 84–85. Rome: Food and Agriculture Organization. http://www.fao.org/docrep/v8180t/v8180t07.htm#livestock%20%20%20a%20driving%20force%20for%20food%20security%20and%20sustainable%20development.

Sandburg, Carl. 1936. *The people, yes.* New York: Harcourt, Brace.

Sang-Hun, Choe. 2008. North Korea to widen access for aid workers; U.S. ship arrives. *New York Times,* July 1.

Santos-Villanueva, P. 1966. The value of rural roads. In *Selected readings to accompany getting agriculture moving,* ed. Raymond E. Borton, 775–795. New York: Agricultural Development Council.

Sazawal, S., et al. 1995. Zinc supplementation in young children with acute diarrhea in India. *New England Journal of Medicine* 333, no. 13 (September):839–844.

Scandizzo, Pasquale L., and Colin Bruce. 1980. *Methodologies for measuring agricultural price intervention effects.* Working Paper no. 394. Washington, D.C.: World Bank, June.

Scandizzo, Pasquale L., and I. Tsakok. 1985. Food price policies and nutrition in developing countries. In *Nutrition and development,* ed. Margaret Biswas and Per Pinstrup-Andersen, 60–76. Oxford: Oxford University Press.

Schaal, B. 2003. Biotechnology, biodiversity, and the environment. Paper presented at the Conference on Biodiversity and Biotechnology and the Protection of Traditional Knowledge, Washington University School of Law, St. Louis, April. http://ls.wustl.edu/centeris/papers/biodiverity/pdfwrddoc/schaalfinal.pdf.

Schemann, J. F., et al. 2003. National immunisation days and vitamin A distribution in Mali: Has the vitamin A status of pre-school children improved?" *Public Health and Nutrition* 6, no. 3:233–244.

Scherr, Sara J., and Satya Yadav. 1997. Land degradation in the developing world: Issues and policy options for 2020. *IFPRI 2020 Brief* no. 44 (June). Washington, D.C.: International Food Policy Research Institute.

Schiff, Maurice, and Alberto Valdés. 1995. The plundering of agriculture in developing countries. *Finance & Development* (March). Washington, D.C.: World Bank.

Schnepf, Randall D., Erik Dohlman, and Christine Bolling. 2001. *Agriculture in Brazil and Argentina: Developments and prospects for major field crops.* Agriculture and Trade Report no. WRS-01-3. Washington, D.C.: US Department of Agriculture, Market and Trade Economics Division, Economic Research Service.

Schroeder, D. G., and K. H. Brown. 1994. Nutritional status as a predictor of child survival: Summarizing the association and quantifying its global impact. *Bulletin of the World Health Organization* 72, no. 4:569–579.

Schuh, G. Edward. 1988. Some issues associated with exchange rate realignments in developing countries. In *Macroeconomics, agriculture, and exchange rates,* ed. Philip L. Paarlberg and Robert G. Chambers, 231–240. Boulder: Westview.

Schultz, T. P. 1999. Health and schooling investments in Africa. *Journal of Economic Perspectives* 13, no. 3:67–88.

Schultz, Theodore W. 1979. *The economics of research and agricultural productivity.* Occasional paper. Arlington, Va.: International Agricultural Development Services.

As quoted in *Agricultural development in the third world,* ed. Carl K. Eicher and John M. Staatz, 335–347. Baltimore: Johns Hopkins University Press, 1984.

Scobie, Grant M. 1983. *Food subsidies in Egypt: Their impact on foreign exchange and trade.* Research Report no. 40. Washington, D.C.: International Food Policy Research Institute, August.

Scobie, Grant M., and Rafael Posada T. 1984. The impact of technical change on income distribution: The case of rice in Colombia. In *Agricultural development in the third world,* ed. Carl K. Eicher and John M. Staatz, 378–388. Baltimore: Johns Hopkins University Press.

Scrimshaw, Nevin S. 1988. Completing the food chain: From production to consumption. Remarks presented at the second annual World Food Price Celebration, Washington, D.C., September 30. An edited version of this paper is available in *Completing the food chain: Strategies for combating hunger and malnutrition,* ed. Paula M. Hirschoff and Neil G. Kolter, 1–17. Washington, D.C.: Smithsonian, 1989.

Scrimshaw, Nevin S., Carl Taylor, and John Gordon. 1968. *Interactions of nutrition and infection.* Geneva: World Health Organization.

Scrimshaw, Nevin S., and Vernon R. Young. 1976. The requirements of human nutrition. In *Food and agriculture,* ed. Scientific American, 26–40. San Francisco: Freeman.

Scrimshaw, Susan. 1978. Infant mortality and behavior in the regulation of family size. *Population and Development Review* 4:383–403.

———. 1984. Infanticide in human populations: Societal and individual concerns. In *Infanticide: Comparative and evolutionary perspectives,* ed. Glen Hausfater and Sarah B. Hardy, 439–462. New York: Aldine.

Seckler, David. 1982. Small but healthy: A basic hypothesis in the theory, measurement, and policy of malnutrition. In *Newer concepts in nutrition and their implications for policy,* ed. P. V. Kukhtame, 127–137. Pune, India: Maharastra Association for the Cultivation of Science Research Institute.

Semba, R., and M. Bloem, eds. 2008. *Nutrition and health in developing countries.* Totowa, N.J.: Humana.

Semba, R. D., et al. 1994. Maternal vitamin A deficiency and mother-to-child transmission of HIV-1. *Lancet* 343, no. 8913 (June 25):1593–1597.

Sen, Amartya K. 1964. Size of holdings and productivity. *Economic and Political Weekly* (February).

———. 1966. Peasants and dualism with or without surplus labor. *Journal of Political Economy* 74:425–450.

———. 1973. *On economic inequality.* London: Oxford University Press.

———. 1981. *Poverty and famines: An essay on entitlement and deprivation.* Oxford: Clarendon.

———. 1987. *Hunger and entitlements.* Helsinki: World Institute for Development Economics Research.

———. 1990. Public action to remedy hunger: Tanco memorial lecture. London: Hunger Project, August. http://www.thp.org/reports/sen/sen890.htm#n1.

Senauer, Benjamin, et al. 1988. Determinants of the intrahousehold allocation of food in the rural Philippines. *American Journal of Agricultural Economics* 70:170–180.

Shakir, A. 1975. The surveillance of protein-calorie malnutrition by simple and economic means. Report to UNICEF. *Journal of Tropical Pediatrics and Environmental Child Health* 21:69–85.

Shapiro, F. 2004. Plural of anecdote is data. Email from Raymond Wolfinger, editor of *The Yale Dictionary of Quotations,* to the American Dialect Society, July 6. http://listserv.linguistlist.org/cgi-bin/wa?a2=ind0407a&l=ads-l&p=8874.

Shaw, A. 2000. Police, troops impose uneasy calm after food riots. Associated Press, October 19.

Shekar, Meera. 1991. *The Tamil Nadu integrated nutrition project: A review of the project with special emphasis on the monitoring and information system.* Working Paper no. 14. Ithaca: Cornell Food and Nutrition Policy Program.

Sherman, Adria R. 1986. Alterations in immunity related to nutritional status. *Nutrition Today* (July–August):7–13.

Shorto, R. 2008. No babies. *New York Times Sunday Magazine,* June 29.

Sicat, Gerardo P. 1983. Toward a flexible interest rate policy, or losing interest in the usury law. In *Rural financial markets in developing countries: Their use and abuse,* ed. J. D. Von Pische et al., 373–386. Baltimore: Johns Hopkins University Press.

Simmons, George B., and Robert J. Lapham. 1987. The determinants of family planning program effectiveness. In *Organizing for effective family planning programs,* ed. G. Lampham and R. Simmons, 683–706. Washington, D.C.: National Academy Press.

Simon, Julian L. 1986. *Theory of population and economic growth.* New York: Blackwell.

———. 1996. *The ultimate resource.* Princeton: Princeton University Press. http://www.rhsmith.umd.edu/faculty/jsimon/ultimate_resource.

Simon, N., M. Cropper, A. Alberini, and S. Arora. 1999. Valuing mortality reductions in India: A study of compensating-wage differentials. Working Paper no. 2978. Washington, D.C.: World Bank.

Sinaga, R. S., and B. M. Sinaga. 1978. Comments on shares of farm earnings from rice production. In *Economic consequences of the new rice technology,* ed. R. Barker and Y. Hayami, 105–109. Los Banos, Philippines: International Rice Research Institute.

Singer, P. 1972. Famine, affluence, and morality. *Philosophy and Public Affairs* 1, no. 3 (Spring):229–243.

Singh, K. S. 1975. *The Indian famine, 1967: A study of crisis and change.* New Delhi: People's Publishing House.

Singh, R. B., P. Kumar, and T. Woodhead. 2002. *Smallholder farmers in India: Food security and agricultural policy.* RAP Publication no. 2002/03. Bangkok: FAO Regional Office for Asia and the Pacific.

Smale, Melinda. 1997. The green revolution and wheat genetic diversity: Some unfounded assumptions. *World Development* 25, no. 8:1257–1269.

Snow, A. 2002. Transgenic crops: Why gene flow matters. *Nature Biotechnology* 20: 542.

Snyder, J. D., and M. H. Merson. 1982. The magnitude of the global problem of acute diarrhoeal disease: A review of active surveillance data. *Bulletin of the World Health Organization* 60:605–613.

Soliman, Ibrahim, and Shahla Shapouri. 1984. See USDA 1984.

Sommer, Alfred, et al. 1986. Impact of vitamin A supplementation on childhood mortality: A randomized controlled community trial. *Lancet* (May 24):1169–1173.

Southgate, D., D. Graham, and L. Tweeten. 2007. *The world food economy.* Malden, Mass.: Blackwell.

Specter, M. 2006. The last drop. *New Yorker,* October 23, p. 34.

Sperduto, R. D., et al. 1993. The Linxian cataract studies: Two nutrition intervention trials. *Archives of Ophthalmology* 830, no. 111:1246–1253.

Spurr, G. B., M. Barac-Nieto, and M. G. Maksud. 1976. See US Department of State 1976.

Stackman, E. C., Richard Bradfield, and Paul C. Mangelsdorf. 1967. *Campaigns against hunger.* Boston: Belknap.

Stanbury, John, ed. 1994. *The damaged brain of iodine deficiency: Cognitive, behavioral, neuromotor, educative aspects.* Port Washington, N.Y.: Cognizant Communication.

Steckel, R. H. 1995. Stature and the standard of living. *Journal of Economic Literature* 33:1903–1940.

Stein, A., H. Barnhart, M. Hickey, U. Ramakrishnan, D. Schroeder, and R. Martorell. 2003. Prospective study of protein-energy supplementation early in life and of growth in the subsequent generation in Guatemala. *American Journal of Clinical Nutrition* 78, no. 1:162–167.

Stein, Z. 1975. *Famine and human development: The Dutch honger winter of 1944–45.* New York: Oxford University Press.

Steindl, Josef. 1987. Pareto distribution. In *The new Palgrave: A dictionary of economics,* vol. 3., ed. John Eatwell et al., 809–810. New York: Stockton.

Steinfeld, H., P. Gerber, T. Wassenaar, V. Castel, M. Rosales, and C. de Haan. 2006. *Livestock's long shadow: Environmental issues and options.* Rome: Food and Agriculture Organization. ftp://ftp.fao.org/docrep/fao/010/a0701e/a0701e04.pdf.

Stephenson, Lani S., M. C. Latham, and A. Jansen. 1983. *A comparison of growth standards: Similarities between NCHS, Harvard, Denver, and privileged African children and differences with Kenyan rural children.* International Nutrition Monograph Series no. 12. Ithaca: Cornell University Press.

Stevens, Robert D., and Cathy L. Jabara. 1988. *Agricultural development principles: Economic theory and empirical evidence.* Baltimore: Johns Hopkins University Press.

Stevenson, B., and J. Wolfers. 2008. *Economic growth and subjective well-being: Reassessing the Easterlin paradox.* Philadelphia: University of Pennsylvania Press. http://bpp.wharton.upenn.edu/betseys/papers/happiness.pdf.

Steyn, M. 2006. Salute Danna Vale. *The Australian,* February 15.

Stiglitz, J. 1986. *Economics of the public sector.* New York: Norton.

———. 2002. *Globalization and its discontents.* New York: Norton.

Stone, Bruce. 1985. Fertilizer pricing policy and foodgrain production strategy. *IFPRI Report* 7 (May):1, 4. Washington, D.C.: International Food Policy Research Institute.

Stout, B. A. 1998. Energy for agriculture in the twenty-first century. In *Feeding a world population of more than eight billion people: A challenge to science,* ed. J. C. Waterlow et al. Oxford: Oxford University Press.

Strategy Page. 2004. Korea, January 21. http://www.strategypage.com//fyeo/qndguide/default.asp?target=korea.htm.

Strauss, J. 1968. Estimating the determinants of food consumption and caloric availability in rural Sierra Leone. In *Agricultural household models: Extensions, applications, and policy,* ed. Inderjit Singh et al., 116–152. Baltimore: Johns Hopkins University Press.

Strauss, John, and Duncan Thomas. 1998. Health, nutrition, and economic development. *Journal of Economic Literature* 36:766–818.

Streeten, Paul. 1987. *What price food? Agricultural price policies in developing countries.* London: Macmillan.

Struck, Doug. 2001. North Korea food crisis intensifies. *Washington Post,* May 16, p. A20.

Stuart, H., and S. Stevenson. 1950. Physical growth and development. In *Textbook of pediatrics,* 5th ed., ed. W. Nelson. Philadelphia: Saunders.

Subramanian, S., and A. Deaton. 1996. The demand for food and calories. *Journal of Political Economy* 104, no. 1:133–162.

Subramanian, U., and M. Cropper. 1995. *Public choices between lifesaving programs: How important are lives saved?* Working Paper no. 1497. Washington, D.C.: World Bank.

Summers, L. 1991. GEP [Global Environmental Policy]. Internal World Bank memo, December 12. http://www.whirledbank.org/ourwords/summers.html.

Susser, E., et al. 1996. Schizophrenia after prenatal famine: Further evidence. *Archives of General Psychiatry* 53, no. 1 (January):25–31.

Sutton, John. 1989. See USDA 1989a.

Svedberg, P. 1999. 841 million undernourished? *World Development* 27:2081–2098.

Swindale, L. D. 1997. The globalization of agricultural research: A study of the control of the cassava mealybug in Africa. Consultative Group for International Agricultural Research. ftp://ftp.cgiar.org/isnar/publicat/pdf/vision/swindale.pdf.

Tang, A. M., et al. 1993. Dietary micronutrient intake and risk of progression to acquired immunodeficiency syndrome (AIDS) in human immunodeficiency virus type 1 (HIV-1)-infected homosexual men. *American Journal of Epidemiology* 138, no. 11 (December 1):937–951.

Tanner, J. M. 1977. Human growth and constitution. In *Human biology: An introduction to human evolution, variation, growth and ecology,* by G. A. Harrison et al., 301–385. Oxford: Oxford University Press.

Tanner, J., R. Whitehouse, and M. Takaishi. 1966. Standards from birth to maturity for height, height velocity, and weight velocity: British children I. *Archives of Disease in Childhood* 41:454.

Tanzi, V., and H. Zee. 2000. *Tax policy for emerging markets: Developing countries.* Working Paper no. 00/35. Washington, D.C.: International Monetary Fund.

Telford, J., J. Cosgrave, and R. Houghton. 2006. *Joint evaluation of the international response to the Indian Ocean tsunami: Synthesis report.* London: Tsunami Evaluation Coalition. http://www.tsunami-evaluation.org/nr/rdonlyres/2e8a3262-0320-4656-bc81-ee0b46b54caa/0/synthrep.pdf.

Teshima, R., H. Akiyama, H. Okunuki, J. Sakushima, Y. Goda, H. Onodera, J. Sawada, and M. Toyoda. 2000. Effect of GM and non-GM soybeans on the immune system of BN rats and B10A mice. *Journal of the Food Hygienic Society of Japan* 41, no. 3:188–193.

Thielke, Thilo. 2006. Kenya's deadly dependency on food aid. *Der Spiegel,* January 19. http://service.spiegel.de/cache/international/spiegel/0,1518,396031,00.html.

Thomas, F., F. Renaud, E. Benefice, T. de Meeus, and J.-F. Guegan. 2001. International variability of ages of menarche and menopause: Patterns and main determinants. *Human Biology* 73, no. 2 (April):271–290.

Thomson-Wadsworth. 2003. 2002 dietary reference intakes (DRI). http://www.newtexts.com/newtexts/nutrition%20tables.pdf.

Timberg, C. 2005. In Zimbabwe, withholding of food magnifies the hunger for change. *Washington Post,* March 30, p. A1.

Timmer, C. Peter. 1984. Choice of technique in rice milling on Java. In *Agricultural development in the third world,* ed. Carl K. Eicher and John M. Staatz, 278–288. Baltimore: Johns Hopkins University Press.

Timmer, C. Peter, Walter P. Falcon, and Scott R. Pearson. 1983. *Food policy analysis.* Baltimore: Johns Hopkins University Press.

Tinker, Anne, et al. 1994. *Women's health and nutrition: Making a difference.* Washington, D.C.: World Bank.

Todaro, Michael P. 1980. Internal migration in developing countries: A survey. In *Population and economic change in developing countries,* ed. Richard A. Easterlin, 361–402. Chicago: University of Chicago Press.

Transport for London. 2003. Congestion charging: Summary of week 6. Press release, April 1. http://www.tfl.gov.uk/tfl/press_cc_news_latest.shtml.

Traub, James. 1988. Into the mouths of babes. *New York Times Magazine,* July 24, p. 18.

Trumbo, P., S. Schlicker, A. Yates, and M. Poos. 2002. Dietary reference intakes for energy, carbohydrate, fiber, fat, fatty acids, cholesterol, protein, and amino acids. *Journal of the American Dietetic Association* 102, no. 11 (November).

Tupasi, T. E. 1985. Nutritional and acute respiratory infection. In *Acute respiratory infections in childhood: Proceedings of an international workshop,* ed. R. Douglas and E. Kerby-Eaton. Adelaide, Australia: University of Adelaide Press.

UN Office of Drug and Crime. 2007. *World drug report 2007.* New York: United Nations. http://www.unodc.org/pdf/research/wdr07/WDR_2007.pdf.

UN Population Division. 2002a. World population prospects: Assumptions underlying the results of the 2002 revisions of the world population prospects. http://esa.un.org/unpp/assumptions.html.

———. 2002b. World population prospects population database. http://esa.un.org/unpp.

UN Population Information Network. 1995. *Population and land degradation.* Vol. 2 of *Population and the environment: A review of issues and concepts for population programmes staff.* New York.

UNAIDS (Joint United Nations Programme on HIV/AIDS). 1994. Women and AIDS: A growing challenge. Fact Sheet 23/11/2004. New York: UNAIDS. http://data.unaids.org/publications/fact-sheets04/fs_women_en.pdf.

———. 2002. Global reports: Estimates, end of 2001. http://www.unaids.org/html/pub/global-reports/barcelona/tableestimatesend2001_en_xls.xls.

———. 2008. *Report on the global AIDS epidemic, 2008.* New York. http://www.unaids.org/en/knowledgecentre/hivdata/globalreport/2008/2008_global_report.asp.

UNAIDS and WHO (World Health Organization), Reference Group for Estimates, Modelling, and Projections. 2002. Improving estimates and projections of HIV/AIDS. http://www.epidem.org/publications/madrid%20report.pdf.

UNDP (United Nations Development Programme). 2003. *Human development report, 2003.* New York. http://www.undp.org/hdr2003/index_indicators.html.

———. 2006. *Human development report, 2006: Beyond scarcity—power, poverty, and the global water crisis.* New York. http://hdr.undp.org/hdr2006/pdfs/report/hdr06-complete.pdf.

———. 2007. *Human development report 2007/08: Fighting climate change: Human solidarity in a divided world.* New York: UNDP. http://hdr.undp.org/en/media/hdr_20072008_en_complete.pdf.

UNEP (United Nations Environment Programme). 1990. *The impacts of climate change on agriculture.* Fact Sheet no. 101. Nairobi: Information Unit for Climate Change.

UNESCO (United Nations Educational, Scientific, and Cultural Organization). 2006. *The 2nd UN world water development report.* New York. http://www.unesco.org/water/wwap/wwdr/wwdr2/table_contents.shtml.

UNICEF (United Nations Children's Fund). 1982. *News,* no. 113:9.

———. 1987. ORT and much more: Developing whole CDD programmes. Memo to all field offices, January 15.

———. 1988. *State of the world's children, 1988.* New York: Oxford University Press.

———. 1996. *Progress of nations.* New York. http://www.unicef.org/pon96.

———. 1998. *State of the world's children, 1998.* New York: Oxford University Press.

———. 2001. *State of the world's children, 2001.* New York: Oxford University Press.

———. 2003. *State of the world's children, 2003.* New York: Oxford University Press.

———. 2006. Progress for children: A report card on nutrition, no. 4, May 2006. New York: United Nations. http://www.unicef.org/progressforchildren/2006n4/index.html.

———. 2007. *Progress for children: A world fit for children—statistical review.* New York. http://www.unicef.org/publications/files/progress_for_children_no_6.pdf.

———. 2008. *State of the world's children, 2008.* New York: Oxford University Press.
———. n.d. Immunization: The big picture. New York: UNICEF. http://www.unicef.org/immunization/index_bigpicture.html.
United Nations. 1991. *World population prospects, 1990.* Population Study no. 120. New York: Department of International Economic and Social Affairs.
———. 2008. *The millennium development goals report, 2008.* New York. http://mdgs.un.org/unsd/mdg/resources/static/products/progress2008/mdg_report_2008_en.pdf.
University of California, Food Task Force. 1974. *A hungry world: The challenge to agriculture—summary report.* Berkeley: Division of Agricultural Sciences.
Urban, Francis, and Arthur J. Dommen. 1989. See USDA 1989b.
US Bureau of the Census. 1961. *Historical statistics of the US from colonial times to 1956.* Washington, D.C.
———. 1987. *Statistical abstract of the United States: 1988.* Washington, D.C.
———. 2008a. Website. http://www.census.gov.
———. 2008b. *American housing survey for the United States, 2007.* Current Housing Report no. H150/07. Washington, D.C.
US Congress. 1974. *National nutrition policy study, 1974: Hearings before the Select Committee on Nutrition and Human Needs,* pt. 3, *Nutrition and special groups.* 93rd Congress. Washington, D.C.: US Government Printing Office.
———, Office of Technology Assessment. 1986. *Technology, public policy, and the changing structure of American agriculture.* Washington, D.C.: US Government Printing Office..
USAID (US Agency for International Development), West African Trade Hub. 2007. *Report on the first results of the improved road-transport governance (IRTG) initiative.* Accra, Ghana. http://www.watradehub.com/images/stories/downloads/studies/report%20on%20first%20irtg%20results,%20english,%20jw.pdf.
USDA (US Department of Agriculture). 1917. *Geography of the world's agriculture.* By V. C. Finch and O. E. Baker. Washington, D.C.: Office of Farm Management.
———. 1963. *Composition of foods: Raw, processed, prepared.* Agricultural Handbook no. 8. Washington, D.C.
———. 1970. *Feed situation report.* Washington, D.C., November.
———. 1984. *The impact of wheat price policy change on nutritional status in Egypt.* By Ibrahim Soliman and Shahla Shapouri. Washington, D.C.: Economic Research Service, International Economics Division, February.
———. 1985. *US demand for food: A complete system of price and income effects.* By Kuw W. Huang. Technical Bulletin no. 1714. Washington, D.C.
———. 1988. *World food needs and availabilities, 1988–89: Summer.* Washington, D.C.: Economic Research Service, August.
———. 1989a. Environmental degradation and agriculture. By John Sutton. *World Agriculture Situation and Outlook Report* no. WAS-55 (June):35–41. Washington, D.C.: Economic Research Service.
———. 1989b. *World agriculture.* By Francis Urban and Arthur J. Dommen. Washington, D.C., June.
———. 1990. *US government concessional exports: Commodity by country and fiscal year.* Unpublished database. Washington, D.C.
———. 1991. *Food cost reviews.* Washington, D.C.
———. 1998. *Agricultural baseline projections to 2007.* Washington, D.C. http://www.econ.ag.gov/briefing/baseline/index98.htm.
———. 2008a. *U.S. biobased products: Market potential and projections through 2025.* Report no. OCE-2008-1. Washington, D.C. http://www.usda.gov/oce/reports/energy/biobasedreport2008.pdf.

———. 2008b. *USDA agricultural projections to 2017.* Report no. OCE-2008-1. Washington, D.C., February. http://www.ers.usda.gov/publications/oce081.

———, Agricultural Research Service. 2003. Research Q&A: Bt corn and monarch butterflies. Washington, D.C. http://www.ars.usda.gov/is/br/btcorn.

———, Foreign Agricultural Service. 2006. *McGovern-Dole international food for education and child nutrition program.* Fact Sheet. Washington, D.C., March.

US Department of Energy. 2003. Residential energy consumption survey. Washington, D.C. http://www.eia.doe.gov/emeu/recs/recs2001/detail_tables.html.

———. 2007. *Research advances in cellulosic ethanol.* Publication no. NREL/BR-510-40742. Washington, D.C.: National Renewable Energy Laboratory, March. http://www.nrel.gov/biomass/pdfs/40742.pdf.

US Department of Health and Human Resources. 1981. *Height and weight of adults ages 18–74 years by socioeconomic and geographic variables, United States.* Publication no. 81-1674 (Public Health Service). Data from National Health Survey Series 11, no. 224. Hyattsville, Md.: National Center for Health Statistics, August.

US Department of Health and Human Services. 1987. *Anthropometric reference data and prevalence of overweight: United States, 1976–80.* Publication no. 87-1688 (Public Health Service). Hyattsville, Md.: National Center for Health Statistics, October.

———. 1988. *The Surgeon General's report on nutrition and health, 1988.* Publication no. 88-50210 (Public Health Service). Washington, D.C.

US Department of Health, Education, and Welfare. 1976. NCHS growth charts. *Monthly Vital Statistics Report* 25, no. 3, supp. 76-1120 (Human Resources Administration). Rockville, Md.: National Center for Health Statistics, June.

———. 1979. *Weight by height and age for adults 18–74 years: United States, 1971–74.* Data from National Health Survey Series 11, no. 208. Hyattsville, Md.: National Center for Health Statistics.

US Department of State. 1976. *Clinical and subclinical malnutrition and their influence on the capacity to do work.* By G. B. Spurr, M. Barac-Nieto, and M. G. Maksud. Project no. AID/CSD 2943, final report. Washington, D.C.

———. 1978. *Agricultural policies and rural malnutrition.* By Phillips Foster. Occasional Paper no. 8. Washington, D.C.: USAID Economics and Sector Planning Division, Office of Agriculture, Technical Assistance Bureau.

———. 1986a. *Development and spread of high yielding rice varieties in developing countries.* By Dana G. Dalrymple. Washington, D.C.: US Agency for International Development.

———. 1986b. *Development and spread of high yielding wheat varieties in developing countries.* By Dana G. Dalrymple. Washington, D.C.: US Agency for International Development.

US Department of Transportation, Federal Highway Administration. 2002. *1999 status of the nation's highways, bridges, and transit: Conditions and performance report.* Washington, D.C. http://www.fhwa.dot.gov/policy/1999cpr/ch_10/cpm10_3.htm.

US Department of Treasury. 2007. *Income mobility in the U.S. from 1996 to 2005.* Washington, D.C. http://www.ustreas.gov/offices/tax-policy/library/incomemobilitystudy03-08revise.pdf.

US Energy Information Administration. 2001. *Residential energy consumption survey, 2001.* Data tables by housing characteristics. http://www.eia.doe.gov/emeu.

———. 2008. *Weekly gasoline prices.* http://tonto.eia.doe.gov/dnav/pet/hist/mg_rt_usw.htm.

US Environmental Protection Agency. 1999. *The benefits and costs of the Clean Air Act, 1990 to 2010.* Washington, D.C. http://www.epa.gov/airprogm/oar/sect812/1990-2010/fullrept.pdf.

US Federal Reserve System, Board of Governors. 1989. *Balance sheets for the US economy, 1949–88.* Washington, D.C., October.

US National Academy of Sciences. 1974. *Recommended dietary allowances.* Washington, D.C.

US Water News Online. 2003. Massive groundwater supply found in arid northwestern China. February. http://www.uswaternews.com/archives/arcglobal/3masgro2.html.

US White House, Council of Economic Advisers. 1989. *Economic indicators.* Washington, D.C., December.

———, President's Science Advisory Committee. 1967. *The world food problem.* Vols. 2–3, *Report of the panel on the world food supply.* Washington, D.C.

Usher, R. 1996. The Cadillac that moos. *Time,* April 1, p. 25.

Uvin, Peter. 1993. State of world hunger. In *The hunger report, 1993.* New York: Gordon and Breach.

van der Zee, H. 1998. *The hunger winter: Occupied Holland, 1944–45.* Lincoln: University of Nebraska Press.

Vergara, Benito S. 1979. *A farmer's primer on growing rice.* Los Banos, Philippines: International Rice Research Institute.

Vertical Farm. 2008. Website. http://www.verticalfarm.com.

Viscusi, W. Kip. 1993. The value of risks to life and health. *Journal of Economic Literature* 31:1912–1946.

Viscusi, W. K., and T. Gayer. 2002. Safety at any price? *Regulation* 25, no. 3. http://www.cato.org/pubs/regulation/regv25n3/regv25n3.html.

Von Braun, Joachim, and Eileen Kennedy. 1986. *Commercialization of subsistence agriculture: Income and nutritional effects in developing countries.* Working Paper on Commercialization of Agriculture and Nutrition, no. 1. Washington, D.C.: International Food Policy Research Institute.

———, eds. 1994. *Agricultural commercialization, economic development, and nutrition.* Baltimore: Johns Hopkins University Press.

Von Braun, Joachim, et al. 1989. *Nontraditional export crops in Guatemala: Effects on production, income, and nutrition.* Research Report no. 73. Washington, D.C.: International Food Policy Research Institute.

Walinsky, Louis. 1962. *Economic development in Burma, 1951–1960.* New York: Twentieth Century Fund.

Walker, Alexander, and Harry Stein. 1985. Growth of third world children. In *Dietary fibre, fibre-depleted foods, and disease,* ed. H. Trowell et al., 331–344. London: Academic Press.

Wallis, J. A. N. 1997. Brazil: The cerrados region. In *Intensified systems of farming in the tropics and subtropics,* 85–103. Discussion Paper no. 364. Washington, D.C.: World Bank.

Walsh, B. 2008. The farmer's bank. *Time,* January 31.

Waterlow, J., D. Armstrong, L. Fowden, and R. Riley, eds. 1998. *Feeding a world population of more than eight billion people.* Oxford: Oxford University Press.

Waterlow, J., R. Buzina, W. Keller, J. Lane, M. Nichaman, and J. Tanner. 1977. The presentation and use of height and weight data for comparing the nutritional status of groups of children under the age of ten years. *Bulletin of the World Health Organization* 55:489–498.

Watts, J. 2007. Riots and hunger feared as demand for grain sends food costs soaring. *The Guardian,* December 4. http://www.guardian.co.uk/world/2007/dec/04/chinabusiness.

Weiner, J. S. 1977. Nutritional ecology. In *Human biology: An introduction to human evolution, variation, growth, and ecology,* by A. G. Harrison et al., 400–423. Oxford: Oxford University Press.

Weiner, T. 2000. A farmer learns about Mexico's lack of the rule of law. *New York Times,* October 27.

WFP (World Food Programme). 1989. *Food aid works.* Rome: Food and Agriculture Organization.

———. 2008. http://www.wfp.org/operations/current-shortfalls/2008.

———. n.d. Zambia food security overview. http://www.wfp.org/countries/zambia.

———. n.d. Where we work—Zimbabwe: Food security overview. http://www.wfp.org/country_brief/indexcountry.asp?country=716.

Whitney, Eleanor N., and Eva Hamilton. 1977. *Understanding nutrition.* St. Paul, Minn.: West Publishing.

WHO (World Health Organization). 1985a. *Energy and protein requirements.* Technical Report no. 724. Geneva.

———. 1985b. *Fourth programme report for control of diarrheal diseases, 1983–1984.* Geneva: Program for Control of Diarrheal Diseases.

———. 1985c. *The management of diarrhoea and use of oral rehydration therapy.* Geneva: WHO and UNICEF.

———. 1989. Report on world health. Press release. Washington, D.C.: Regional Office for the Americas, September 25.

———. 1995a. *Global prevalence of vitamin A deficiency.* Geneva. http://www.who.int/nutrition/publications/vad_intro_background.pdf.

———. 1995b. *Physical status: The use and interpretation of anthropometry.* Technical Report no. 854. Geneva.

———. 1995c. *World health report, 1995: Bridging the gaps.* Geneva.

———. 1996a. *Investing in health research and development: Ad hoc committee on health research relating to future intervention options.* New York: United Nations.

———. 1996b. *State of the world's vaccines and immunization.* New York: United Nations. http://www.who.ch/gpv/tenglish/avail/sowvi.htm.

———. 1996c. *World health report, 1996.* Geneva.

———. 1997. *World health report, 1997.* Geneva.

———. 2001. *Iron deficiency anaemia: Assessment, prevention, and control.* Geneva. http://www.who.int/vmnis/anaemia/prevalence/anaemia_data_status_prevalence/en/index.html.

———. 2002. *Diet, nutrition, and the prevention of chronic diseases.* Technical Report no. 916. Geneva.

———. 2003a. Progress towards global immunization goals, 2001. Geneva. http://www.who.int/vaccines-surveillance/documents/slidesglobalimmunization_2002update.

———. 2003b. Water supply, sanitation, and hygiene development. http://www.who.int/water_sanitation_health/hygiene/en.

———. 2007. *Assessment of iodine deficiency disorders and monitoring their elimination.* 3rd ed. Geneva. http://whqlibdoc.who.int/publications/2007/9789241595827_eng.pdf.

———. 2008. *Worldwide prevalence of anaemia, 1993–2005.* Edited by B. de Benoist, E. McLean, I. Egli, and M. Cogswell. Geneva. http://whqlibdoc.who.int/publications/2008/9789241596657_eng.pdf.

WHO and United Nations. 1997. Micronutrient and trace element deficiencies: General information. http://www.who.org/nut/micr/micrgen.htm.

Wild, Alan. 2003. *Soils, land, and food: Managing the land during the twenty-first century.* Cambridge: Cambridge University Press.

Williamson, J. 1990. What Washington means by policy reform. In *Latin American adjustment: How much has happened?* ed. J. Williamson. Washington, D.C.: Institute for International Economics.

Wines, M. 2007. Caps on prices only deepen Zimbabweans' misery. *New York Times,* August 2.

Winick, M., K. Meyer, and R. Harris. 1973. Malnutrition and environmental enrichment by early adoption. *Science* 190:1173–1175.

Winter, Roger P. 1988. In Sudan, both sides use food as a weapon. *Washington Post,* November 19, p. A25.

Wittwer, Sylvan. 1995. *Food, climate, and carbon dioxide: The global environment and world food production.* New York: Lewis.

Wolfenbarger, L., and P. Phifer. 2000. The ecological risks and benefits of genetically engineered plants. *Science* 290:2088–2093.

World Bank. 1975. *Land reform sector.* Policy paper. Washington, D.C., May.

———. 1982. *World development report: Agriculture and economic development.* Washington, D.C.: Oxford University Press.

———. 1994. *Enriching lives: Overcoming vitamin and mineral malnutrition in developing countries.* Washington, D.C.

———. 1997. Does better nutrition improve academic achievement? Yes. *World Bank Policy and Research Bulletin* 8, no. 2 (April–June).

———. 2001. Private-sector development strategy: Issues and options. http://rru.worldbank.org/strategy/psdstrategy-june1.pdf.

———. 2003a. Globalization, growth, and poverty: Building an inclusive world economy. http://econ.worldbank.org/prr/globalization/text-2857.

———. 2003b. *World development indicators.* Washington, D.C. http://www.worldbank.org/data/wdi2003/worldview.htm.

———. 2007. *World development report, 2008.* Washington, D.C. http://econ.worldbank.org/wbsite/external/extdec/extresearch/extwdrs/extwdr2008/0,,menupk:2795178~pagepk:64167702~pipk:64167676~thesitepk:2795143,00.html.

———. 2008a. *China quarterly update.* Washington, D.C., February. http://siteresources.worldbank.org/intchina/resources/318862-1121421293578/cqu_jan_08_en.pdf.

———. 2008b. *World development report, 2009.* Washington, D.C. http://econ.worldbank.org/wbsite/external/extdec/extresearch/extwdrs/extwdr2009/0,,contentmdk:21955654~pagepk:64167689~pipk:64167673~thesitepk:4231059,00.html.

———. Various dates. *Monthly commodity price report sheet.* Washington, D.C. http://siteresources.worldbank.org/intdailyprospects/resources.

———. Various years. *World development report.* New York: Oxford University Press.

World Bank and UNDP (United Nations Development Programme). 1990. *A proposal for an internationally supported programme to enhance research in irrigation and drainage technology in developing countries.* Vol. 2. Washington, D.C.

World Food Council. 1988. *The global state of hunger and malnutrition, 1988 report.* Nicosia, Cyprus, May.

World Resources Institute. 1997. *World resources, 1996/97: The urban environment.* Washington, D.C. http://www.wri.org/wr-96-97.

———. 2000. *World resources, 2000–01: People and ecosystems.* Washington, D.C. http://pdf.wri.org/world_resources_2000-2001_people_and_ecosystems.pdf.

Worldwatch Institute. 1996a. Cropland losses threaten world food supplies. Press release, July 27. http://www.worldwatch.org/press/news/1996/07/27.

———. 1996b. Worldwatch Institute urges World Bank and FAO to overhaul misleading food supply projections. Press release, May 1. http://www.worldwatch.org/node/1594.

Yang, Jisheng. N.d. Tombstone. Not published in English. See Applebaum 2008.
Yao, Shujie. 1999. A note on causal factors of China's famine in 1959–61. *Journal of Political Economy* 107, no. 6:1365–1372.
Ying, Yvonne. 1996. *Poverty and inequality in China.* Washington, D.C.: World Bank. http://www.worldbank.org/html/prddr/trans/ja96/art2.htm.
Zaidi, S., and M. C. Fawzi. 1995. Health of Baghdad's children [letter]. *Lancet* 346, no. 8988 (December 2):1485.

Index

About the Book

Why do millions of people in the less developed countries go hungry—while there is an abundance of food in the world? What can be done about it? These are the issues explored in this accessible and comprehensive text.

In addition to incorporating updated data throughout, this new edition includes:

- a comprehensive description and analysis of the 2008 food crisis
- an expanded discussion of the impact of using food crops to produce biofuels
- new case studies and recent examples to illustrate key points
- examples of successful and unsuccessful policy approaches
- reference to the latest research findings (with more than 150 new citations)

The result is the best available analysis of the current world food problem, as well as a provocative assessment of prospects for the future.

Howard D. Leathers is associate professor of agricultural and resource economics, and **Phillips Foster** is professor emeritus of agricultural and resource economics, both at the University of Maryland, College Park.